BLACK TIO$_2$ NANOMATERIALS FOR ENERGY APPLICATIONS

BLACK TiO₂ NANOMATERIALS FOR ENERGY APPLICATIONS

Editors

Xiaobo Chen
University of Missouri-Kansas City, USA

Yi Cui
Stanford University, USA

World Scientific

NEW JERSEY · LONDON · SINGAPORE · BEIJING · SHANGHAI · HONG KONG · TAIPEI · CHENNAI · TOKYO

Published by

World Scientific Publishing Europe Ltd.

57 Shelton Street, Covent Garden, London WC2H 9HE

Head office: 5 Toh Tuck Link, Singapore 596224

USA office: 27 Warren Street, Suite 401-402, Hackensack, NJ 07601

Library of Congress Cataloging-in-Publication Data

Names: Chen, Xiaobo, 1976–   author. | Cui, Yi, 1976–   author.
Title: Black TiO$_2$ nanomaterials for energy applications / Xiaobo Chen
   (University of Missouri-Kansas City, USA), Yi Cui (Stanford).
Description: [Hackensack?] New Jersey : World Scientific, [2016] |
   Includes bibliographical references.
Identifiers: LCCN 2016033508 | ISBN 9781786341655 (hc : alk. paper)
Subjects: LCSH: Fuel cells--Materials. | Titanium dioxide--Industrial applications. |
   Nanostructured materials--Industrial applications. | Hydrolysis--Industrial applications.
Classification: LCC TK2931 .C465 2016 | DDC 621.3028/4--dc23
LC record available at https://lccn.loc.gov/2016033508

British Library Cataloguing-in-Publication Data
A catalogue record for this book is available from the British Library.

Desk Editors: Herbert Moses/Mary Simpson

Typeset by Stallion Press
Email: enquiries@stallionpress.com

Printed in Singapore

Preface

My Sunday copy of The New York Times shows on front page this title in bold face:

Global Warming's Mark: Coastal Inundation

and the subtitle in smaller print says:

Decades of Warnings by Scientists Are No Longer Theoretical.

(September 4, 2016 edition of The New York Times)

The time for scientists to sound the alarm of global warming has already passed. It is now time for scientists to come up with the solutions. Scientists have so far thought that there is no "silver bullet" for this global problem. Instead, many tools together may help to solve this problem eventually one day. However, this book shows us what may be considered the closest approximation to a "silver bullet". It involves a special form of titanium dioxide known commonly as "black titanium dioxide" in the recent literature.

Titanium dioxide (TiO_2 or titania) is an inexpensive and widely available material. It occurs in nature as common minerals known as rutile and anatase. It has already been used in household products such as white paint. It is chemically stable and yet is a powerful photo-catalyst because of its special electronic properties. It has been studied for years as an ideal photo-catalyst in water splitting and pollution remediation. In spite of all its advantages, TiO_2 has a serious disadvantage in harvesting solar energy. It is transparent to visible light and starts absorbing light strongly only in the near ultraviolet region. This means that most of the solar spectrum cannot be harvested by TiO_2.

There have been many attempts to lower the absorption onset of TiO_2 but success has limited to making TiO_2 colored. Then in 2011 a breakthrough was announced by Xiaobo Chen, one of the editors of the present book, and his collaborators from the Lawrence Berkeley National Laboratory in Berkeley, CA. By hydrogenating nanometer (nm) size TiO_2 under pressure, this group succeeded in creating a new form of nm TiO_2 which is black in color. Furthermore, this "black TiO_2" was demonstrated to be chemically stable and to enhance the photo-catalytic efficiency of the usual "white TiO_2" by orders of magnitude under solar illumination. After this surprising result was reproduced around the world, the race was on to produce black TiO_2 with other techniques. These techniques have been applied to other forms of nanostructured TiO_2 and extensive theoretical effort was devoted to understand the reason why these techniques have worked where others have failed. It turns out that black TiO_2 is a true product of nanotechnology. The technique would not have worked if the sample is not nm in size. New applications of black TiO_2 related to energy have also been discovered such as in batteries. The worldwide interest in black TiO_2 has been increasing exponentially. In a short span of five years, the original paper of Chen *et al.* in Science has been cited over 2000 times and has been classified by Citation Index as one of the highly cited papers. Judging from the feverish rate at which progress has been made in finding applications for black TiO_2 in renewable energy generation and storage, it is likely that it will become an important tool in our "combat" against global warming.

Given the highly competitive nature of the field of renewable energy, the present volume edited by Xiaobo Chen and Yi Cui is highly timely. It covers nearly all the important developments in the field within the past five years and involves contributions by a roster of international experts in this nascent field. It will be an invaluable guide to both students and experienced researchers interested in learning about this new and exciting form of an old material which forms the base of white paint!

Peter Y. Yu
Berkeley, CA

About the Editors

Xiaobo Chen is an Assistant Professor and the Director of Materials Modification and Characterization Laboratory at the Department of Chemistry, University of Missouri–Kansas City (UMKC). His research interests include nanomaterials' synthesis, characterization, modifications, and applications in renewable energies, such as photocatalysis, battery, electrochemistry, and environmental pollution removal. He has published around 100 peer-reviewed articles with over 26,000 citations. During his Ph.D. study from 2001 to 2005 at Case Western Reserve University, he developed sol-gel and solid-state methods in synthesizing nitrogen, sulfur and carbon-doped TiO_2 nanoparticles. In his post-doctoral research in University of California, Berkeley, and Lawrence Berkeley National Laboratory, he invented the black TiO_2 nanoparticles and a new self-cleaning building technology (2011 R&D 100 Award). Since his independent research at UMKC from 2011, he further developed a series of new materials to improve the performance of lithium-ion battery electrodes, photocatalytic and electrocatalytic materials for water splitting and hydrogen generation (one electrolysis technology is on the way for large-scale demonstration and commercialization), and proposed new concepts for microwave absorbing materials with the black TiO_2 nanoparticles.

Yi Cui went to University of Science and Technology of China (USTC), where he received a Bachelor's degree in Chemistry in 1998. He attended graduate school from 1998 to 2002 at Harvard University, where he worked under supervision of Professor Charles M. Lieber. His Ph.D thesis concerned semiconductor nanowires for nanotechnology including synthesis, nanoelectronics and nanosensor applications. After that, he went on to work as a Miller Postdoctoral Fellow with Professor Paul Alivisatos at University of California, Berkeley. His postdoctoral work was mainly on electronics and assembly using colloidal nanocrystals. In 2005, he became an Assistant Professor in Department of Materials Science and Engineering at Stanford University. In 2010, he was promoted to an Associate Professor with tenure and named as David Filo and Jerry Yang Faculty Scholar. In 2010, he was promoted to be a Full Professor. In 2011, he started a joint appointment in Photon Science Faculty, SLAC National Accelerator Laboratory. His current research is on nanomaterials for energy storage, photovoltaics, topological insulators, biology and environment.

He is a highly proliferate materials scientist and has published ~330 research papers, filed more than 40 patent applications and has given ~300 plenary/keynote/invited talks. His works have generated a very large impact and he is among top most cited scientists in the world (Google Scholar, H-index 123). In 2014, he was ranked No.1 in Materials Science by Thomson Reuters as "The World's Most Influential Scientific Minds". He is an Associate Editor of *Nano Letters*. He is a Co-Director of the Bay Area Photovoltaics Consortium and a Co-Director of Battery 500 Consortium. He founded Amprius Inc. in 2008, a company to commercialize the high-energy battery technology. Now the high-energy batteries invented by him have started to be used in commercial market, which could potentially revolutionize portable electronics and transportation applications. In 2015, he and Professor Steve Chu co-founded 4C

Air Inc., to commercialize their invented breakthrough technology to remove particle pollutants from the air.

He has won numerous awards recognizing his scientific contributions in these research areas, including MRS Fellow (2016), MRS Fred Kavli Distinguished Lectureship in Nanoscience (2015), Fellow of Royal Society of Chemistry (2015), Small Young Innovator Award (2015), Resonate Award for Sustainability (2015), Blavatnik National Award Finalist (2015), Inorganic Chemistry Frontiers Award for Young Scientist (2015), Inaugural Schlumberger Chemistry Lectureship (University of Cambridge, 2015), Top 10 World Changing Ideas for His Work on Batteries to Capture Low-Grade Waste Heat (Scientific American, 2014), No. 1 "Hottest Researchers of Today" in Materials Science (Thomas Reuters, 2014), Closs Lectureship (University of Chicago, 2014), Inaugural Nano Energy Award (2014), Bau Family Awards in Inorganic Chemistry (2014), Blavatnik National Award Finalist (2014), the IUPAC Distinguished Award for Novel Materials and Synthesis (2013), Scientist in Residence of University of Duisburg-Essen (2013), Next Power Visiting Chair Professorship (National Tsinghua University, 2013), the Wilson Prize (2011), the David Filo and Jerry Yang Faculty Scholar (2010), the Sloan Research Fellowship (2010), the Global Climate and Energy Project Distinguished Lecturer (2009), KAUST Investigator Award (2008), ONR Young Investigator Award (2008), MDV Innovators Award (2007), Terman Fellowship (2005), the Technology Review World Top Young Innovator Award (2004), Miller Research Fellowship (2003), Distinguished Graduate Student Award in Nanotechnology (Foresight Institute, 2002), and Gold Medal of Graduate Student Award (Material Research Society, 2001).

Contents

3.   Black TiO$_2$ Nanomaterials Through Electrochemical and Mechanical Methods

33

Haidong Bian, Chris Lee, Hui Li, Jian Lu
and Yang Yang Li

4.   The Effect of Points Defects and Ordered/ Disordered Morphology on the Electronic and Structural Features in Black TiO$_2$ Nanomaterials

49

Vladimiro Dal Santo and Alberto Naldoni

7. Black Titania Coatings: Fabrication Process and Photoelectrochemical/Photocatalytic Properties

Tomohiko Nakajima and Tetsuo Tsuchiya

8. Hydrogen-Treated TiO_2 Nanowires for Charge Storage and Photoelectrochemical Water Splitting

Gongming Wang, Xihong Lu and Yat Li

9. Rationalizing the Efficiency of Hydrogen-Treated TiO$_2$ Nanomaterials in Light Driven Water-Splitting Applications

215

Mark Forster and Alexander J. Cowan

10. Black TiO$_2$ Nanomaterials for Lithium-Ion Batteries

249

Laifa Shen, Shengyang Dong, Xiaogang Zhang and Guozhong Cao

11. Black TiO$_2$ Nanomaterials for Lithium–Sulfur Batteries

275

Zheng Liang, Xinyong Tao and Yi Cui

CHAPTER ONE

Introduction

***Xiaobo Chen**[*] **and Yi Cui**[†]

[*]*Department of Chemistry, University of Missouri–Kansas City,
Kansas City, Missouri, USA*

[†]*Department of Materials Science and Engineering,
Stanford University, Stanford, CA*

Titanium dioxide (TiO_2) has been widely studied for various photocatalytic, photoelectrochemical, and electrical energy applications. One of its most important properties is its optical absorption, as its photocatalytic and photoelectrochemical activity and performance largely depends on the amount of light it absorbs. As a large bandgap semiconductor with bandgap of 3.0–3.2 eV, TiO_2 only absorbs light in the ultraviolet (UV) region, which limits its efficiency in utilizing the solar spectrum where the UV light only accounts for less than 5%. Thus, it is highly desirable to make TiO_2 absorb larger amount of sunlight in order to improve its solar-related applications.

The discovery of black TiO_2 by hydrogenation has triggered worldwide research effort in this new material, as it absorbs light from UV to visible and near-infrared and displays superior photocatalytic and photoelectrochemical performances. Distinct from traditional metal and non-metal doping techniques, hydrogenation induces a thin layer of disordered shell on the outside of crystalline TiO_2 nanoparticles, which causes the alteration of its electronic structure

1

and the optical properties. This book presents the most recent progresses on the various aspects of black TiO$_2$ nanomaterials.

In Chapter Two by Dr. Xiaobo Chen from the University of Missouri–Kansas City, the synthesis of black TiO$_2$ nanomaterials by hydrogenation has been presented under various hydrogenation conditions, and the properties of the resulting black TiO$_2$ nanomaterials are discussed, along with their structural, chemical, electronic, and optical properties. Thus, this chapter will give a good overview on the hydrogenated black TiO$_2$ nanomaterials. In Chapter Three by Dr. Yangyang Li from the City University of Hong Kong, the preparation of black TiO$_2$ nanomaterials by electrochemical and mechanical methods is presented, with the introduction of strain engineering through mechanical processing, ball milling of free TiO$_2$ nanostructures, surface attrition treatment, and doping/surface energy creation and effects.

In Chapter Four, Dr. Vladimiro Dal Santo and Dr. Alberto Naldoni from the Institute of Molecular Science and Technologies, Italy, discuss the effects of points defects and ordered/disordered morphology on the electronic and structural features in black TiO$_2$ nanomaterials. They elucidate the electronic singularities in black TiO$_2$ nanomaterials: valence band edge modification, electronic transitions due to intragap states, and H,N doping, along with the structural properties in various black TiO$_2$ nanostructures. In Chapter Five, the black and white issue of TiO$_2$ nanomaterials are presented by Dr. Fuqiang Huang and Dr. Guilian Zhu at Shanghai Institute of Ceramics, Chinese Academy of Sciences, and their colleague Dr. Tao Xu at Northern Illinois University. They address the structural and chemical reasons of black TiO$_2$ nanomaterials, their properties and synthetic methods, and applications in photocatalysis, photothermal, photochemical sensors, etc. In Chapter Six by Dr. Lei Liu at Changchun Institute of Optics, Fine Mechanics, and Physics, Chinese Academy of Sciences, the theoretical efforts on understanding the black TiO$_2$ nanomaterials are presented, with consideration on the intrinsic defects, doping, surface/interface effects, amorphous/disordered phases, lattice strain effects, and nanosize effects. This chapter thus provides a theoretical understanding of the properties of black TiO_2 nanomaterials.

Chapter Seven by Dr. Tomohiko Nakajima and Dr. Tetsuo Tsuchiya from National Institute of Advanced Industrial Science and Technology, Japan, presents the photoelectrochemical properties of black TiO_2 photoelectrode coatings, with an introduction of the overview of the black TiO_2 coating process, the hydrogenated coating, and the oxygen-deficient photoelectrode coating. Chapter Eight by Dr. Yat Li and his colleagues from University of California, Santa Cruz, discusses hydrogenated black TiO_2 nanowires for photo-electrochemical water splitting and supercapacitor. The discussion includes hydrogen-treated rutile TiO_2 nanowire arrays, improved efficiency of charge separation in the hydrogen-treated TiO_2, and the synergistic effects between hydrogen treatment and elemental doping, along with hydrogenated TiO_2 nanowire as supercapacitor electrode and as supports for supercapacitor electrodes. Chapter Nine by Dr. Alexander J Cowan from the University of Liverpool, United Kingdom, rationalizes the efficiency of hydrogen-treated TiO_2 nanomaterials in light driven water splitting applications, including discussion on the roles of oxygen vacancies in hydrogen-treated TiO_2 and surface reaction kinetics in oxygen-deficient TiO_2.

Chapter Ten by Dr. Laifa Shen from Nanjing University of Aeronautics and Astronautics, China and Dr. Guozhong Cao at University of Washington, Seattle along with their colleagues, summarizes black TiO_2 nanomaterials for lithium-ion batteries, including the mechanism of lithium storage in TiO_2, the advantage of using black TiO_2, and vacuum-assisted, hydrogenated, nitridated, and sulfureted synthesis of black TiO_2 nanomaterials for lithium-ion batteries. Chapter Eleven by Dr. Yi Cui and his colleagues from Stanford University lays out the fundamentals of the usage of TiO_2 nanomaterials in Li–S rechargeable batteries. They discuss the fabrication of black TiO_2 nanomaterials and the composite electrode, their properties, and four examples.

Overall, this book presents the cutting-edge research progress on black TiO_2 nanomaterials, one of the most promising photo-catalysts discovered recently. It covers various synthetic methods and approaches, experimental and theoretical understanding of the chemical, physical, and electronic properties of black TiO_2 nanomaterials and their applications in catalysis, photocatalysis, photothermal, photoelectrochemical water splitting, lithium-ion batteries, and supercapacitors. Thus, this book not only provides an excellent introduction but also an in-depth analysis of black TiO_2 nanomaterials, and may inspire new thoughts in related fields for readers with various knowledge levels and research backgrounds.

CHAPTER TWO

Synthesis and Properties of Hydrogenated Black TiO$_2$ Nanomaterials

Xiaodong Yan[*], *Lihong Tian*[*,†] *and Xiaobo Chen*[*,‡]

[*]*Department of Chemistry, University of Missouri–Kansas City,*
Kansas City, MO 64110, USA

[†]*Hubei Collaborative Innovation Center for Advanced*
Organochemical Materials, Hubei University,
Wuhan, Hubei 430062, China

[‡]*chenxiaobo@umkc.edu*

2.1 Introduction

Climate change, environmental pollution, energy crisis, and disease are the most challenging issues facing humanity today. Titanium dioxide (TiO$_2$) as a multifunctional material has thus drawn intense attention and effort worldwide owing to its potential applications in energy harvesting and storage, pollutant removal, and biomedical applications.[1–4] Its performance in most of these applications is highly dependent on its morphological, structural, surface, electronic and/or optical properties.[5–7] So far, much effort has been devoted to tailoring the morphology and facet of TiO$_2$ nanomaterials. TiO$_2$ nanomaterials with different dimensions and/or selective facets demonstrated highly enhanced performance in many applications.[8–11] However, morphology and facet control are far from satisfactory for the most desired applications in photocatalysis

---

[‡]Corresponding author.

due to the intrinsic wide bandgap of TiO_2. Its wide bandgap (3.0–3.2 eV) restricts its optical response only to the ultraviolet (UV) radiation, accounting for ∼5% of the solar radiation on earth and thus resulting in inefficient photocatalytic properties.[1,3,12] The inefficiency originates from its intrinsic electronic band structure.[12] Hence, it is imperative to extend its optical activity from UV light to visible light by engineering its electronic structure.

Structural modification through incorporation of heteroatoms into the lattice of TiO_2 is a simple and effective strategy to engineer the electronic structure. The corporation of alien metal ions can raise the valence band (VB) by forming a solid solution of two or more semiconductors, while the introduction of nitrogen can create intraband states above the VB.[13] Both can extend the optical absorption of TiO_2 nanomaterials to visible light if the dopants are introduced appropriately. Recently, codoping of TiO_2 nanomaterials with different elements has received considerable attention in order to further improve their photocatalytic activity.[14,15] In addition to heteroatom doping, another effective strategy is self-structural modification, that is, the bandgap narrowing is achieved just by introducing intrinsic defects (e.g. oxygen vacancy and Ti–OH) into TiO_2 nanomaterials.[15] However, the mechanism of self-structural modification in improving the visible-light activity is still under debate. For example, Gu *et al.* reported that bulk vacancy defects tended to act as trap states and charge carrier recombination centers, which were detrimental to the photocatalytic properties.[16] Naldoni *et al.* found that bulk oxygen vacancy in black TiO_2 nanoparticles played an important role in bandgap narrowing.[17]

Since its discovery in 2011,[18] defect-engineered black TiO_2 nanomaterials have become the focus of intense interest.[19] Hydrogenation treatment is the most commonly used approach to synthesize black TiO_2 nanomaterials. Hence, we would like to offer a comprehensive review on hydrogenated black TiO_2 nanomaterials with emphasis on their synthesis and properties. We try to summarize all the research works on hydrogenated black TiO_2 nanomaterials, and collect accumulated information to provide a better understanding on the inner workings of hydrogenated black TiO_2 nanomaterials.

2.2 Synthesis of Hydrogenated Black TiO$_2$ Nanomaterials

Hydrogenation is defined as a chemical reaction between molecular hydrogen (H$_2$) and another compound. Usually, hydrogenation refers to the reduction or saturation of organic compounds often in the presence of a catalyst. Herein, it is employed specially representing the mild reduction reaction at the interface between hydrogen and TiO$_2$ at elevated temperatures, but no metallic titanium will be formed. It is reasonable to make the assumption that hydrogen atoms will be attached onto the surfaces of TiO$_2$ crystals through the formation of O–H bond during the hydrogenation process. If the reaction further goes forward, oxygen vacancy, Ti–H bond, and even Ti^{3+} interstitial tend to be generated.

It is natural to claim that the defects will be generated on the surface of the TiO$_2$ crystals as the hydrogenation reaction occurs at the interface. Indeed, surface disorder was commonly observed in hydrogenated black TiO$_2$ nanomaterials.[18, 20, 21] However, bulk defects in a few cases were detected maybe owing to the high hydrogenation temperature.[16, 17] On the other hand, the types of defects in hydrogenated black TiO$_2$ nanomaterials were also different with varying preparation conditions. Therefore, the synthesis of hydrogenated black TiO$_2$ nanomaterials plays a decisive role in their structures and properties. In detail, the structures and properties of hydrogenated black TiO$_2$ nanomaterials are highly dependent on the inherent physicochemical properties of the pristine TiO$_2$ nanocrystals (e.g. size, shape, morphology, crystal facet, and defect content), hydrogenation time and temperature, and hydrogen pressure and purity.[19] Herein, we focus on the synthesis processes of hydrogenated black TiO$_2$ nanomaterials, along with a brief introduction to the remarkable structural evolution of the TiO$_2$ nanomaterials after hydrogenation.

2.2.1 *Hydrogenation in High-Pressure Hydrogen*

Chen *et al.* first reported the black TiO$_2$ nanoparticles by hydrogenating the pristine white anatase TiO$_2$ nanoparticles in a vacuum

(a)   (b)

Figure 2.1. HRTEM images of TiO_2 nanocrystals (a) before and (b) after hydrogenation (the insets in (a) and (b) are the digital images of the corresponding samples).[18] Reproduced from Ref. [18]. © The American Association for the Advancement of Science, 2011.

for 1 h and then in 20.0-bar H_2 atmosphere at about 200°C for 5 days.[18] The pristine TiO_2 nanoparticles were obtained by heating the reaction solution containing titanium tetraisopropoxide, Pluronic F127, hydrochloric acid, deionized (DI) water, and ethanol with the molar ratio of 1:0.005:0.5:15:40 at 40°C for 24 h, followed by evaporating and drying at 110°C for 24 h and then calcinating at 500°C for 6 h.[18] Figure 2.1 shows the high resolution transmission electron microscopy (HRTEM) images of (a) pristine white TiO_2 nanoparticles and (b) hydrogenated black TiO_2 nanoparticles.[18] The size of the individual TiO_2 nanoparticle is $\sim$8 nm in diameter.[18] The black TiO_2 nanoparticle featured a core–shell structure, where the core presented ordered crystal lattice and the shell was lattice-disordered.[18] The disordered shell was believed to host the possible hydrogen dopant in the form of Ti–H and Ti–O–H bonds, bringing about the mid-gap states and the black color of the hydrogenated TiO_2 nanoparticles.[18] Further research showed that hydrogenation greatly affected the crystal facets and volume of the TiO_2 crystals.[22] Hydrogenated black TiO_2 nanoparticles experienced a 17.0% decrease in surface area and a 20.2% decrease in volume compared to the pristine TiO$_2$ crystals, along with the disappearance of (102) facet and the birth of (215) facet.[22]

Sun *et al.* investigated the hydrogen incorporation and storage in well-defined nanocrystals of anatase TiO$_2$ nanocrystals during which black TiO$_2$ nanocrystals were generated.[23] For (001)-faceted anatase TiO$_2$, 5 mL of titanate isopropoxide or tetrabutyl titanate and 0.4–0.6 mL of HF (48% w/w) were added to a Teflon-lined stainless steel autoclave with a capacity of 30–50 mL and then kept at 180°C for 24 h.[23] For (101)-faceted anatase TiO$_2$, 0.73 mL of hydrochloric acid (10% w/w) was added to 30 mL of titanium tetrachloride aqueous solution (5.33 mM), and then hydrothermal reaction was carried out in a Teflon-lined autoclave at 180°C for 14 h.[23] The anatase TiO$_2$ nanocrystals were collected by centrifugation accompanied by being washed with ethanol and DI water three times, and dried in vacuum overnight.[23] The thickness of the (001)-faceted TiO$_2$ nanosheets was less than 10 nm, with the percentage of (001) facet being >80%.[23] The diameter of the (101)-faceted TiO$_2$ nanocrystals was in the range of 10–30 nm, with the percentage of (001) being around 2%.[23] The (101)-faceted anatase TiO$_2$ nanocrystals turned black after hydrogenation at 450°C for 24 h with an initial hydrogen pressure of 7 MPa.[23] Experimental data and density functional theory (DFT) calculations showed that hydrogen incorporation through (101) facet was more favorable than through (001) facet.[23] However, surface disorder was not observed, probably owing to the high hydrogenation temperature.[23]

Lu *et al.* synthesized hydrogenated black TiO$_2$ nanocrystals from commercial P25 TiO$_2$ at room temperature.[24] In a typical synthesis, 0.5 g of P25 TiO$_2$ (Degussa) with an anatase/rutile ratio of 80/20 was first maintained in a vacuum for 24 h and then hydrogenated in a 35-bar hydrogen atmosphere at room temperature ($\sim$15°C) in a homemade stainless steel cell (500 mL).[24] Black TiO$_2$ was obtained after reacting for 20 days.[24] A typical crystalline-disordered core–shell structure was observed in the hydrogenated black TiO$_2$ nanocrystals.[24] Noticeably, a small amount of Ti$_2$O$_3$ was generated as evidenced by the X-ray diffraction (XRD) pattern.[24]

Leshuk *et al.* studied the photocatalytic activity of hydrogenated porous nanocrystalline anatase TiO$_2$ shells on silica spheres.[16] Typically, tetraethyl orthosilicate (TEOS) was added to a water/ethanol mixture containing NH$_4$OH under vigorous stirring (the concentrations of reagents were approximately 3.33 M H$_2$O, 1.4 M NH$_4$OH, and 0.2 M TEOS).[16] After reacting for at least 10 h under vigorous stirring, the silica particles were collected by centrifugation, washed thrice with ethanol, and dried at 70°C in air.[16] Dried silica particles were dispersed in ethanol by sonication, to which DI water and hydroxypropyl cellulose (HPC) were added.[16] After the HPC was completely dissolved under stirring, titanium(IV) butoxide (TB) in ethanol (40% TB, v/v) was then added to the mixture dropwise over 1 h with a syringe pump, such that the final concentrations of reagents in the solution were 2.5 g L^{-1} SiO$_2$, 3 g L^{-1} HPC, 0.45 M H$_2$O, and 0.12 M TB.[16] After reacting at 85°C for 1.5 h under stirring and cooling down, the product was centrifuged, washed thrice with ethanol, and dispersed in DI water at a SiO$_2$ concentration of 10 g L^{-1} for subsequent hydrothermal treatments.[16] Hydrothermal treatment of the SiO$_2$–TiO$_2$ particles was performed in a PTFE-lined stainless steel autoclave at 180°C for 1.5 h.[16] After cooling down naturally, the precipitate was centrifuged, washed thoroughly with DI water, and dried at 70°C in air.[16] The dried powders were calcined in air at 500°C for 3 h with a ramp rate of 60°C h^{-1}, and cooled down to room temperature over 4 h.[16] Hydrogenation (30–40 mg) was carried out at 250°C, 350°C, or 450°C for 24 h in a 20-bar H$_2$ atmosphere, and the reactor was rapidly cooled by means of compressed air.[16] Gray and black TiO$_2$ were achieved by hydrogenation at 350°C and 450°C, respectively, but no significant change in the structure and morphology of the gray and black TiO$_2$ was observed.[16] In another research, Leshuk *et al.* proposed that that H$_2$ would more likely react with preexistent reactive sites in TiO$_2$ crystals, possibly dangling bonds or other lattice disorder, wherein some samples might become black as a result of hydrogenation, while others turned different colors or even remained unchanged.[25] In addition, hydrogenation pressure and temperature were disclosed to be fundamentally important parameters in controlling the coloration of hydrogenated TiO$_2$ nanomaterials.[25]

Wang *et al.* reported on the black rutile TiO$_2$ nanowire arrays on fluorine-doped tin oxide (FTO) glass substrate prepared by hydrothermal treatment and subsequent thermal treatments.[26] In a typical synthesis, 0.5 mL of titanium *n*-butoxide was added to a solution containing 15 mL of concentrated hydrochloric acid and 15 mL of DI water in a 100 mL beaker.[26] The clear solution was transferred to a Teflon-lined stainless steel autoclave (40 mL volume), where a clean FTO glass substrate was submerged in the solution.[26] Hydrothermal reaction was carried out in an electric oven at 150°C for 5 h, and then cooled down to room temperature slowly.[26] The FTO glass substrate covered with white TiO$_2$ nanowire film was thoroughly washed with DI water, air dried, and annealed in air at 550°C for 3 h.[26] Hydrogenation was carried out at various temperatures in a range of 200–550°C for 30 min in a home-built tube furnace filled with pure hydrogen gas.[26] Figure 2.2(a) shows the scanning electron microscope (SEM) image of the vertically aligned TiO$_2$ nanowire arrays on FTO substrate. The nanowires had diameters of 100–200 nm.[26] The color of the hydrogenated rutile TiO$_2$ nanowires turned yellowish green at 350°C and black at $\geq$ 450°C (Figure 2.2(b)).[26]

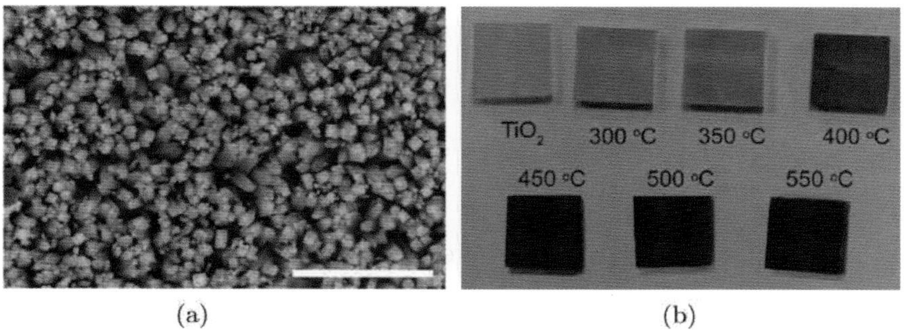

(a)   (b)

Figure 2.2. (a) SEM image of vertically aligned rutile TiO$_2$ nanowire arrays on a FTO substrate. Digital pictures of pristine and hydrogenated TiO$_2$ nanowires on FTO substrates annealed in a hydrogen atmosphere at various temperatures (300°C, 350°C, 400°C, 450°C, 500°C, and 550°C).[26] Reprinted from Ref. [26]. © American Chemical Society, 2011.

Eom *et al.* synthesized black anatase TiO$_2$ nanotube arrays on Ti foil through electrochemical oxidation and subsequent hydrogenation.[27] Typically, a commercially titanium foil (99.5% purity) with a thickness of 0.05 mm was degreased with acetone and then rinsed with DI water and dried in a nitrogen stream.[27] Electrochemical anodization was conducted at a constant potential with a DC power supply in a two-electrode system with Ti foil (1 cm^2) as a working electrode and platinum as a counter electrode.[27] The electrolyte was 0.05 wt.% NH$_4$F in ethylene glycol (99.8%, anhydrous).[27] The as-prepared TiO$_2$ nanotubes were thoroughly washed with ethanol and DI water and then dried at 80°C overnight.[27] The as-prepared TiO$_2$ nanotubes were 30–50 nm in diameter and 5–10 nm for wall thickness.[27] Black TiO$_2$ nanotubes were obtained via hydrogenation treatment at 450°C for 2 h in a hydrogen atmosphere.[27] Liu *et al.* reported on a series of hydrogenated anodic TiO$_2$ (anatase) nanotube arrays with variable colors under various conditions: (i) pristine anatase TiO$_2$ nanotube arrays annealed in air at 450°C for 1 h; (ii) TiO$_2$ nanotube arrays treated in Ar or H$_2$/Ar at 500°C for 1 h under atmospheric pressure; (iii) high-pressure hydrogenated TiO$_2$ nanotube arrays in pure hydrogen (20 bar, 500°C for 1 h); and (iv) high-pressure but mild-temperature hydrogenated TiO$_2$ nanotube arrays (pure H$_2$, 20 bar, 200°C for 5 days).[28] The color of the TiO$_2$ nanotubes turned from white into black in H$_2$/Ar, deep purple in Ar, light blue in high-pressure H$_2$ at 500°C, and gray in high-pressure H$_2$ at 200°C.[28]

In summary, high-pressure hydrogenation is a powerful technique to achieve black TiO$_2$ nanomaterials. As clearly shown in Table 2.1, the coloration of hydrogenated TiO$_2$ nanomaterials is highly dependent on the crystalline structure, morphology, hydrogenation time, and temperature. The surface disorder was observed only in black TiO$_2$ nanoparticles that were hydrogenated at 200°C for 5 days and at room temperature for 20 days, indicating that low temperature (e.g. ≤200°C), long reaction time (e.g. ≥5 day), and high pressure (e.g. ≥20 bar) may be favorable for the generation of surface-disordered thin layers.

Table 2.1. Key parameters of black TiO$_2$ nanomaterials by high-pressure hydrogenation.

Starting TiO$_2$	Reaction temperature	Reaction time	Hydrogen pressure	Color	Surface disorder	Ref.
Anatase nanoparticles	200°C	120 h	20 bar	Black	√	18
Anatase nanoparticles	250°C	24 h	20 bar	Pale yellow	×	16
Anatase nanoparticles	350°C	24 h	20 bar	Gray	×	16
Anatase nanoparticles	450°C	24 h	20 bar	Black	×	16
Anatase nanoparticles	450°C	24 h	7 MPa	Black	×	23
P25	15°C	480 h	35 bar	Black	√	24
Rutile nanowires	350°C	0.5 h	N/S	Yellowish green	×	26
Rutile nanowires	≥450°C	0.5 h	N/S	Black	×	26
Anatase nanotubes	450°C	2 h	N/S	Black	×	27
Anatase nanotubes	200°C	120 h	20 bar	Gray	×	28
Anatase nanotubes	500°C	1 h	20 bar	Black	×	28

2.2.2 *Hydrogenation in Atmospheric-Pressure Hydrogen*

Naldoni *et al.* synthesized black TiO$_2$ nanoparticles through hydrogenation of commercial amorphous TiO$_2$.[17] In a typical synthesis, ~300 mg of commercial amorphous TiO$_2$ (NanoActive®, NanoScale Corporation, USA) was initially treated in vacuum (10^{-5} mbar) and then heated at 200°C under an oxygen flow for 1 h in order to oxidize and favor the desorption of molecular species adsorbed onto the surface of TiO$_2$ nanoparticles.[17] After cooling down to room temperature, the pre-treated TiO$_2$ nanoparticles were heated at 500°C in air for 1 h to obtain crystalized white TiO$_2$ nanoparticles, or hydrogenated at 500°C for 1 h under a hydrogen flow to achieve black TiO$_2$ nanoparticles.[17] In order to study the influence of precursor crystallinity, crystalline P25 Degussa and white TiO$_2$ nanoparticles were hydrogenated at 500°C for 1 h under a hydrogen flow.[17] All thermal treatments were carried out using a heating rate of 10°C min^{-1}.[17] It was found that fast cooling (cooling rate: 50°C min^{-1}) in inert environment until room temperature resulted in black TiO$_2$ nanoparticles, while very slow cooling rate (2°C min^{-1}) or instantaneous exposure to air resulted in a gray coloration.[17] In addition, it was found that hydrogenated TiO$_2$ nanoparticles from crystalline white TiO$_2$ and P25 only presented pale blue color. These results to some extent confirmed the possibly preferred reaction of H$_2$ with pre-existent reactive sites (dangling bonds or other lattice disorder) in TiO$_2$ crystals.[25] On the other hand, hydrogen reduction triggered the phase change below the usual phase change temperature, that is, black TiO$_2$ nanoparticles obtained at 500°C presented 81% anatase and 19% rutile phases, whereas white TiO$_2$ prepared in flowing O$_2$ at 500°C for 1 h was 100% anatase.[17] However, reduction of TiO$_2$ at 400°C and 450°C generated samples that were 100% anatase.[17] It was believed that the interaction between TiO$_2$ host matrix and hot H$_2$ molecule gave rise to oxygen vacancies that overcame the activation energy of TiO$_2$ lattice rearrangement and accelerated the phase transformation from anatase to rutile.[17,29] As shown in Figure 2.3, surface disorder was observed on black TiO$_2$ nanocrystals.

Zhou *et al.* reported on the synthesis of ordered mesoporous black anatase TiO$_2$ fabricated via an evaporation-induced self-assembly method combined with an ethylenediamine encircling process, followed by hydrogenation.[20] Figure 2.4 schematically shows the synthesis process. Typically, 1.0 g of triblock copolymer Pluronic P123 was dissolved in 15 mL of anhydrous ethanol by being stirred for 0.5 h at room temperature and then sonicated for 10 min.[20] 3.2 mL concentrated hydrochloric acid was added dropwise to 4.0 g of tetrabutyl titanate solution, and the solution was stirred for 0.5 h.[20] Then the P123/ethanol solution was added to the above solution with vigorous stirring for at least 3 h.[20] The resulting sol solution was gelled in an open petri dish at room temperature and 50–60% relative

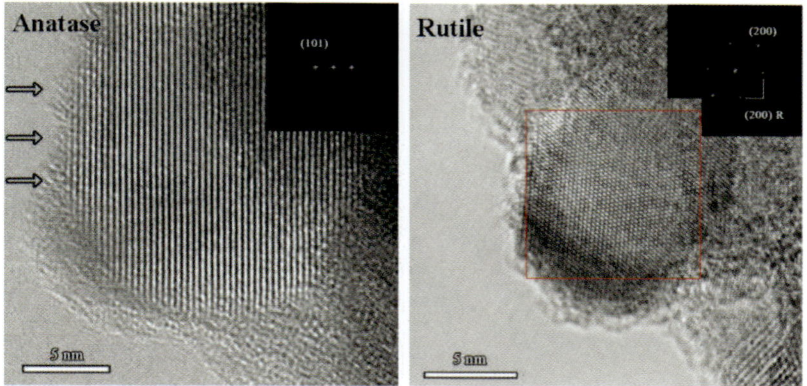

Figure 2.3. HRTEM images of black TiO$_2$ nanocrystals in the form of anatase and rutile.[17] Reprinted from Ref. [17]. © American Chemical Society.

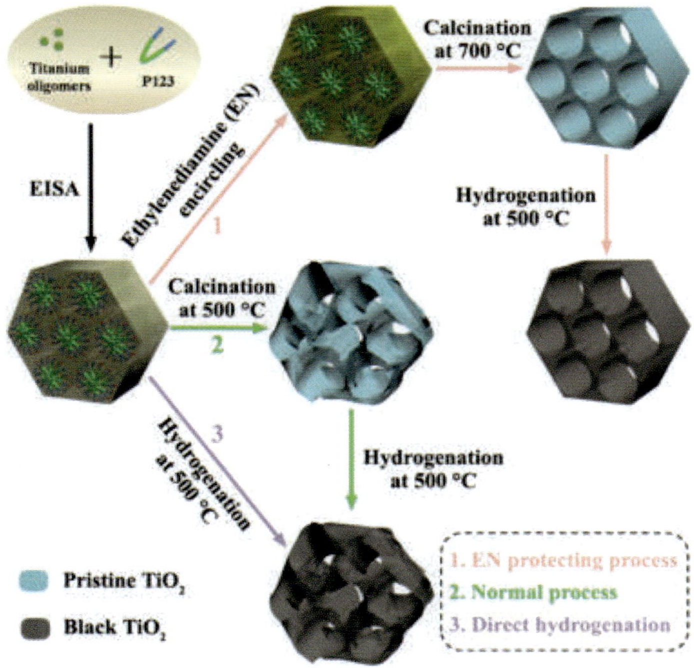

Figure 2.4. Illustration of the synthesis process for the ordered mesoporous black TiO$_2$ materials.[20] Reprinted from Ref. [20]. © American Chemical Society.

humidity (RH) for several weeks.[20] The precursor was calcined at 200°C for 4 h in order to stabilize the network, then refluxed with ethylenediamine aqueous solution for 48 h at 90–100°C (the pH was in the range of 11–12), and finally washed by DI water and dried at 60°C overnight.[20] The resulting samples were first calcined at 350°C under N$_2$ atmosphere for 3 h to stabilize the mesostructure network and then calcined at 700°C in air for 2 h to remove the organic template and improve crystallinity.[20] After cooling down to 200°C, it was deaerated under inert gas flow (N$_2$) for 0.5 h, and then hydrogenated in a H$_2$ flow at 500°C for 3 h with a constant heating rate of 5°C min^{-1}.[20] The mesoporous black TiO$_2$ maintained the well-defined anatase phase after hydrogenation, and a surface-disordered thin layer on a crystalline core was clearly observed by HRTEM.[20]

Lu *et al.* reported the hydrogenated black anatase TiO$_2$ nanotube arrays by anodic oxidation and subsequent thermal treatments.[30] Typically, titanium foils with a thickness of 0.25 mm were cleaned by sonication in ethanol for 5 min, rinsed with DI water, and finally dried in an oven at 100°C for 10 min.[30] The cleaned Ti foil was anodized in the electrolyte of ethylene glycol (99.8) dissolved with 0.5 wt.% ammonium fluoride (99+%) and 4–6 vol% DI water using platinum foil as the counter electrode at a constant potential of 60 V at room temperature for 0.5 h.[30] The as-prepared TiO$_2$ nanomaterials were annealed in air in an oven at 500°C for 3 h to fully remove the organic residues.[30] Hydrogenation was carried out in a tube furnace at 450°C for 1 h with a ramping rate of 10°C min^{-1} under a 5% H$_2$/Ar flow (100 sccm), and the samples were naturally cooled down to room temperature under the same atmosphere.[30] Black anatase TiO$_2$ nanotube arrays were obtained after hydrogenation.[30] Interestingly, an obvious amorphous layer with a thickness of ~2 nm on the surface of the pristine TiO$_2$ nanotube was observed, whereas the black nanotube had no typical surface-disordered thin layer.[30] Lu *et al.* also prepared black rutile TiO$_2$ nanorod arrays. Pristine rutile TiO$_2$ nanorod arrays were prepared according to the preparation process in Ref. [27].[30] After hydrogenation at 450°C for 1 h with a ramping rate of 10°C min^{-1} under a 5% H$_2$/Ar flow (100 sccm), black rutile

Table 2.2. Key parameters of black TiO$_2$ by atmospheric-pressure hydrogenation.

Starting TiO$_2$	Reaction temperature	Reaction time	Atmosphere	Color	Surface disorder	Ref.
Anatase/rutile nanoparticles	500°C	1 h	H$_2$ flow	Black	✓	17
Mesoporous anatase	500°C	3 h	H$_2$ flow	Black	✓	20
Anatase nanotubes and nanorods	450°C	1 h	5% H$_2$/ Ar flow	Black	✗	30

TiO$_2$ nanorod arrays were obtained.[30] Table 2.2 summarizes the key features of the black TiO$_2$ nanomaterials by atmospheric-pressure hydrogenation.

Atmospheric-pressure hydrogenation was less used than high-pressure hydrogenation. Nevertheless, surface disorder was more likely to form during atmospheric-pressure hydrogenation, though the hydrogenation temperature was very high (e.g. 500°C).

2.2.3 *Hydrogen Plasma Treatment*

Hydrogen plasma technology has attracted increasing interest owing to its effectiveness in engineering surface-disordered TiO$_2$ nanomaterials.[21, 31–33] Wang *et al.* reported H-doped black TiO$_2$ nanoparticles by localized surface plasmon resonance.[21] Typically, the hydrogenation of TiO$_2$ (Degussa P25) was performed in a thermal plasma furnace by hydrogen plasma for 8 h at 500°C with the plasma input power being 200 W.[21] As shown in Figure 2.5, the hydrogen-plasma-reduced black TiO$_2$ nanoparticle feathers a typical crystalline-amorphous core–shell structure.

Yan *et al.* investigated the effect of reaction time on the coloration and surface disorder of hydrogen-plasma-reduced TiO$_2$ nanoparticles.[31] In a typical synthesis, 0.1 g of commercial TiO$_2$ nanoparticles (Degussa P25) were dispersed in 50 mL ethanol with ultrasonication for 10 min, and then the suspension was drop-cast onto a 6-inch Si wafer.[31] The drop casting process was repeated

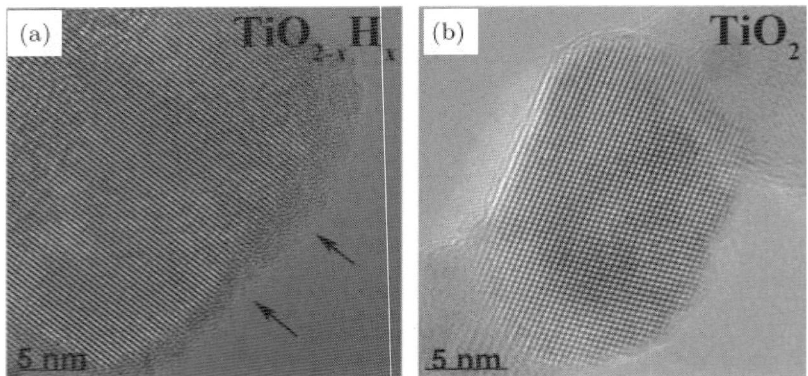

Figure 2.5. HRTEM images of (a) black $TiO_{2-x}H_x$ and (b) pristine TiO_2 nanoparticles.[21] Reprinted from Ref. [21]. © Wiley-VCH.

several times to achieve a TiO_2 mass loading of 0.5–0.6 mg cm^{-2}.[31] Then, the wafer covered with TiO_2 was transferred into a chamber for plasma-enhanced hydrogenation treatment, where an inductively coupled plasma (ICP) was used.[31] The ICP power was 3000 W, the chamber pressure was 25.8–27.1 mTorr, and the hydrogen flow rate was 50 sccm.[31] The hydrogen plasma treatment was performed at 150°C for 30 s, 1 min, 3 min, 5 min, and 20 min to obtain H–TiO_2-30s, H–TiO_2-1min, H–TiO_2-3min, H–TiO_2-5min, and H–TiO_2-20min, respectively.[31] As shown in Figure 2.6, with increasing the reaction time, the TiO_2 nanoparticles turned from white to gray then to black, and as well the surface-disordered domain increased obviously in thickness. No phase transformation occurred in all these hydrogenated TiO_2 nanomaterials.[31]

Teng *et al.* reported on the synthesis of black TiO_2 via hydrogen plasma assisted by hot-filament chemical vapor deposition (HFCVD).[33] In a typical synthesis, 2 g of commercial TiO_2 nanoparticles (Degussa P25) were dispersed in 50 mL of 10 M NaOH solution in a 100 mL beaker.[33] The suspension was transferred into a Teflon-lined stainless steel autoclave, and then hydrothermal reaction was performed in an oven at 120°C for 12 h.[33] The resultant precipitate was collected, washed with 1 M HCl aqueous solution and then water until the pH value of the washing solution reached

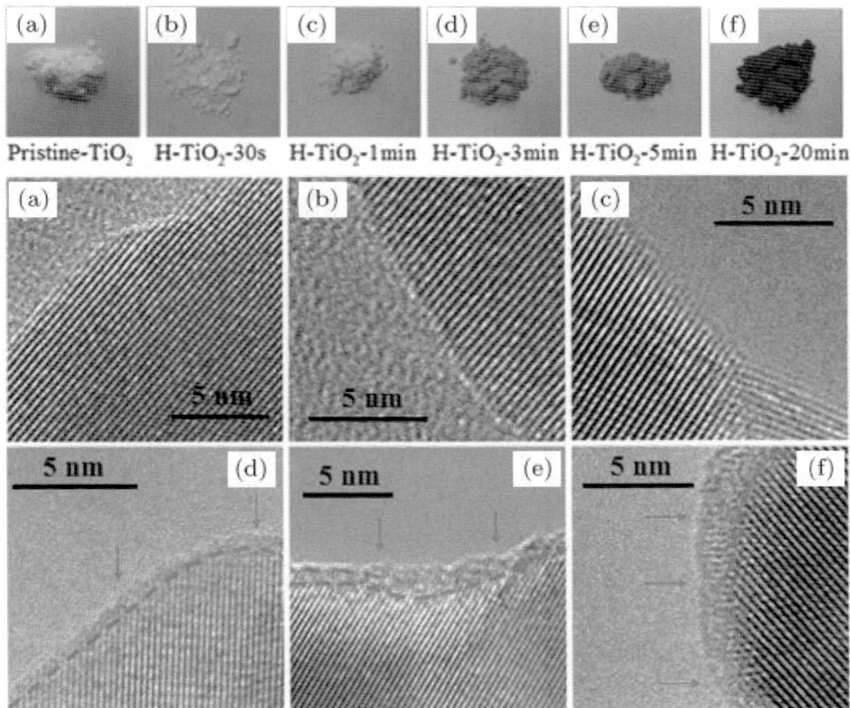

Figure 2.6. Digital and HRTEM images of the pristine TiO$_2$ and hydrogenated TiO$_2$ by hydrogen plasma treatment with different reaction time: (a) pristine TiO$_2$; (b) H–TiO$_2$-30s; (c) H–TiO$_2$-1min; (d) H–TiO$_2$-3min; (e) H–TiO$_2$-5min; (f) H–TiO$_2$-20min.[31] Reprinted from Ref. [31]. © The Royal Society of Chemistry.

6–7, and dried at 110°C.[33] The hydrogenation was performed in an HFCVD apparatus under a hydrogen flow,[33] as schematically shown in Figure 2.7. The temperature of the filament was maintained at 2000°C, and the black TiO$_2$ was achieved through hydrogenation at 350°C or 500°C for 3 h.[33] The black TiO$_2$ nanoparticles had a disorder amorphous layer surrounding a crystalline core to form an amorphous shell/crystalline core structure.[33]

Table 2.3 summarizes the key features of the black TiO$_2$ nanomaterials by hydrogen plasma treatment. In summary, hydrogen plasma treatment is a powerful tool that can synthesize black TiO$_2$ nanomaterials with a typical crystalline-amorphous core–shell

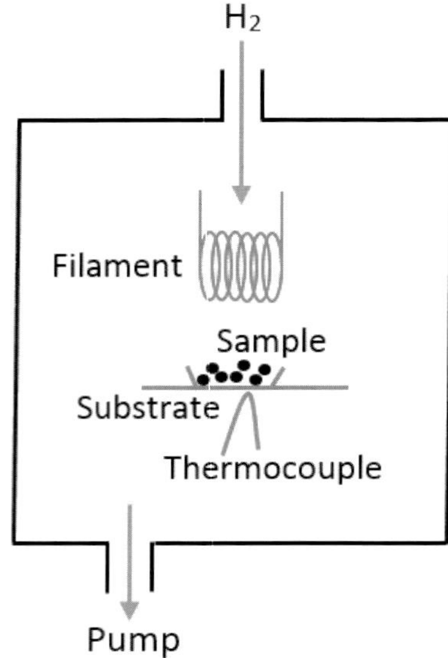

Figure 2.7. Schematic illustration of synthesis process by the hydrogen plasma assisted by HFCVD.

Table 2.3. Key parameters of black TiO$_2$ by hydrogen plasma treatment.

Starting TiO$_2$	Reaction temperature	Reaction time	ICP power	Color	Surface disorder	Ref.
Degussa P25	500°C	8 h	200 W	Black	√	21
Degussa P25	150°C	5 min	3000 W	Gray	√	31
Degussa P25	150°C	20 min	3000 W	Black	√	31, 32
Degussa P25	350°C or 500°C	3 h	N/S	Black	√	33

structure. More importantly, the thickness of the amorphous shell is tunable through controlling the reaction parameters.

2.3 Properties of Black TiO$_2$ Nanomaterials

2.3.1 *Surface Structure and Chemical Bonding*

2.3.1.1 Surface disorder

Surface disorder on hydrogenated black TiO$_2$ nanomaterials is highly dependent on the reaction parameters, such as hydrogenation temperature, time, and pressure. Also, the morphology or crystal growth direction may affect the surface disorder. For example, surface-disordered thin layer was not observed, so far, in hydrogenated one-dimensional TiO$_2$ nanomaterials.[26–28,30]

Surface-disordered thin layer was not commonly observed in black TiO$_2$ nanomaterials synthesized by common hydrogenation treatment. Only in a few cases was the surface disorder observed. Chen *et al.*[18] and Lu *et al.*[24] clearly revealed the surface-disordered thin layer on the hydrogenated black TiO$_2$ nanoparticle from high-pressure hydrogenation. As shown in Figure 2.8, HRTEM and

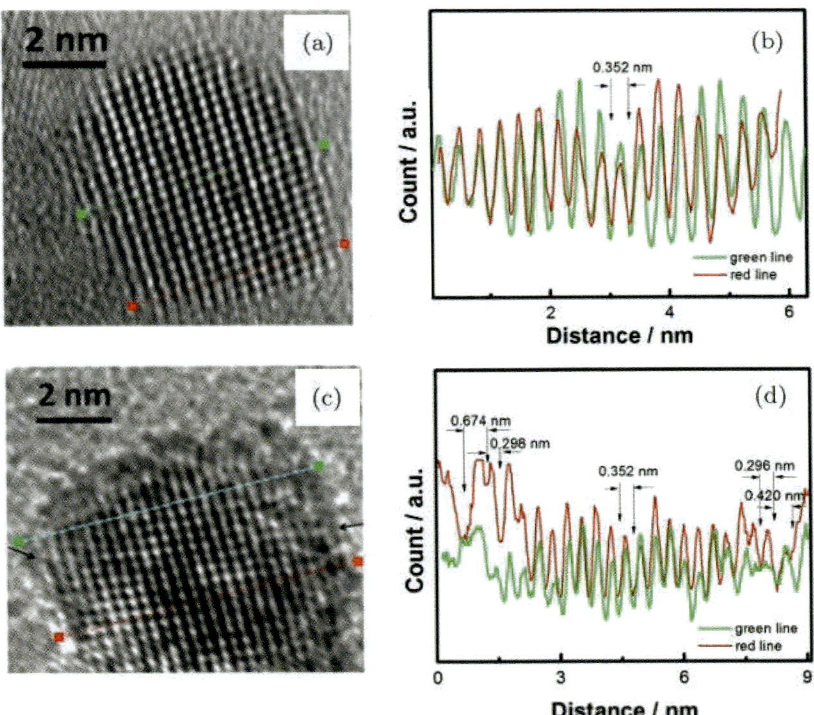

Figure 2.8. HRTEM (a) and line analyses (b) of one white TiO$_2$ nanoparticle, HRTEM (c) and line analyses (d) of one black TiO$_2$ nanoparticle. The zeros of the axis in (b) and (d) correspond to the left ends of the lines in (a) and (c). The red and green curves in (b) and (d) correspond to the red and green lines in (a) and (c).[34] Reprinted from Ref. [34]. © Nature Publishing Group.

line analyses were employed to comparatively study the structures of the pristine white TiO$_2$ nanoparticle and the hydrogenated black TiO$_2$ nanoparticle.[34] The pristine white TiO$_2$ nanoparticle displayed clearly-resolved and well-defined lattice fringes throughout the whole nanocrystal (Figure 2.8(a)), and the distance between the adjacent lattice planes was typical for anatase (0.352 nm) (Figure 2.8(b)).[34] In sharp contrast, the black TiO2 nanoparticle had a crystalline-disordered core–shell structure (Figure 2.8(c)), and a structural deviation from the standard crystalline anatase was readily seen at the outer layer, where the straight lattice line was bent at the edge of the nanoparticle and the plane distance was no longer uniform (Figure 2.8(d)).[34] Naldoni *et al.* showed that only black TiO$_2$ nanoparticle from atmospheric-pressure hydrogenation presented a unique core–shell morphology characterized by a disordered surface layer with a thickness of $\sim$1.5 nm.[17] Zhou *et al.* also observed a thin disordered surface layer coating on a crystalline core of black mesoporous TiO$_2$ prepared by atmospheric-pressure hydrogenation.[20]

Hydrogen plasma treatment is a more powerful tool that can synthesize black TiO$_2$ nanomaterials with a typical crystalline-amorphous core–shell structure. So far, all the black TiO$_2$ nanomaterials fabricated by hydrogen plasma treatment were found to possess a typical crystalline-amorphous core–shell structure.[21,31–33] For example, the HRTEM image clearly showed the crystalline-amorphous core–shell structure of the black TiO$_2$ nanoparticle (Figure 2.5(a)).[21]

2.3.1.2   Oxygen vacancy

Oxygen vacancies are the dominant defects in many transition metal oxides and play a crucial role in the physical and chemical properties of transition metal oxides.[35,36] Oxygen vacancies are of significant importance in the photocatalytic process of TiO$_2$. For example, Car–Parrinello molecular dynamic simulations on the rutile (110) surface showed that the water dissociation occurred only in the presence of surface oxygen vacancies.[37] The mechanism of the formation of oxygen vacancies by hydrogen reduction at elevated temperatures has been investigated long before the discovery of black TiO$_2$. In

1993, Khader *et al.* proposed that the hydrogen reduction of rutile TiO$_2$ could lead to the oxygen loss and the formation of oxygen vacancy[38]:

$$\text{TiO}_2 + x\text{H}_2(\text{g}) \rightarrow \text{TiO}_{2-x} + x\text{H}_2\text{O}(\text{g}).$$

This process involved three steps: (1) hydrogen adsorption on the defects; (2) O^{2-} ion reduction on the surface; and (3) oxygen diffusion outward.[38] The newly generated oxygen vacancies were the active sites for H$_2$ adsorption.[38] In 1997, Rekoske *et al.* found that the initial reduction of anatase and rutile TiO$_2$ with hydrogen was slow due to the limited number of available surface H atoms and then was accelerated when more sites (i.e. reduced Ti sites) capable of H$_2$ dissociation were produced.[39]

Electron spin resonance (ESR) technique is commonly used to detect the oxygen vacancy.[28,33] For example, Figure 2.9 shows the ESR spectrum of the black TiO$_2$ synthesized in Ar or H$_2$/Ar showed a pronounced oxygen vacancy signal ($g = 2.002$).[28] The oxygen

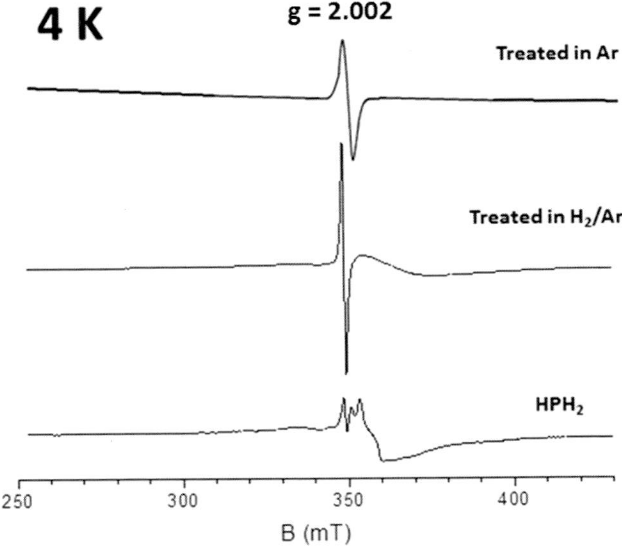

Figure 2.9. ERS spectrum of the various TiO$_2$ synthesized at different conditions.[28] Reprinted from Ref. [28]. © American Chemical Society.

vacancy signal is generally located at around $g = 2.003$ in the ESR spectrum.[28,33]

2.3.1.3   Ti–OH

Ti–OH groups are commonly found on the surfaces of TiO$_2$ nanocrystals. Owing to the partial reduction of TiO$_2$ in hydrogen atmosphere, hydrogenated TiO$_2$ nanomaterials will theoretically bear more −OH groups. X-ray photoelectron spectroscope (XPS) is frequently employed to detect the surface −OH groups.[16–18,20,21,24,26,27] As an example shown in Figure 2.10(a), the spectrum of hydrogenated black TiO$_2$ nanocrystals shows an additional peak centered at ∼530.9 eV, which is attributed to the Ti–OH species.[18] Fourier transform infrared (FTIR) spectroscopy is another important technique to reveal the change in surface −OH groups by comparing the magnitude of the intensity of the peak corresponding to the −OH vibrational band.[21,24,34,40] Wang *et al.* found that hydrogenation led to another new absorbance at $3710 \, \text{cm}^{-1}$ in the FTIR spectrum

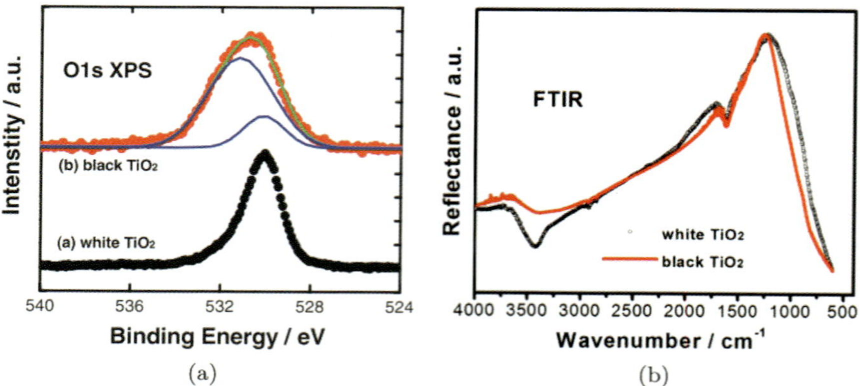

Figure 2.10. (a) O 1s XPS spectra of the white and black TiO$_2$ nanocrystals. The red and black circles are XPS data.[18] The green curve is the fitting of experimental data for black TiO$_2$ nanocrystals, which can be decomposed into a superposition of two peaks shown as blue curves.[18] Reproduced from Ref. [18]. © The American Association for the Advancement of Science, 2011. (b) FTIR reflectance spectra of hydrogenated black TiO$_2$ nanocrystals compared with white TiO$_2$ nanocrystals.[34] Reprinted from Ref. [34]. © Nature Publishing Group.

corresponding to terminal OH groups,[21] whereas hydrogenated black TiO$_2$ nanocrystals by Chen *et al.* showed reduced terminal OH groups (Figure 2.10(b)) as evident from the weakened peaks centered at 3400 cm^{-1} and 3700 cm^{-1}.[34] Lu *et al.* also found that the intensity of the OH peak (*ca.* 3400 cm^{-1}) in the FTIR spectrum of black TiO$_2$ was much lower than that of the pristine TiO$_2$.[24] Therefore, hydrogenation will not necessarily bring more –OH groups in the black TiO$_2$ nanomaterials.

2.3.1.4   Ti–H

A few research groups have found the Ti–H groups in the black TiO$_2$ by ^{1}H NMR measurement.[21,28,34] Figure 2.11 shows the typical ^{1}H magic angle NMR spectra of hydrogenated black TiO$_2$ nanocrystals compared with white TiO$_2$ nanocrystals.[34] Both the black and white TiO$_2$ show a large peak at chemical shift of 15.7 ppm. The slightly larger linewidth in black TiO$_2$ may be caused by the incorporation of H at bridging sites at the disordered phase produced during the hydrogenation process, or may be due to the bridging sites located on different crystallographic planes on the surface.[34] Similar ^{1}H magic

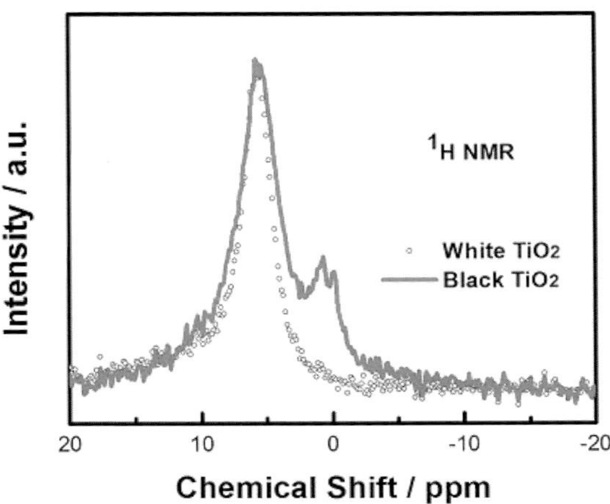

Figure 2.11. ^{1}H magic angle NMR spectra of hydrogenated black TiO$_2$ nanocrystals compared with white TiO$_2$ nanocrystals.[34] Reprinted from Ref. [34]. © Nature Publishing Group.

angle NMR spectra of hydrogenated black TiO$_2$ nanocrystals were obtained by Wang *et al.*[21] and Liu *et al.*[28] Wang *et al.* also observed a broad peak centered at 457.1 eV in the XPS spectrum of black TiO$_2$, which was ascribed to the surface Ti–H groups.[21]

2.3.1.5 Ti^{3+}

Ti^{3+} sites were found only in a few cases.[17,27,28,31] Based on the XPS measurements, Eom *et al.*[27] and Yan *et al.*[31] found the Ti^{3+} ions in the black TiO$_2$. As an example shown in Figure 2.12, the XPS spectrum of black TiO$_{2-x}$ experiences a right shift. The peaks centered at $\sim$458.6 eV and $\sim$464.3 eV were ascribed to the Ti 2p$_{3/2}$ and Ti 2p$_{1/2}$ peaks of Ti^{3+}.[27] However, the XPS technique failed to detect the Ti^{3+} ions in a few papers.[16,30,34] Naldoni *et al.* found the Ti^{3+} ions in the black TiO$_2$ nanoparticles using synchrotron X-ray powder diffraction and cathodoluminescence techniques.[17] Liu *et al.* adopted ESR measurement to detect the Ti^{3+}, and Ti^{3+} ions were found in the black TiO$_2$ nanotubes prepared in H$_2$/Ar atmosphere.[28] Some authors argued that any Ti^{3+} which may have been present at the particle surface would have been readily oxidized on exposure of the black TiO$_2$ to air.[16] Therefore, the existence of Ti^{3+} ions is still under debate, and the current evidence may not be enough as trace

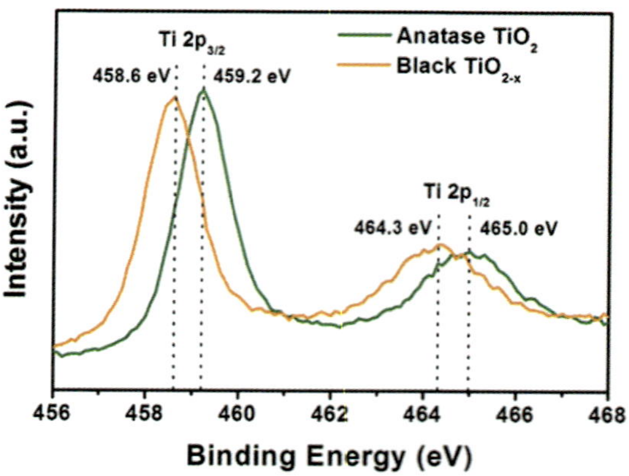

Figure 2.12. XPS spectra of the pristine TiO$_2$ and hydrogenated black TiO$_{2-x}$.[27] Reprinted with permission from Ref. [27]. © The Royal Society of Chemistry.

amount of Ti$_2$O$_3$ may be generated during the hydrogen reduction of TiO$_2$. For example, even a small amount of Ti$_2$O$_3$ was found in the black TiO$_2$ nanoparticles prepared at room temperature.[24]

2.3.2 Electronic Properties

Chen *et al.* found that the pristine white TiO$_2$ nanocrystals displayed typical VB density of states (DOS) characteristics of TiO$_2$, with the edge of the maximum energy at about 1.26 eV from the VB XPS spectrum (Figure 2.13(a)).[18] For the black TiO$_2$ nanocrystals, the VB maximum energy blue-shifted toward the vacuum level at approximately -0.92 eV (Figure 2.13(a)).[18] Conduction band tail states arising from the surface disorder of black TiO$_2$ were presumably proposed.[18] Further, Chen *et al.* revealed that the Ti^{3+} did not contribute to the additional VB-edge states in black TiO$_2$ as the appearance of Ti^{3+} was accompanied with the disappearance of the additional VB-edge states.[34] Zhou *et al.* considered that the bandgap narrowing from 3.15 eV to 2.82 eV was ascribed to the formation of Ti^{3+} in the surface disordered layer.[20] Naldoni *et al.* reported that VB XPS of black TiO$_2$ showed a main absorption onset located at 0.6 eV, whereas the maximum energy associated with

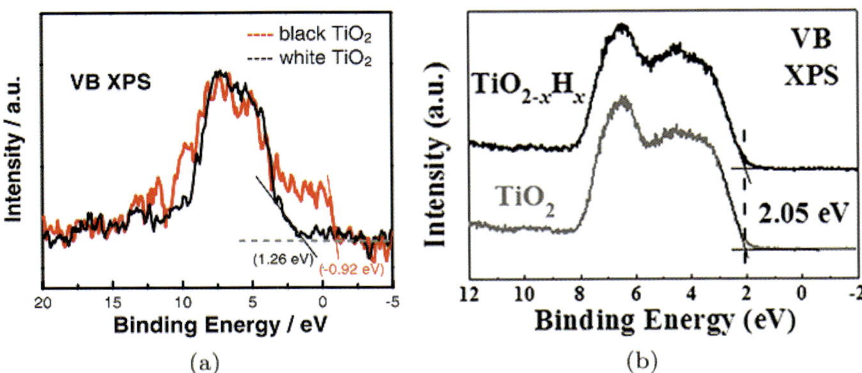

Figure 2.13. (a) VB XPS spectra of the white and black TiO$_2$ nanocrystals.[18] Reproduced from Ref. [18]. © The American Association for the Advancement of Science, 2011. (b) VB XPS spectra of pristine TiO$_2$ and black TiO$_{2-x}$H$_x$.[21] Reprinted from Ref. [21]. © Wiley-VCH.

the band tail blue shifted further toward the vacuum level at about -0.3 eV.[17] And the remarkable bandgap narrowing (a bandgap of $\sim$1.85 eV) was attributed to the synergistic effect of oxygen vacancies and surface disorder.[17] Lu *et al.* also found that oxygen vacancies could narrow the bandgap through VB-edge shifting from 0.11 eV to 0.25 eV after hydrogenation treatment.[30] Teng *et al.* thought that the bandgap narrowing of black TiO$_2$ was attributed to the synergistic effect of oxygen vacancies and Ti–H bonds.[17]

On the other hand, Wang *et al.* found that the VB XPS spectra of pristine TiO$_2$ and black TiO$_{2-x}$H$_x$ were similar (Figure 2.13(b)), indicating that hydrogen plasma treatment had negligible effect on the VB position of the surface disorder.[21] Leshuk *et al.*[16] and Wang *et al.*[26] also did not observe the VB change for black TiO$_2$ using VB XPS measurement.

2.3.3 *Optical Properties*

The large bandgap of TiO$_2$ restricts its light absorption to the UV light, which only accounts for $\sim$5% of the solar light. This explains the low efficiency of TiO$_2$ nanomaterials in photocatalytic water splitting and environmental pollution removal. Therefore, bandgap engineering aiming to narrow the bandgap is now of high interest. Introduction of surface defects to TiO$_2$ nanomaterials by hydrogenation treatment is one of the most efficient approaches to engineer the VB and thus narrow the bandgap.[17,18,20,30,34] The narrowed bandgap extended the light absorption of TiO$_2$ to visible light. For example, diffusive reflectance and absorbance spectroscopy (Figure 2.14(a)) revealed that the light absorption of the unmodified white TiO$_2$ nanocrystals was limited to UV light, whereas the onset of optical absorption of the hydrogenated black TiO$_2$ nanocrystals was $\sim$1200 nm.[18] However, there are exceptions. For instance, Wang *et al.* found that the black TiO$_{2-x}$H$_x$ possessed a remarkably high solar absorption of $\sim$83% (Figure 2.14(b)), though its VB was not narrowed by hydrogen plasma treatment.[21] The H-doped amorphous shell was proposed to play the same role as Ag or Pt loaded on TiO$_2$ nanocrystals.[21]

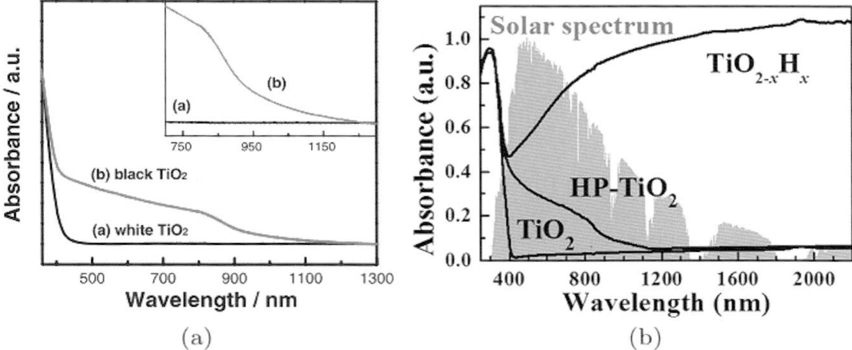

Figure 2.14. (a) Spectral absorbance of the white and black TiO$_2$ nanocrystals. The inset enlarges the absorption spectrum in the range from approximately 750 nm to 1200 nm.[18] Reproduced from Ref. [18]. © The American Association for the Advancement of Science, 2011. (b) Diffuse reflectance spectra of hydrogen plasma treated black TiO$_{2-x}$H$_x$, the high-pressure hydrogenated black titania (HP–TiO$_2$), and pristine TiO$_2$. The background is the total solar spectrum.[21] Reprinted from Ref. [21]. ©Wiley-VCH.

2.4 Summary and Prospective

In conclusion, hydrogenation including hydrogen plasma treatment is an effective approach to achieve black TiO$_2$. However, reaction parameters (e.g. the crystalline structure and morphology of pristine TiO$_2$, hydrogenation temperature and time, and hydrogen pressure) highly influence the surface structure and colorization of hydrogenated TiO$_2$. Low temperature (e.g. $\leq$200°C), long reaction time (e.g. $\geq$5 day), and high pressure (e.g. $\geq$20 bar) may favor the generation of surface-disordered thin layer; nevertheless, high temperature (e.g. 500°C) still has a chance to produce the surface-disordered thin layer under atmospheric-pressure hydrogenation. Hydrogen plasma treatment can controllably synthesize the crystalline-amorphous core–shell TiO$_2$ nanostructure with tunable surface amorphous layer. On the other hand, hydrogenation will induce changes in surface chemical bonding. For instance, oxygen vacancies are intrinsic defects for TiO$_2$ and tend to be enhanced after hydrogenation treatment; the increase of surface −OH groups in hydrogenated black TiO$_2$ is frequently reported; Ti–H bonds and Ti^{3+} interstitials are not commonly detected and are still under debate. ESR, HRTEM, XPS, FTIR, ^{1}H NMR, CL techniques can be used to detect the surface structure, defects, and chemical bonding. Often, the surface defects of hydrogenated black TiO$_2$ are a collective of different types of defects, making it difficult to investigate the role of each single type of defect. Therefore, the inner workings of the hydrogenated black TiO$_2$ in enhancing its photocatalysis are still under debate.

Another important consideration of hydrogenated black TiO$_2$ is its narrowed bandgap, which is believed to extend its light absorption from UV to visible light. These outstanding electronic and optical features are desired to exploit the applications of TiO$_2$ in different fields. It should, however, be noted that there are still arguments on what is the key factor leading to the bandgap narrowing. In addition, bandgap narrowing phenomena are only limited to a few cases. Therefore, understanding the bandgap narrowing in the aspects of materials science and engineering is highly important for the future well-controlled synthesis of bandgap-narrowed black TiO$_2$ and its practical applications.

Acknowledgments

X. C. thanks the support from the College of Arts and Sciences, the University of Missouri–Kansas City and the University of Missouri Research Board. L. T. thanks the National Natural Science Foundation of China (No. 51302072) and China Scholarship Council for their financial supports.

References

(1) Chen, X.; Mao, S. S. *Chem. Rev.* **2007**, *107*, 2891–2959.
(2) Yan, X.; Wang, Z.; He, M.; Hou, Z.; Xia, T.; Liu, G., Chen, X. *Energy Technol.* **2015**, *3*, 801–814.
(3) Pelaez, M.; Nolan, N. T.; Pillai, S. C.; Seery, M. K.; Falaras, P.; Kontos, A. G.; Dunlop, P. S. M.; Hamilton, J. W. J.; Byrne, J. A. *Appl. Catal. B: Environ.* **2012**, *125*, 331–349.
(4) Wu, S.; Weng, Z.; Liu, X.; Yeung, K. W. K.; Chu, P. K. *Adv. Funct. Mater.* **2014**, *24*, 5464–5481.

(5) Burda, C.; Chen, X.; Narayanan, R.; El-Sayed, M. A. *Chem. Rev.* **2005**, *105*, 1025–1102.

(6) Chen, X.; Li, C.; Grätzel, M.; Kostecki, R.; Mao, S. S. *Chem. Soc. Rev.* **2012**, *41*, 7909–7937.

(7) Chen, X.; Shen, S.; Guo, L.; Mao, S. S. *Chem. Rev.* **2010**, *110*, 6503–6570.

(8) Wang, X.; Li, Z.; Shi, J.; Yu, Y. *Chem. Rev.* **2014**, *119*, 9346–9384.

(9) Liu, G.; Yang, H. G.; Pan, J.; Yang, Y. Q.; Lu, G. Q.; Cheng, H.-M. *Chem. Rev.* **2014**, *114*, 9559–9612.

(10) Fattakhova-Rohlfing, D.; Zaleska, A.; Bein, T. *Chem. Rev.* **2014**, *114*, 9487–9558.

(11) Lee, K.; Mazare, A.; Schmuki, P. *Chem. Rev.* **2014**, *114*, 9385–9454.

(12) Liu, L.; Yu, P. P.; Chen, X.; Mao, S. S.; Shen, D. Z. *Phys. Rev. Lett.* **2013**, *111*, 065505.

(13) Habisreutinger, S. N.; Schmidt-Mende, L.; Stolarczyk, J. K. *Angew. Chem. Int. Ed.* **2013**, *52*, 7372–7408.

(14) Schneider, J.; Matsuoka, M.; Takeuchi, M.; Zhang, J.; Horiuchi, Y.; Anpo, M.; Bahnemann, D. W. *Chem. Rev.* **2014**, *114*, 9919–9986.

(15) Liu, L.; Chen, X. *Chem. Rev.* **2014**, *114*, 9890–9918.

(16) Leshuk, T.; Parviz, R.; Everett, P.; Krishnakumar, H.; Varin, R. A.; Gu, F. *ACS Appl. Mater. Interfaces* **2013**, *5*, 1892–1895.

(17) Naldoni, A.; Allieta, M.; Santangelo, S.; Marelli, M.; Fabbri, F.; Cappelli, S.; Bianchi, C. L.; Psaro, R.; Dal Santo, V. *J. Am. Chem. Soc.* **2012**, *134*, 7600–7603.

(18) Chen, X.; Liu, L.; Yu, P. Y.; Mao, S. S. *Science* **2011**, *331*, 746–750.

(19) Chen, X.; Liu, L.; Huang, F. *Chem. Soc. Rev.* **2014**, *44*, 1861–1885.

(20) Zhou, W.; Li, W.; Wang, J.-Q.; Qu, Y.; Yang, Y.; Xie, Y.; Zhang, K.; Wang, L.; Fu, H.; Zhao, D. *J. Am. Chem. Soc.* **2014**, *136*, 9280–9283.

(21) Wang, Z.; Yang, C.; Lin, T.; Yin, H.; Chen, P.; Wan, D.; Xu, F.; Huang, F.; Lin, J.; Xie, X.; Jiang, M. *Adv. Funct. Mater.* **2013**, *23*, 5444–5450.

(22) Xia, T.; Chen, X. *J. Mater. Chem. A* **2013**, *1*, 2983–2989.

(23) Sun, C.; Jia, Y.; Yang, X.-H.; Yang, H.-G.; Yao, X.; (Max) Lu G. Q.; Selloni, A.; Smith, S. C. *J. Phys. Chem. C* **2011**, *115*, 25590–25594.

(24) Lu, H.; Zhao, B.; Pan, R.; Yao, J.; Qiu, J.; Luo, L.; Liu, Y. *RSC Adv.* **2014**, *4*, 1128–1132.

(25) Leshuk, T.; Linley, S.; Gu, F. *Can. J. Chem. Eng.* **2013**, *91*, 799–807.

(26) Wang, G.; Wang, H.; Ling, Y.; Tang, Y.; Yang, X.; Fitzmorris, R. C.; Wang, C.; Zhang, J. Z.; Li, Y. *Nano Lett.* **2011**, *11*, 3026–3033.

(27) Eom, J.-Y.; Lim, S.-J.; Lee, S.-M.; Ryu, W.-H.; Kwon, H.-S. *J. Mater. Chem. A* **2015**, *3*, 11183–11188.

(28) Liu, N.; Schneider, C.; Freitag, D.; Hartmann, M.; Venkatesan, U.; Muller, J.; Spiecker, E.; Schmuki, P. *Nano Lett.* **2014**, *14*, 3309–3313.

(29) Salari, M.; Konstantinov, K.; Liu, H. K. *J. Mater. Chem.* **2011**, *21*, 5128–5133.

(30) Lu, Z.; Yip, C.-T., Wang, L.; Huang, H.; Zhou, L. *ChemPlusChem* **2012**, *77*, 991–1000.

(31) Yan, Y.; Han, M.; Konkin, A.; Koppe, T.; Wang, D.; Andreu, T.; Chen, G.; Vetter, U.; Morante, J. R.; Schaaf, P. *J. Mater. Chem. A* **2014**, *2*, 12708–12716.

(32) Ren, W.; Yan, Y.; Zeng, L.; Shi, Z.; Gong, A.; Schaaf, P.; Wang, D.; Zhao, J.; Zou, B.; Yu, H.; Chen, G.; Brown, E. M. B.; Wu, A. *Adv. Healthcare Mater.* **2015**, *4*, 1526–1536.

(33) Teng, F.; Li, M.; Gao, C.; Zhang, G.; Zhang, P.; Wang, Y.; Chen, L.; Xie, E. *Appl. Catal. B* **2014**, *148–149*, 339–343.

(34) Chen, X.; Liu, L.; Liu, Z.; Marcus, M. A.; Wang, W.-C.; Oyler, N. A.; Grass, M. E.; Mao, B.; Glans, P.-A.; Yu, P. Y.; Guo, J.; Mao, S. S. *Sci. Rep.* **2013**, *3*, 1510.

(35) Schaub, R.; Wahlström, E.; Rønnau, A.; Lægsgaard, E.; Stensgaard, I.; Besenbacher, F. *Science* **2003**, *299*, 377–379.

(36) Pacchioni, G. *ChemPhysChem* **2003**, *4*, 1041–1047.

(37) Tilocca, A.; Selloni, A. *J. Phys. Chem. B* **2004**, *108*, 4743–4751.

(38) Khader, M. M.; Kheiri, F. M.-N.; El-Anadouli, B. E.; Ateya, B. G. *J. Phys. Chem.* **1993**, *97*, 6074–6077.

(39) Rekoske, J. E.; Barteau, M. A. *J. Phys. Chem. B* **1997**, *101*, 1113–1124.

(40) Xia, T.; Zhang, C.; Oyler, N. A.; Chen, X. *Adv. Mater.* **2013**, *25*, 6905–6910.

CHAPTER THREE

Black TiO$_2$ Nanomaterials Through Electrochemical and Mechanical Methods

Haidong Bian[*,¶], *Chris Lee*[*,†,¶], *Hui Li*[*,†], *Jian Lu*[‡]
and Yang Yang Li[*,†,§]

[*]*Center of Super-Diamond and Advanced Films (COSDAF),*
City University of Hong Kong, Kowloon, Hong Kong SAR, China

[†]*Department of Physics and Materials Science,*
City University of Hong Kong, Kowloon, Hong Kong SAR, China

[‡]*Department of Mechanical and Biomedical Engineering,*
City University of Hong Kong, Kowloon, Hong Kong SAR, China

[§]*City University of Hong Kong Shenzhen Research Institute,*
8 Yuexing 1st Road, Shenzhen Hi-Tech Industrial Park,
Nanshan District, Shenzhen, China

3.1 Black TiO$_2$ Nanomaterials Through Electrochemical Method

TiO$_2$, as an important semiconductor for photocatalyst and charge storage material, offers attractive technological advantages, e.g. low toxicity, high stability, and wide commercial availability. It is intensively studied for various applications in energy and environment areas, such as light-induced water splitting devices, lithium-ion batteries, lithium–sulfur batteries, supercapacitors, and solar cells.[1–8] However, the native forms of TiO$_2$ present considerable limitations, particularly a wide electronic bandgap (e.g. 3.0/3.2 eV for rutile/anatase) and a high resistivity.[9,10] The wide bandgap of

---

[¶]These authors contributed equally.

TiO$_2$ limits its light-harvesting ability from solar light that finally restricts its application in photocatalysis and light-induced water splitting. Meanwhile, the high resistivity of TiO$_2$ leads to high possibilities for electrons and holes to recombine, and results in high internal resistance of the charge-storage devices when it is applied as an electro-active material. To address these limitations, focused efforts have been devoted to modify native TiO$_2$.[9–14] As discussed in depth in other chapters of this book, encouraging progress has been achieved using different strategies. For example, a common one is to dope TiO$_2$ with impurity species (e.g. N), e.g. by heating native TiO$_2$ in a dopant-abundant atmosphere (e.g. NH$_3$ or N$_2$).[9,15–17] Interestingly, crystallinity of TiO$_2$ was found also tailorable from thermal treatment and showed a strong impact on the electronic structure of TiO$_2$.[10] Self-doping of TiO$_2$, with intentionally introduced Ti^{III} or oxygen vacancies, usually through annealing TiO$_2$ in a reductive/inert atmosphere or vacuum,[18–21] has proved to be another efficient strategy for modifying the electronic structure of TiO$_2$.

Here in the first session of this chapter, we introduce the solution-based electrochemical method for modifying TiO$_2$, which can be done at ambient temperature and pressure. This method is closely related to the electrochromatic effects of TiO$_2$. It is well known that some semiconductors (such as Ni oxide, W oxide, and Ti oxide) can exhibit reversible color changes under electric field.[22–24] This phenomenon is known as electrochromism. The solution-immersed semiconductor electrode undergoes redox reactions on the electrode surface, involving transportation of electrons and ions, exhibiting changeable light adsorption from UV to near-infrared. Schmuki *et al.* studied the transparency of thin layers of mesoporous anodic TiO$_2$ in the visible and near-infrared region (Figure 3.1), and found that they exhibited high electrochromic contrast for switchable electrochromic devices.[25]

Similarly, under cathodic bias in aqueous electrolyte, a simple electrochemical doping method was found to improve the photocatalytic efficiency and capacitance of pure TiO$_2$. Meekins and Kamat[26] found that cations such as H$^+$ and Li$^+$ can be intercalated into

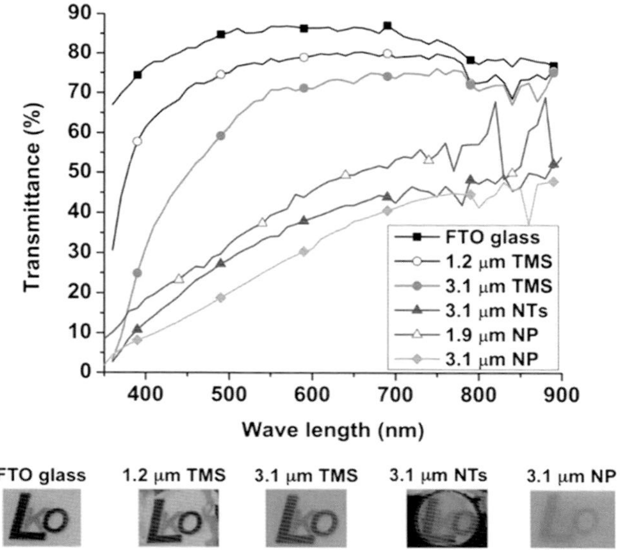

Figure 3.1. Optical transmittance spectra and optical images of TiO_2 mesoporous structures (TMSs), nanotubes (NTs), and nanoparticles (NPs) on FTO glass. Rectangles show optimal images of the different structures after being formed or deposited on FTO and then annealed at 450°C. Reprinted with permission from Ref. [25]. Copyright 2012 The Royal Society of Chemistry.

TiO_2 NT arrays under controlled potentials ($< -1.0\,V$ vs. Ag/AgCl), accompanied by the reduction of Ti^{4+} to Ti^{3+} and a nearly three-fold increase in both photocurrent and photoconversion efficiency (IPCE) (Figure 3.2). A similar method for Li-ion insertion into TiO_2 NT arrays was fulfilled by biasing the electrodes under potential pulses (-1.0 to $-1.7\,V$ vs. saturated calomel electrode, SCE) for 1–11 s in 1 M aqueous $LiClO_4$ solution.[27] Li-ion insertion was found able to enhance the photocurrent generation and the IPCE of TNTs by $\sim$70% and $\sim$2.5 times. Xu *et al.*[28] used electrochemical reductive doping process to produce hydrogenated-TiO_2 photoanodes. At a $-5\,V$ bias voltage, with 10 s of processing time, the photocurrent density was enhanced from $0.29\,mA/cm^2$ to $0.65\,mA/cm^2$, possibly due to the oxygen vacancies induced by the electrochemical process. It was found that the reductive electrochemical doping of TiO_2 significantly improves its performance as photoelectrocatalysts and

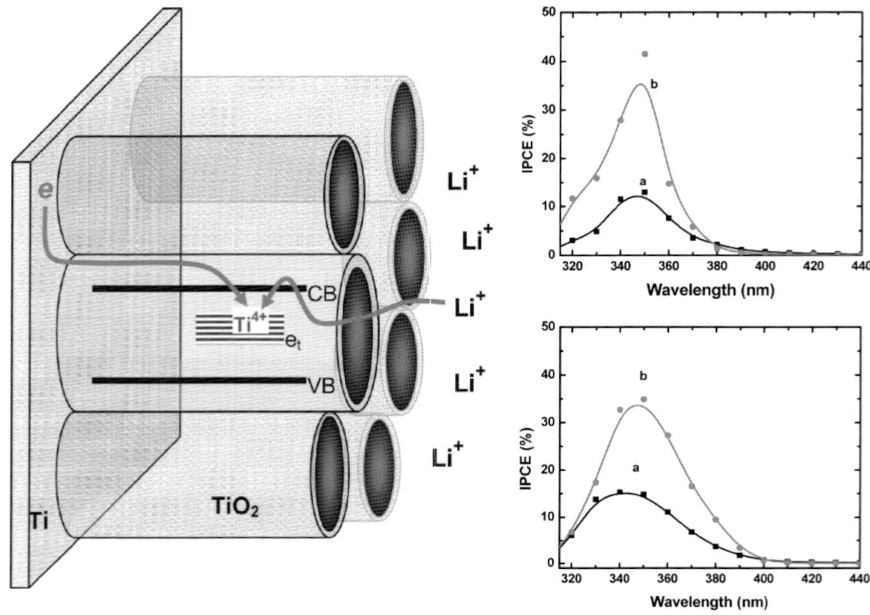

Figure 3.2. (Left) Scheme of electrochemical conversion of Ti^{4+} to Ti^{3+} and simultaneous incorporation of Li$^+$ cations in TiO$_2$ NT array electrode. (Right top) IPCE (%) of the TiO$_2$ NT array electrode: before (a) and after (b) Li$^+$ intercalation. (Right bottom) IPCE (%) of the TiO$_2$ NT electrode before (a) and after (b) H$^+$ intercalation. Electrolyte in all experiments was 1 M KOH. Reprinted with permission from Ref. [26]. Copyright 2009 American Chemical Society.

dye-sensitized solar cells (Figure 3.3).[29] The improvement can be attributed to the easier charge transport with reduced recombination rate and improved electron injection.

Electrochemically self-doped TiO$_2$ NT arrays were investigated for high-performance supercapacitors. Zhang *et al.*[6] demonstrate a novel method for fabricating self-doped TiO$_2$ NT arrays through cathodic polarization of pristine TiO$_2$ to achieve improved conductivity and capacitive properties (Figure 3.4). After polarized at -1.4 V (vs. SCE), the TiO$_2$ sample exhibited 5 orders of magnitude improvement on carrier density and 39 times enhancement in capacitance (1.84 mF cm^{-2} at a scan rate of 5 mV s^{-1}), with 93.1% capacitance retention after 200 cycles. Grimes *et al.*[30] found that the TiO$_2$ NT

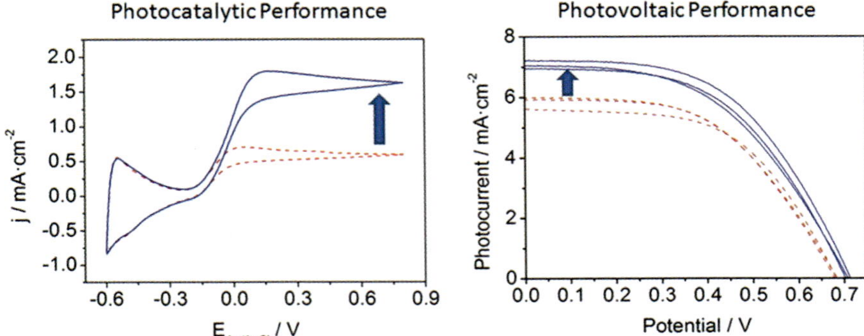

Figure 3.3. (Left) CVs under illumination for undoped (dashed line) and doped (solid lines) samples. (Right) Current–Voltage characteristics under AM 1.5 illumination of DSC test devices based on undoped (dashed line) and doped (solid lines) electrodes. Reprinted with permission from Ref. [29]. Copyright 2013 American Chemical Society.

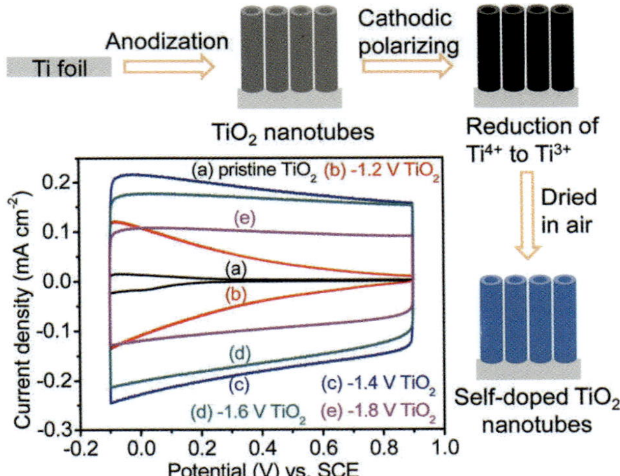

Figure 3.4. Schematic diagram showing the fabrication of electrochemically self-doped TiO$_2$ NT arrays and CV curves of the samples (different potential used during polarization process) obtained at a scan rate of $100\,\mathrm{mV\,s^{-1}}$. Reprinted with permission from Ref. [6]. Copyright 2014 American Chemical Society.

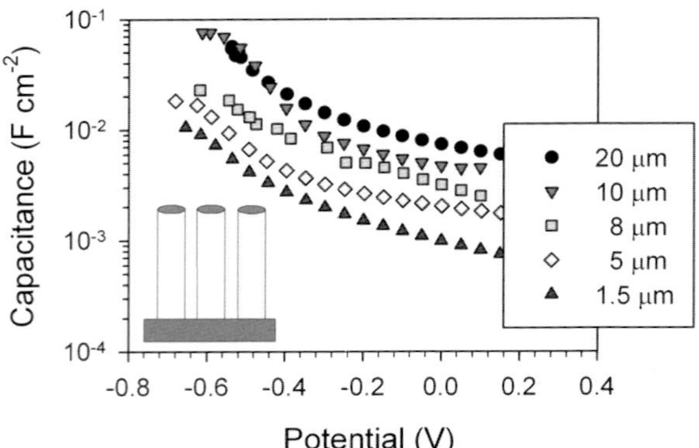

Figure 3.5. The exponential relationship between the bias potential and capacitance of TiO$_2$ NT arrays with different length under 1.5–20 μm. Reprinted with permission from Ref. [30]. Copyright 2008 American Chemical Society.

arrays display the standard behavior of nanoparticulate TiO$_2$ films in the basic media, and their chemical capacitance is exponentially dependent on the bias voltage applied (Figure 3.5). The capacitance improvement was ascribed to the proton intercalation, indicating potential supercapacitor applications. In general, the cathodically treated TiO$_2$ suffers from low stability with the treatment effects fading away quickly upon removal of the bias (e.g. within minutes or hours).[26,31] This low stability is possibly because the cathodic treatments were generally carried out in the low-viscosity aqueous electrolytes, which is in good agreement with a very recent study revealing that the dynamics of ion intercalation/release into/from TiO$_2$ are highly dependent on the electrolyte viscidity.[32]

Recently, Li *et al.*[33] reported a facile electrochemical method to modify anatase TiO$_2$ by cathodically biasing TiO$_2$ in a more viscous ethylene glycol electrolyte, leading to enhanced photocatalytic and electrochemical performance. The doping process is shown in Figure 3.6. The resulting black TiO$_2$ (Figure 3.7) was highly stable with a significantly narrower bandgap and higher electrical conductivity. Furthermore, largely improved photoconversion efficiency (increased from 48% to 72% in the visible region, and from nearly 0% to 7% in the UV region), photocatalytic efficiency, and charge-storage capability (42-fold increase) were achieved in the

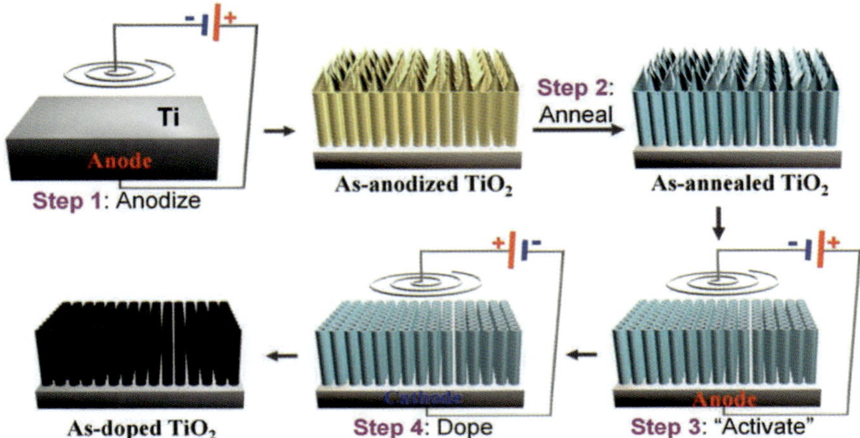

Figure 3.6. Fabrication procedures for generating doped black TiO₂ NTs in ethylene glycol solution. Reprinted with permission from Ref. [33]. Copyright 2014 The Royal Society of Chemistry.

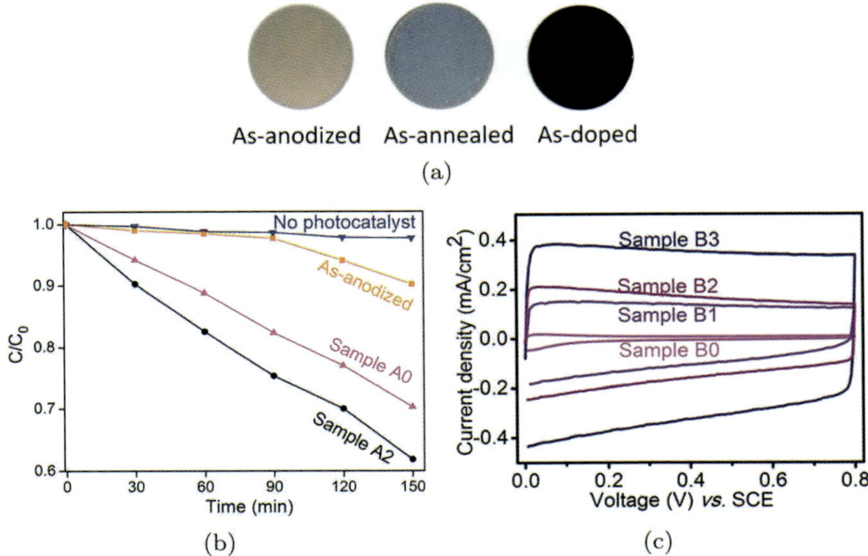

Figure 3.7. (a) Optical photographs of the TiO₂ films obtained after different steps shown in Figure 3.7: "as-anodized" (after step 1), "as-annealed" (after step 2), and "as-doped" (after step 4). (b) Photocatalytic properties of the as-anodized and as-doped samples under UV light illumination. (c) CV curves for the anodized and doped samples at a scan rate of $100\,\mathrm{mV\,s^{-1}}$. Reprinted with permission from Ref. [33]. Copyright 2014 The Royal Society of Chemistry.

modified TiO_2.[33] It is noteworthy that the electrochemical doping methods discussed here can be potentially applied to modify other anodic metal oxides (e.g. ZrO_2, Nb_2O_5, WO_3, and Fe_2O_3) for improved properties.

3.2 Black TiO₂ Nanomaterials Through Mechanical Methods

Previous theoretical study has shown that strain created along the soft axial directions can modify the effective bandgap of both the rutile and anatase TiO_2 (Figure 3.8),[34] indicating the potential interest in using strain engineering for functional modification of TiO_2. Strain along the soft axes in latticed TiO_2 not only has shown to lead to improved electrical (pseudocapacitive, and energy storage) performance of the material, but also been linked to bandgap modification and photoactive performance.[34] Furthermore, the potential of defect introduction, simultaneous doping, and phase

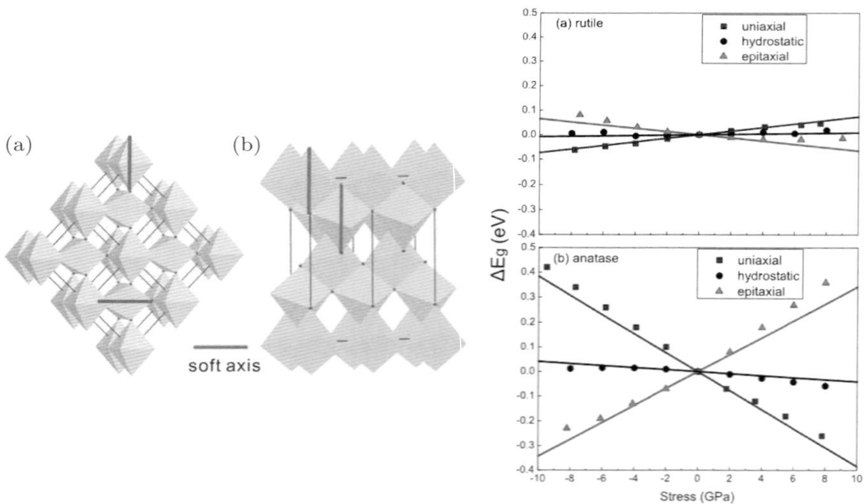

Figure 3.8. (Left) The octahedral packing style of both (a) rutile and (b) anatase TiO_2, with the soft axis directions shown. (Right)The bandgap variation as calculated by first principles in (a) rutile and (b) anatase when under uniaxial, hydrostatic, and epitaxial stress. Reprinted with permission from Ref. [34]. Copyright 2010 AIP Publishing LLC.

transformation enabled by strain engineering has opened a new avenue of producing black TiO$_2$.[34] Here several key methods for strain-engineering TiO$_2$ will be introduced.

3.2.1 *Ball Milling for Treating Powdered TiO$_2$*

To date, TiO$_2$ modification via strain introduction has mainly focused on ball milling. Ball milling can be employed to create strain on TiO$_2$ powders, create and modify TiO$_2$-based nanocomposites and induce crystal transformation. A recent study has reported that the ball milling treatment of TiO$_2$ powders can lead to significant photocatalytic improvement, which was mainly ascribed to the particle size reduction resulted from ball milling.[35] Besides particle size reduction, another attractive benefit of the ball milling is the addition of dopants that can be introduced by this treatment. Previous research has shown that both anions and cations can be introduced into the sol–gel fabricated TiO$_2$ NPs by ball milling.[36] By including energy driven dopants (e.g. nitrogen or aluminum) to the lattice of the NPs, other structural defects can potentially also be introduced to TiO$_2$, leading to bandgap and light absorption modifications. In addition to the defect and dopant effects made possible by ball milling, phase transformation in TiO$_2$ (from anatase to rutile) was reportedly observed with extended ball milling treatment time.[37,38] With the demonstrated improvements through powder/grain size reduction,[39,40] defect/dopant introduction,[41] and bandgap modifications,[42] ball milling appears to be a new promising method for modifying photoactive TiO$_2$ particles.

3.2.2 *Surface Mechanical Attrition Treatment for Treating TiO$_2$ Nanostructures*

On one hand, the ball milling technology is well suited mainly for treating powders, while difficult for processing bulky materials or on-substrate nanostructures. On the other hand, the major issue with strain engineering of delicate TiO$_2$ nanomaterials (e.g. thin films, NTs, or mesoporous TiO$_2$) is that the strain created must not exceed the strain to failure, or the material will simply get destroyed. Thus, there exists an intricate balance to being able to create the most-effective strain on the material, while ensuring that the strain input does not lead to material destruction or loss. Therefore, there is a pressing need for more innovative methods of creating small-scale stress in materials. Here we introduce a novel strain-creation method of great flexibility particularly useful for treating on-substrate nanostructures: the Surface Mechanical Attrition Treatment (SMAT).

The SMAT technology has long been used for processing bulky metal parts to create defects (e.g. to create nanograins and nanotwinings) and high surface energy.[43–45] It entails using a vibrational source to propel projectiles towards a target, allowing the projectiles to mechanically collide with the target, inducing localized strain and stress in the target, allowing for structural refinement at a nanoscale level.[46] It is also an effective way of boosting the surface energy of the treated materials, largely facilitating for instance surface chemical reactions or diffusion. The SMAT method is known for high efficiency and high flexibility in tuning the treating parameters. It is similar to ball milling, but different in some key features: SMAT has much higher efficiency — on the scale of a magnitude higher than ball milling,[43] leading to cheaper and faster material processing. SMAT features a direct processing chamber, suitable for treating on-substrate films.[44] Thus, with SMAT, a novel method for conveniently treating delicate nanostructures is enabled. These key differences allow SMAT to become a better strain engineering method for nanomaterials, and with the ease of tuning the SMAT processing machine at different operational parameters, SMAT can be tailored to treat the delicate and brittle on-substrate nanomaterials, e.g. TiO$_2$ NT arrays grown using the anodization procedures.[3, 47]

SMAT can be easily applied for treating on-substrate TiO$_2$, either as films or as nanostructures (NTs, nanorods, etc.). Taking the anodic TiO$_2$ NT arrays as an example, the SMAT-treated TiO$_2$ nanostructures showed significantly improved photocatalytic performance (Figure 3.9), which is ascribed to the variety dislocations (e.g. slips and twinning). From the XRD patterns of the films, the peaks underwent broadening, indicating grain subdivision and/or strain-related

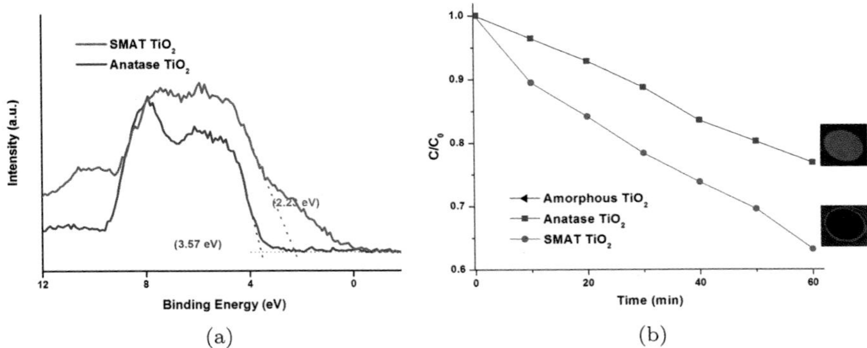

Figure 3.9. Valence band XPS spectra (a) and the methylene blue-photodegradation measurement under the simulated solar radiation (b) of pristine and SMAT-treated anatase TiO$_2$ NT samples. The insets in (b) show the corresponding photographs of the samples.

by-products (defect, vacancies, slips, or twinning). Notably, dramatic color change from tan to black with increased SMAT treatment time was observed, indicating its modified electric structure. It should be pointed out that SMAT is able to induce strain-facilitated doping as well. This doping modification is facilitated by the high surface energy and rich defects created by the continuous and repeated collisions of the tiny SMAT balls with the TiO$_2$ nanostructures. With strain and doping performed simultaneously, the SMAT method is capable of creating a "hybridized" effect to create more effective TiO$_2$ materials for photocatalysis or energy storage.

Furthermore, the great flexibility of SMAT can be embodied through the design and selection of the projectile substances. For example, TiO$_2$ can be SMAT-treated with a small amount MoS$_2$ powder present in the chamber, where the constant movement of balls within the SMAT chamber continuously suspend the MoS$_2$ powder allowing it to be deposited onto the surface of the substrate either through direct attachment, or by mechanical friction. This facile method of energetically mixing two separate entities at once — a process that is conventional for powder processing in ball milling, but challenging for on substrate bulk processing — has led to enhanced photocatalytic performance. Furthermore, this process is highly efficient, entailing only very little introduction of MoS$_2$ powder (e.g. a few milligrams) for treating a large batch of TiO$_2$ NT array samples.

3.2.3 *Other Strain-Related Methods*

Other innovative methods of creating small-scale stress in materials range from pulsed or modified deposition.[48,49] to ion bombardment.[50,51] and mechanical compression.[52] and tension. For example, previous research indicates that strain in the pulsed laser deposition (PLD) fabricated TiO$_2$ can affect its UV/visible light absorption and enhance its photocatalytic performance.[52,53]

3.3 Summary and Prospective

This chapter introduces two kinds of TiO$_2$ modification methods through electrochemical and mechanical means. Although both are relatively less explored compared with other mainstream efforts, both these kinds of methods possess great potential of enabling low-cost and high-performance TiO$_2$ materials, particularly TiO$_2$ nanomaterials, for photocatalysis and energy-storage applications.

Acknowledgments

The authors would like to thank the Research Grants Council of Hong Kong (Project 9042231 (CityU 11302515)), Shenzhen Basic Research (Project No. JCYJ20150630140546704), and Innovation and Technology Commission via the Hong Kong Branch of National Precious Metals Material Engineering Research Center, the Centre for Functional Photonics at the City University of Hong Kong for financial support.

References

(1) Liu, N.; Häublein, V.; Zhou, X.; Venkatesan, U.; Hartmann, M.; Mačković, M.; Nakajima, T.; Spiecker, E.; Osvet, A.; Frey, L.; Schmuki, P. *Nano Lett.* **2015**, *15*, 6815.

(2) Chen, X.; Mao, S. S. *Chem. Rev.* **2007**, *107*, 2891.

(3) Paramasivam, I.; Jha, H.; Liu, N.; Schmuki, P. *Small* **2012**, *8*, 3073.

(4) Wei Seh, Z.; Li, W.; Cha, J. J.; Zheng, G.; Yang, Y.; McDowell, M. T.; Hsu, P.-C.; Cui, Y. *Nat. Commun.* **2013**, *4*, 1331.

(5) Hong, Z.; Wei, M.; Lan, T.; Jiang, L.; Cao, G. *Energy Environ. Sci.* **2012**, *5*, 5408.

(6) Zhou, H.; Zhang, Y. *J. Phys. Chem. C* **2014**, *118*, 5626.

(7) Liu, N.; Schneider, C.; Freitag, D.; Venkatesan, U.; Marthala, V. R. R.; Hartmann, M.; Winter, B.; Spiecker, E.; Osvet, A.; Zolnhofer, E. M.; Meyer, K.; Nakajima, T.; Zhou, X.; Schmuki, P. *Angew. Chem. Int. Ed.* **2014**, *53*, 14201.

(8) Nakajima, T.; Nakamura, T.; Shinoda, K.; Tsuchiya, T. *J. Mater. Chem. A* **2014**, *2*, 6762.

(9) Asahi, R.; Morikawa, T.; Ohwaki, T.; Aoki, K.; Taga, Y. *Science* **2001**, *293*, 269.

(10) Chen, X.; Liu, L.; Yu, P. Y.; Mao, S. S. *Science* **2011**, *331*, 746.

(11) Zuo, F.; Wang, L.; Wu, T.; Zhang, Z.; Borchardt, D.; Feng, P. *J. Am. Chem. Soc.* **2010**, *132*, 11856.

(12) Liu, G.; Yang, H. G.; Wang, X.; Cheng, L.; Lu, H.; Wang, L.; Lu, G. Q.; Cheng, H.-M. *J. Phys. Chem. C* **2009**, *113*, 21784.

(13) Salari, M.; Konstantinov, K.; Liu, H. K. *J. Mater. Chem.* **2011**, *21*, 5128.

(14) Naldoni, A.; Allieta, M.; Santangelo, S.; Marelli, M.; Fabbri, F.; Cappelli, S.; Bianchi, C. L.; Psaro, R.; Dal Santo, V. *J. Am. Chem. Soc.* **2012**, *134*, 7600.

(15) Vitiello, R. P.; Macak, J. M.; Ghicov, A.; Tsuchiya, H.; Dick, L. F. P.; Schmuki, P. *Electrochem. Commun.* **2006**, *8*, 544.

(16) Chen, X.; Burda, C. *J. Am. Chem. Soc.* **2008**, *130*, 5018.

(17) Macak, J. M.; Ghicov, A.; Hahn, R.; Tsuchiya, H.; Schmuki, P. *J. Mater. Res.* **2006**, *21*, 2824.

(18) Wang, G.; Ling, Y.; Li, Y. *Nanoscale* **2012**, *4*, 6682.

(19) Zheng, Z.; Huang, B.; Lu, J.; Wang, Z.; Qin, X.; Zhang, X.; Dai, Y.; Whangbo, M.-H. *Chem. Commun.* **2012**, *48*, 5733.

(20) Wang, G.; Wang, H.; Ling, Y.; Tang, Y.; Yang, X.; Fitzmorris, R. C.; Wang, C.; Zhang, J. Z.; Li, Y. *Nano Lett.* **2011**, *11*, 3026.

(21) Lu, X. H.; Wang, G. M.; Zhai, T.; Yu, M. H.; Gan, J. Y.; Tong, Y. X.; Li, Y. *Nano Lett.* **2012**, *12*, 1690.

(22) Niklasson, G. A.; Granqvist, C. G. *J. Mater. Chem.* **2007**, *17*, 127.

(23) Lin, F.; Nordlund, D.; Weng, T.-C.; Sokaras, D.; Jones, K. M.; Reed, R. B.; Gillaspie, D. T.; Weir, D. G. J.; Moore, R. G.; Dillon, A. C.; Richards, R. M.; Engtrakul, C. *ACS Appl. Mat. Interfaces* **2013**, *5*, 3643.

(24) Sorar, I.; Pehlivan, E.; Niklasson, G. A.; Granqvist, C. G. *Sol. Energy Mater. Sol. Cells* **2013**, *115*, 172.

(25) Lee, K.; Kim, D.; Berger, S.; Kirchgeorg, R.; Schmuki, P. *J. Mater. Chem.* **2012**, *22*, 9821.

(26) Meekins, B. H.; Kamat, P. V. *ACS Nano* **2009**, *3*, 3437.

(27) Kang, U.; Park, H. *Appl. Catal. B* **2013**, *140–141*, 233.

(28) Xu, C.; Song, Y.; Lu, L.; Cheng, C.; Liu, D.; Fang, X.; Chen, X.; Zhu, X.; Li, D. *Nanoscale Res. Lett.* **2013**, *8*, 1.

(29) Idígoras, J.; Berger, T.; Anta, J. A. *J. Phys. Chem. C* **2013**, *117*, 1561.

(30) Fabregat-Santiago, F.; Barea, E. M.; Bisquert, J.; Mor, G. K.; Shankar, K.; Grimes, C. A. *J. Am. Chem. Soc.* **2008**, *130*, 11312.

(31) Idígoras, J.; Berger, T.; Anta, J. A. *J. Phys. Chem. C* **2012**, *117*, 1561.

(32) Song, W.; Luo, H.; Hanson, K.; Concepcion, J. J.; Brennaman, M. K.; Meyer, T. J. *Energy Environ. Sci.* **2013**, *6*, 1240.

(33) Li, H.; Chen, Z.; Tsang, C. K.; Li, Z.; Ran, X.; Lee, C.; Nie, B.; Zheng, L.; Hung, T.; Lu, J.; Pan, B.; Li, Y. Y. *J. Mater. Chem. A* **2014**, *2*, 229.

(34) Yin, W.; Chen, S.; Yang, J.; Gong, X.; Yan, Y.; Wei, S. *Appl. Phys. Lett.* **2010**, *96*, 221901.

(35) Farbod, M.; Khademalrasool, M. *Powder Technol.* **2011**, *214*, 344.

(36) Santos, D. M. D.; Navas, J.; Sanchez-Coronilla, A.; Alcantara, R.; Fernandez-Lorenzo, C.; Martin-Calleja, J. *Mater. Res. Bull.* **2015**, *70*, 704.

(37) Ren, R. M.; Yang, Z. G.; Shaw, L. L. *J. Mater. Sci.* **2000**, *35*, 6015.

(38) Pan, X. Y.; Ma, X. M. *J. Solid State Chem.* **2004**, *177*, 4098.

(39) Dutta, H.; Sahu, P.; Pradhan, S. K.; De, M. *Mater. Chem. Phys.* **2003**, *77*, 153.

(40) Hafizah, N. N.; Musa, M. Z.; Mamat, M. H.; Rusop, M. *Adv. Mater. Eng. Technol.* **2012**, *626*, 425.

(41) Amade, R.; Heitjans, P.; Indris, S.; Finger, M.; Haeger, A.; Hesse, D. *J. Photochem. Photobiol. A* **2009**, *207*, 231.

(42) Chen, S. F.; Chen, L.; Gao, S.; Cao, G. Y. *Chem. Phys. Lett.* **2005**, *413*, 404.

(43) Tao, N.; Wang, Z.; Tong, W.; Sui, M.; Lu, J.; Lu, K. *Acta Mater.* **2002**, *50*, 4603.

(44) Chan, H.; Ruan, H.; Chen, A.; Lu, J. *Acta Mater.* **2010**, *58*, 5086.

(45) Wei, Y. H.; Liu, B. S.; Hou, L. F.; Xu, B. S.; Liu, G. *J. Alloys Compd.* **2008**, *452*, 336.

(46) Chen, A.; Ruan, H.; Zhang, J.; Liu, X.; Lu, J. *Mater. Chem. Phys.* **2011**, *129*, 1096.

(47) Chen, A.; Ruan, H.; Wang, J.; Chan, H.; Wang, Q.; Li, Q.; Lu, J. *Acta Mater.* **2011**, *59*, 3697.

(48) Kamei, M.; Miyagi, T.; Ishigaki, T. *Chem. Phys. Lett.* **2005**, *407*, 209.

(49) Nambara, T.; Yoshida, K.; Miao, L.; Tanemura, S.; Tanaka, N. *Thin Solid Films* **2007**, *515*, 3096.

(50) Diwald, O.; Thompson, T. L.; Goralski, E. G.; Walck, S. D.; Yates, J. T. *J. Phys. Chem. B* **2004**, *108*, 52.

(51) Potapenko, D. V.; Li, Z.; Kysar, J. W.; Osgood, R. M. *Nano Lett.* **2014**, *14*, 6185.

(52) Miyamura, A.; Kaneda, K.; Sato, Y.; Shigesato, Y. *Thin Solid Films* **2008**, *516*, 4603.

(53) Sauthier, G.; György, E.; Figueras, A.; Sánchez, R. S.; Hernando, J. *J. Phys. Chem. C* **2012**, *116*, 14534.

CHAPTER FOUR

The Effect of Points Defects and Ordered/Disordered Morphology on the Electronic and Structural Features in Black TiO$_2$ Nanomaterials

Vladimiro Dal Santo and Alberto Naldoni

CNR-Istituto di Scienze e Tecnologie Molecolari,
Via Golgi 19, 20133 Milano, Italy

So-called black titania is the resulting material of a number of different treatments with reducing agents, including vacuum, hydrogen, Al, CaH$_2$, NaBH$_4$, hydrogen plasmas, etc., performed on stoichiometric titanium dioxide, having, as distinctive feature, a black color resulting from electromagnetic radiation absorption on a wide range of frequencies, from visible light down to infrared (IR) typically. From a physical, more rigorous, point of view, reductive treatments leading to black titania generate points defects and ordered/disordered morphology which, in turn, influences both structural features and electronic states of the resulting materials.

4.1 Electronic Singularities in Black TiO$_2$ Nanomaterials

Black TiO$_2$ nanostructures have attracted enormous interest due to the ability to harvest a great portion of solar radiation through the entire visible and near-infrared (NIR) wavelength range.[1] The unique absorption features of black TiO$_2$ allow to overcome the main limitation of pristine TiO$_2$, which is represented by its wide bandgap (3.2 eV) enabling only the absorption of ultraviolet (UV) photons. Considering that the solar spectrum is composed of 5% UV light, 45% visible light, and 50% of NIR–IR radiation, the discovery of black TiO$_2$ have opened the way to enhance solar energy conversion technologies to unexpected upper limits.

Figure 4.1 shows some examples of optical properties of black TiO$_2$ nanomaterials obtained using different preparation strategies.[2–4] Figure 4.1(a) shows diffusive reflectance absorbance

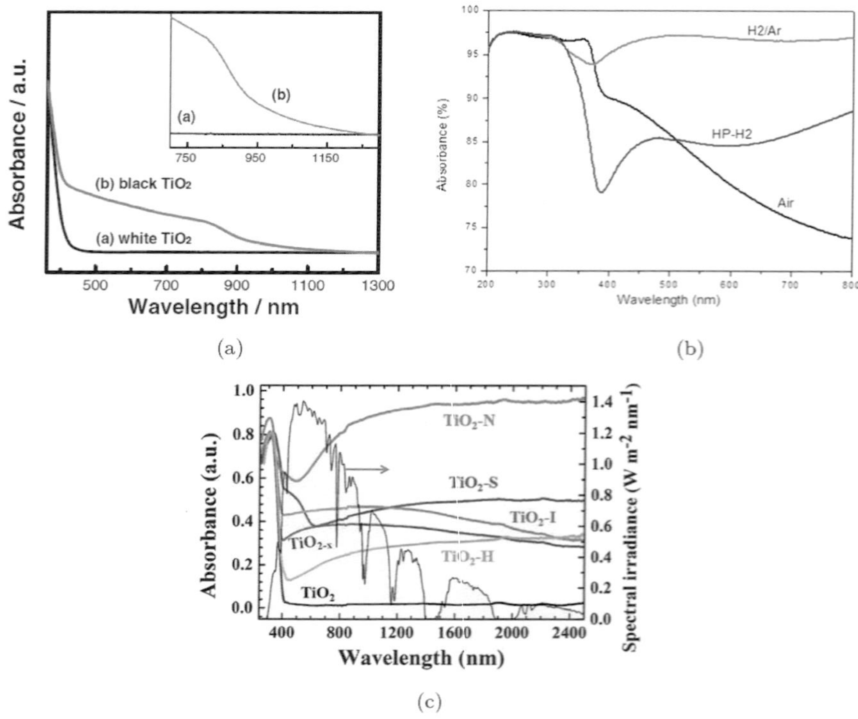

Figure 4.1. (a) Spectral absorbance of the white and black TiO$_2$ nanocrystals. Reprinted from Ref. [2]. © The American Association for the Advancement of Science, 2011. (b) Integrated light reflectance results of TiO$_2$ nanotubes annealed in air, Ar/H$_2$, and high pressure H$_2$ (HP–H$_2$). (Reprinted from Ref. [3]. © American Chemical Society). (c) Diffuse reflectance spectra of TiO$_2$–x (X = H, N, S, I) and solar spectral irradiance (right). (Reprinted from Ref. [4]. © The Royal Society of Chemistry).

spectroscopy of white and black TiO_2 nanocrystals hydrogenated at high H_2 pressure. The bandgap of the unmodified white TiO_2 nanocrystals was approximately $3.30\,eV$, while the onset of optical absorption of the black TiO_2 nanocrystals was lowered to about $1.0\,eV$ ($\sim 1200\,nm$). The abrupt change in absorbance spectra at approximately $1.54\,eV$ ($806.8\,nm$) reflects the narrowing of the black TiO_2 optical gap caused by intraband transitions.[2] Absorption features of TiO_2 nanotubes thin film reduced in different conditions are reported in Figure 4.1(b).[3] In general, the reduction on thin film produces different changes in the absorption of TiO_2 if compared to powdered samples. Differently from the latter, TiO_2 nanotubes annealed in a H_2/Ar mixture absorb over 97% of incoming light from UV to visible wavelength range probably due to scattering effect from the nanostructured surface. Interestingly, if X (H, N, S, I) atoms are introduced into the black TiO_{2-x} during the synthesis step, the spectra reveal an increased absorption (Figure 4.1(c)).[4] For example, the N-doped black titania (TiO_2–N) possesses far larger solar absorption (85% of solar spectral irradiance) than black TiO_{2-x} (65%). This optical effect is due to the combined modifications on TiO_2 electronic structure produced by the concomitant reduction and non-metal-doping.

The extension of light absorption to the visible range in TiO_2 nanomaterials finds origin in the profound modification on its electronic structure, which is represented by the total density of states (DOS), that doping induced. During the past few decades, an intense research activity focused on improving TiO_2 optical properties and photocatalytic properties through metal and non-metal doping. First approach includes the incorporation of ions such as V, Ni, Cr, Mo, Fe, Sn, Mn, and so on.[5] The electronic structures of TiO_2 compounds doped with the 3d transition metals (V, Cr, Mn, Fe, Co, and Ni) were obtained using *ab initio* band calculations.[6] It was found that the 3d metal doping created an occupied level either in the bandgap or in valence band (VB) due to the t_{2g} state of the dopant. The charge-transfer transition between this t_{2g} level and the conduction band (CB) (or VB) of TiO_2 contributed to the photoexcitation under visible light. The second doping approach used to modify UV-light-active photocatalysts is based on non-metal ion doping. Typically, the introduction of non-metal elements such as N, S, and C in the TiO_2 crystalline lattice narrows the bandgap and improves its visible-light-driven photocatalytic activity. Unlike metal-ion dopants, non-metal ion dopants are less likely to form donor levels in the forbidden band but instead shift the VB edge upward (Figure 4.2(a)).[7] Despite metal-ion doping has proved to substantially extend the absorption of TiO_2 to the visible range, many reports have proven that metal centers act also as a recombination center for photogenerated charges thus making the effect of doping not a suitable strategy to design efficient photocatalyst. On the other side, non-metal ion doping, especially nitrogen insertion, is a valid way to increase at the same time absorption and activity of nanostructured materials. However, the bandgap narrowing induced by N-doping is intrinsically limited by energy position of new electronic states/band close to VB that is reflected by a maximum extension of absorbed visible light to the wavelength range of 500–550 nm.

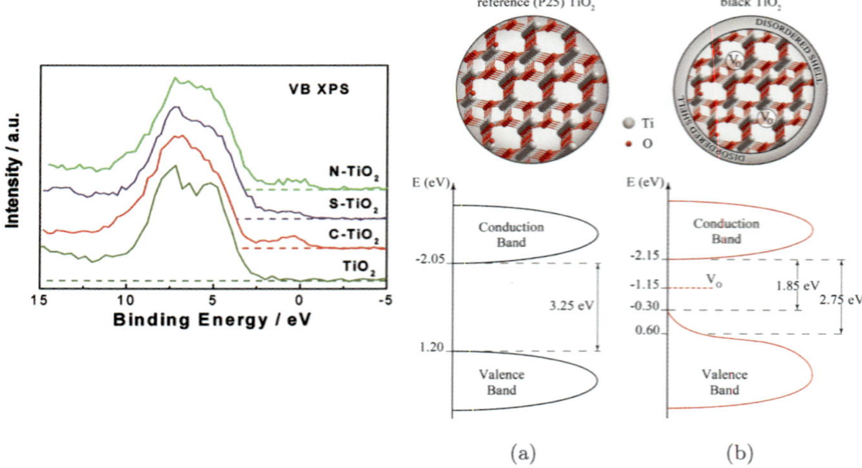

Figure 4.2. (a) VB X-ray photoelectron spectroscope (XPS) spectra of pure TiO_2, C–TiO_2, S–TiO_2, and N–TiO_2. Reprinted from Ref. [7]. © American Chemical Society. (b) Schematic of structure and DOS for TiO_2 P25 Degussa and black TiO_2 nanoparticles. Both representations were built using experimental data various techniques. Reprinted from Ref. [8]. © American Chemical Society.

As shown at the beginning of this paragraph, black TiO_2 instead presents an extended absorption until the NIR wavelength region. This is due to unique electronic feature induced by the so-call self-doping, i.e. introduction of oxygen vacancies (V_{Os}), Ti^{3+} centers, H-doping, and to addition peculiar structural features such as a disordered shell with peculiar electronic properties.[2,8] The advantages of inducing modifications in electronic and thus optical properties in TiO_2 through hydrogenation/reduction is mainly due to the elimination of foreign dopant elements into the crystalline lattice. Importantly, black TiO_2 nanomaterials have shown electronic features that can be considered a combination of metal and non-metal doping effects. The appearance of the black color in TiO_2 is not accompanied by identical modifications on the electronic structure. Depending on the preparation strategy of black TiO_2 nanomaterials (e.g. reduction of powder, reduction of thin film, reduction of nanowires, or superstructures such as inverse opals),[9] the structure, the morphology, and the electronic properties show unique features. For such reasons, the rationalization of electronic properties is quite elusive in the literature since making general statements on structure–electronic properties relationships is very challenging. For instance, the most relevant modification on the TiO_2 electronic structure due to reductive synthetic methods are the introduction of intragap electronic levels below the CB and the rigid shift of the VB-edge (Figure 4.2(b)).

In the following sections, we rationalize the origin of black TiO_2 color through an analysis that correlates the structural modification and electronic singularities observed after TiO_2 reduction. In particular, we will discuss (a) the origin of the VB-edge shift, (b) the origin of the intragap states created in TiO_2 and electronic transitions therefore activated, (c) the coexistence of multiple doping effects such as H and N doping at the same time.

4.1.1 *VB-Edge Modification*

One of the signature of disordered-engineered black TiO_2 nanocrystal is the shift of the VB toward the Fermi Level due to electronic structure modification. Such profound modification has been observed

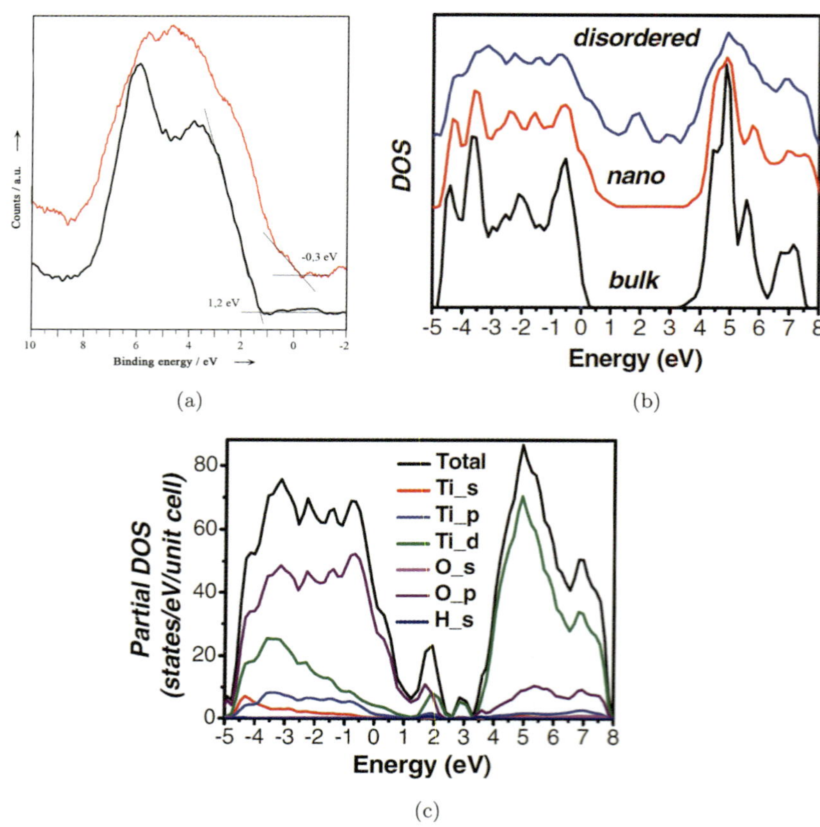

Figure 4.3. (a) VB XPS spectra of P25 Degussa (black line) and black TiO$_2$ (red line). Thin black lines show the linear extrapolation of the curves used for deriving the band edge position of TiO$_2$ samples. Reprinted from Ref. [8]. © American Chemical Society. (b) Calculated DOS of TiO$_2$ in the form of a disorder-engineered nanocrystal, an unmodified nanocrystal, and a bulk crystal. (c) Decomposition of the total DOS of disorder-engineered black TiO$_2$ nanocrystals into partial DOS of the Ti, O, and H orbitals. Reprinted from Ref. [2]. © The American Association for the Advancement of Science, 2011.

and varies depending on synthesis conditions and TiO$_2$ morphology.[2,8,10–14] Figure 4.3(a) shows the VB XPS of stoichiometric (white) and defective (black) TiO$_2$ obtained through the reduction of an amorphous TiO$_2$ powder at 500°C under H$_2$ flow. White TiO$_2$ displayed the characteristic VB DOS of TiO$_2$, with the band edge at

~1.2 eV below the Fermi energy. Since the optical bandgap of TiO_2 is 3.25 eV, the CB minimum would occur at -2.05 eV. On the other hand, VB XPS of black TiO_2 showed notable differences: the main absorption onset was located at 0.6 eV, whereas the maximum energy associated with the band tail blue-shifted further toward the vacuum level at about -0.3 eV.[8]

Only few reports have tried to address the electronic origin of the VB shift in black TiO_2 nanocrystal so far.[2,15] In particular, Chen *et al.* used a first-principles density functional theory (DFT) approach to calculate the energy band structures of black TiO_2. Usually, existing models of modified TiO_2 are focused on point defects, which tend to produce shallow or deep energy levels near the CB minimum with typical Ti^{3+} state characteristics. However, considering on-lattice disorders in TiO_2 nanocrystals in the presence of hydrogen, it was found that, rather than generating levels near the CB minimum, disorder-induced mid-gap-states can upshift the VB edge of TiO_2 nanocrystals. The primary effect of surface reconstruction in TiO_2 nanocrystals is to produce strong band tailing near the VB edge. The disordered TiO_2 nanocrystal model, in which one H atom is bonded to an O atom while another H atom is bonded to a Ti atom, yields electronic band structures consistent with the VB XPS measurements.[2]

Figure 4.3(b) plots the calculated DOS of the disorder-engineered TiO_2 nanocrystals along with those of the bulk and the unmodified TiO_2 nanocrystals. Two groups of mid-gap-states (centered at about 1.8 eV and 3.0 eV) can be observed in the DOS of the disordered TiO_2 nanocrystals, for which the Fermi level was found to locate slightly below 2.0 eV. The different nature of these two groups of mid-gap states is revealed by the calculated partial DOS (Figure 4.3(c)). Whereas the higher-energy mid-gap states (~3.0 eV) are derived from the Ti 3d orbitals only, the lower-energy states (~1.8 eV) are hybridized from both O 2p orbitals and Ti 3d orbitals, and mainly from the VB states as the result of disorders stabilized by hydrogen.[2] The hydrogen 1s orbital coupling to the Ti atom does not make a substantial contribution to either state, which suggests that lattice disorder accounts for the mid-gap states; hydrogen may have stabilized the lattice disorders by passivating their dangling bonds. Because the lower-energy mid-gap states lie below the Fermi level, they can account for a large blue shift of the VB edge.[2]

The electronic singularities that produce the VB shift observed in some cases for black TiO$_2$ nanomaterials should be investigated in more detail in the future to shed more light in fundamental aspects that will help the design of more efficient black TiO$_2$ nanomaterials.

4.1.2   *Electronic Transitions Due to Intragap States*

Stoichiometric TiO$_2$ is a wide bandgap semiconductor with an energy gap of about 3.2 eV. In this material, the photoexcitation involves transitions from the O 2p states in the VB to the 3d states of the Ti^{4+} cations in the CB.[16] Hydrogenation of TiO$_2$ nanomaterials often is accompanied to the formation of substoichiometric oxide form. Defective black TiO$_2$ nanomaterials obtained through different reduction strategies, both for thin films and nanopowders, have shown the presence of V$_{Os}$ (TiO$_{2-x}$) or Ti interstitials (Ti$_{1+x}$O$_2$) into the crystalline lattice. In both cases, (i) Ti is in excess with respect to O, and (ii) reduction is accompanied by the appearance of Ti^{3+} species.[8] Black TiO$_2$ is usually obtained in relative mild conditions such as 500°C under reductive gas mixture at ambient pressure. The presence of interstitial Ti ions in most black TiO$_2$ nanomaterials has been ruled out since they are most probably formed under more severe conditions (i.e. high-temperature vacuum annealing). In addition, in the O-poor regime the formation of V$_{Os}$ should be energetically favored compared to that of interstitial Ti ions.[8]

Therefore, many reports have used the detection of Ti^{3+} as motivation for the black color in reduced TiO$_2$ samples. For instance, the introduction of Ti^{3+} and thus of V$_{Os}$ can be detected using a plethora of electronic techniques (e.g. EPR, XPS, UV–Vis–NIR, and IR absorption spectroscopy) and structural approaches through the detection of variation in Ti–O distance (e.g. XRD, electron diffraction) since the introduction of V$_O$ produces the contraction of bond length. Nevertheless, a clear understanding on the origin of electronic transitions induced by the incorporation of Ti^{3+} defects is still scarce. Only, recently Wang *et al.* have reported a study that, combining two photoemission spectroscopy (2PPE) and DFT, shed some light on this matter.[16] In particular, using 2PPE, an excited resonant state derived from Ti^{3+} species was identified at $2.5\pm0.2\,eV$ above the Fermi level on the reduced TiO_2 surfaces. DFT calculations revealed that this excited state is closely related to the gap state at $\sim 1.0\,eV$ below the Fermi level, as it results from the Jahn–Teller induced splitting of the 3d orbitals of Ti^{3+} ions in reduced TiO_2. Localized excitation of Ti^{3+} ions via $3d \rightarrow 3d$ transitions from the gap state to this empty resonant state significantly increases the TiO_2 photoabsorption and extends the absorbance to the visible region, consistent with the observed enhancement of the visible light induced photocatalytic activity of hydrogenated TiO_2.[16] More fundamental investigation on the physical origin of the Ti^{3+} related photoabsorption and visible light–NIR photocatalytic activity in black TiO_2 are needed in order to provide important details to design new efficient black TiO_2 nanomaterials for advanced solar energy conversion applications.

4.1.3   *H,N Doping*

One of the driving design issues in solar energy conversion applications is the realization of novel nanomaterials with broadband light absorption. This means that new materials should be able to absorb efficiently electromagnetic radiation along the whole solar spectrum wavelength range. An interesting approach to achieve broadband light harvesting based on the use of black TiO_2 nanomaterials is to add foreign non-metal atoms during the hydrogenation step.[4,17] In particular, the most promising approach is to couple the nitrogen doping electronic modifications (i.e. creation of VB upward shift due to N states) and disordered-induced electronic features in black TiO_2. Following this strategy, for example, advanced photocatalysts based on N–H codoping have been realized showing enhanced photocatalytic performances with respect to bare hydrogenated samples.

4.2   Structural Properties of Black TiO$_2$

Treatments inducing the formation of black TiO$_2$ phases often result also in the formation of disordered features, both stoichiometric and non-stoichiometric. Crystalline phases are influenced too and lattice expansion or contraction phenomena are reported. Most commonly encountered structures are core–shell nanoparticles/nanocrystals, usually crystalline core–amorphous (disordered) shell. Depending on the location of defects, stoichiometric core, and defective shell or its opposite are reported. Defects, induced by reductive treatments, include reduced Ti^{3+} sites, V$_{Os}$, while Ti–H sites (H filling V$_{Os}$) or dangling bonds are less common.

Other nanostructures, such as nanowires, nanoribbons, nanobelts, nanotubes or porous thin films, inverse opals are also reported, both as crystalline "core" surrounded by amorphous layers or as fully crystalline structures. Crystalline phases include anatase and, in lesser extent, rutile; Ti suboxides and Magneli phases, Ti$_x$O$_y$ ($y/x < 2$) are also reported. A correlation between treatment conditions/parameters and the resulting structure(s) is hard to draw, however some common structural features related to the formation of black titania can be highlighted.

4.2.1   *Core–Shell Nanoparticles*

Anatase crystalline core/amorphous shell nanocrystals were the first black TiO$_2$ nanostructures reported by Chen *et al.*[18] and Naldoni *et al.*[8, 19] who independently reported on black TiO$_2$ nanocrystals obtained by high pressure and high temperature H$_2$ treatment, respectively. HRTEM measurements show clear structural deviation from the standard crystalline anatase at the outer layer, where the straight lattice line is bent at the edge of the nanoparticle (Figure 4.4). While the core shows well resolved (101) lattice plane with typical anatase plane distance, on the disordered outer layer, the distances between adjacent lattice planes are no longer uniform.[19] The formation of the peculiar defective crystalline core–amorphous stoichiometric shell was ascribed to the combination of amorphous high specific surface area TiO$_2$ precursor and the fast cooling down

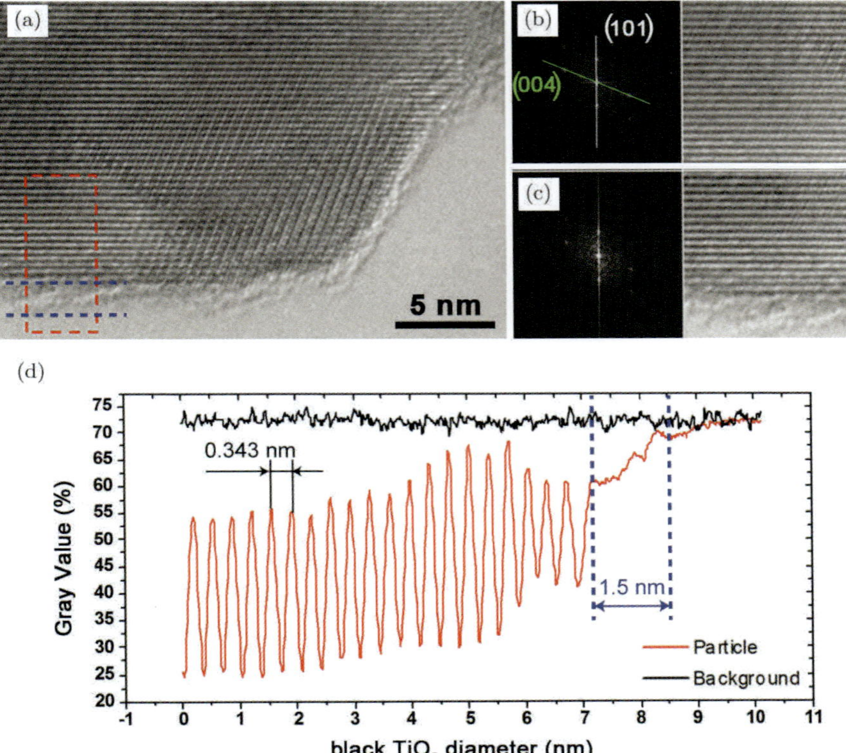

Figure 4.4. (a) HRTEM image of a black anatase TiO_2 nanocrystal. FFT and reconstructed images of (b) the crystalline core and (c) the boundary area including both the core and the disordered shell. (d) The line profile across the area marked in red in panel (a), showing the lattice distance of the core and the disordered shell thickness taken along the (010) direction. Reprinted from Ref. [19]. © The Royal Society of Chemistry.

under inert atmosphere at the end of the high temperature reduction under hydrogen atmosphere that induces the freezing of a metastable defective phase.

In addition Chen *et al.* performed a detailed study on lattice variations and crystalline shapes (i.e. percentage of different crystalline exposed facets) of 20 bar H_2 reduced TiO_2 nanocrystals. Hydrogenation treatment resulted in a contraction of unit cell parameters a and c, of hydrogenated black TiO_2 nanocrystals, of 0.12%, and 0.41%, respectively, while the cell volume contracts 0.64%.

The exposed facets undergo important modifications too: from white to black TiO$_2$ NPs, the percentage of facets (001), (110), decreases from 11.3% to 6.5%, and from 17.5% to 4.7%, respectively. Conversely, facet (100) increases from 3.6% to 5.4%, and (102) disappears from 67.6% to 0% and converts to facet (215) with 83.3%. The surface of the white TiO$_2$ nanocrystal is dominated by facet (102) (66.7%), and it converts to facet (215) on the surface of the hydrogenated TiO$_2$ nanocrystal.

On the other hand, Tian *et al.* reported a detailed study on the structure of core–shell NPs having a crystalline rutile core surrounded by an amorphous shell of stoichiometric Ti$_2$O$_3$ phase.[20] NPs were obtained by crystallization of amorphous ultrasmall nanoparticles synthesized by pulsed laser vaporization, and the formation of the disordered layer occurred only in an oxygen-free environment (pure Ar at 970 K). Exact structure was elucidated by monochromated electron energy-loss spectroscopy (EELS), fifth-order aberration-corrected scanning transmission electron microscopy (STEM), and nanobeam electron diffraction (NBED). EELS confirmed that the core consists of perfectly crystalline rutile and the disordered shell consists of Ti$_2$O$_3$ but was unable to determine the structure of the transition region present between the core and the shell. On the other hand, atomic-resolution high-angle annular dark field (HAADF) imaging identified two types of defective layers both containing interstitial Ti atoms (Figure 4.5). Amorphous TiO$_2$ NPs and crystallization under O$_2$ free atmosphere were identified as the two key parameters leading to the formation of the structural sequence crystalline rutile core — transition layers of highly defective rutile — amorphous stoichiometric Ti$_2$O$_3$ outer shell. The driving force of this mechanism is the creation of a dynamic oxygen deficiency gradient during crystallization that results in V$_{Os}$ accumulating and propagating toward the surface region, creating an effective composition gradient, Ti$_2$O$_3$, the stoichiometric phase of Ti^{3+}, forms on the outermost surface when the concentration of V$_{Os}$ reaches a saturation level beyond which the TiO$_2$ structure is unstable (Figure 4.6).[20]

Liu *et al.*[21] reported a detailed study on the effect of high pressure (20 bar) medium temperature (500°C) reduction of commercial

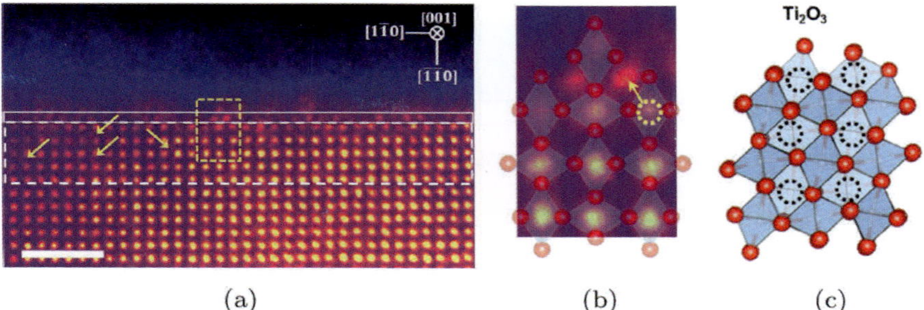

Figure 4.5. Atomic-resolution HAADF image of a rutile nanoparticle viewed in the (001) direction and crystal structure models of connected TiO_6 octahedra for rutile and Ti_2O_3. (a) HAADF image of a rutile nanoparticle viewed in the (001) direction. The surface reconstruction layer consisting of an ordered line of interstitial Ti atoms is located above the rutile (110) plane designated by a solid white rectangle. The disordered Ti_2O_3 layers form on top of the surface reconstruction layer. The yellow arrows designate a few Ti atoms near the surface that are displaced from the rutile basis in a series of defective rutile layers designated by a dashed white rectangle. (b) Magnified image from the area highlighted by the dashed yellow rectangle in (a) and compared with the corundum Ti_2O_3 structure on the right. (c) The Ti atoms in the light blue octahedra in corundum Ti_2O_3 originate from the diffusion of interstitial Ti atoms in rutile, associated with a unit cell distortion. Scale bar is 2 nm. Reprinted from Ref. [20]. © American Chemical Society.

anatase, rutile and mixed phase commercial samples. In agreement with the results of Tian,[20] they did not observe any effect on rutile samples. However, as anatase regards, the pristine nanoparticles show already a crystalline core–amorphous shell structure that is unaffected by the reductive treatment. Interestingly, the observed lattice parameter variation and shrinkage of the average nominal crystallite size was attributed to the formation of voids inside NPs (revealed by TEM taken under defocus, Fresnel contrast, conditions) due to vacancy condensation. Moreover, the void-free portions of nanoparticles did not show any significant change in lattice parameters.

Core–shell black titania nanocrystal (TiO_{2-x}) consisting of a crystalline TiO_2 a core and a highly disordered surface layer (2 nm thick) can be obtained also by CaH_2 reduction at $400°C$[22] Interestingly, CaH_2 can effectively reduce the rutile phase of Degussa P25

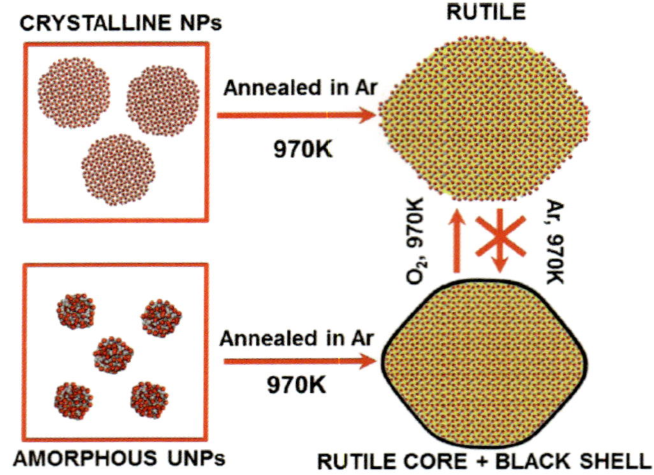

Figure 4.6. Schematic illustration of the steps in black TiO$_2$ core–shell nanoparticle formation and conversion. Both crystalline nanoparticles and amorphous nanoparticles agglomerate into larger particles when annealed in Ar. However, only the amorphous TiO$_2$ nanoparticles result in the formation of a Ti$_2$O$_3$ shell characteristic of black TiO$_2$. Annealing of black TiO$_2$ in O$_2$ at 970 (20 K forms ordinary white rutile NPs. This process cannot be reversed to obtain black TiO$_2$ by repeating the annealing in Ar. Reprinted from Ref. [20]. © American Chemical Society.

samples, which reaches 8% at 400°C, while at 500°C a complete transition of rutile to Magneli phase (Ti$_4$O$_7$) occurs.

Crystalline core disordered defective shell larger nanoparticles were also obtained by Chen *et al.*[11] by laser ablation of commercial TiO$_2$ nanoparticles (Aldrich, 100 nm). Laser treatment induces a narrowing of size distribution and a decrease of mean particles size (120 nm for sample treated for 120 min). X-ray powder diffraction (XRD) and selected area electron diffraction (SAED) measurements revealed that there are at least two crystalline phases coexisting in the ablated nanospheres anatase and rutile, and the SAED for TiO$_2$-120 reveals an almost single-crystal morphology. Surface of nanoparticles show many disordered areas and the presence of V$_{Os}$, around 12% of Ti^{3+}, and the formation of Ti–OH sites.

Even larger black TiO_2 spherical granules (20–30 μm), composed of micrograins with a size of 3–5 μm can be obtained by a flame spraying with a C_2H_2/O_2 flame with Ar as carrier gas (i.e. under a high-temperature, O-poor regime).[23] More complex structures, such as Ti^{3+} self-doped 2 μm yolk–shell particles, were obtained by CaH_2 reduction.[24]

Jang *et al.* reported a two-step fabrication, a novel carbon layer coated T^{i3+}–TiO_2 rod-like structure by carbothermal reduction that results in fully crystalline (anatase phase) structure with a surface interlayer rich in Ti^{3+} and V_{Os} defects just below the thin carbon external layer.[25]

Wei *et al.* reported the preparation of fully crystalline anatase black TiO_2 NPs having non-stoichiometric $TiO_{1.980}$ and $TiO_{1.974}$ outermost surface containing Ti^{3+} and V_{Os} by a simple calcination under N_2 or Ar atmosphere.[10] A highly disordered defective (Ti^{3+} and V_{Os} defects) surface layer surrounding the crystalline TiO_2 (P25) core was introduced into the titania nanocrystal (TiO_{2-x}) can be obtained also by reduction with aluminum powder.[26]

When metal nanoparticles are present, some peculiar composites nanostructures can be obtained: "black" composite TiO_2 nanoparticles consisting of a stoichiometric anatase TiO_2 "core" surrounded by silver nanoparticles covered by titanium suboxides (Ti_3O_5 and Ti_4O_7) originated from SMSI were prepared by one-step continuous flame spray pyrolysis.[27]

Teng *et al.*[28] obtained by hot filament H_2 plasma treatment black TiO_2 NPs with amorphous shell/crystalline core structure consisting of a stoichiometric anatase/rutile crystalline core surrounded by an amorphous defective layer, eventually containing Cr^{3+} dopant, in doped samples. Wang *et al.*[29] reported the preparation, by H_2 plasma treatment, of H-doped black titania with a core/shell structure (TiO_2–$TiO_{2-x}H_x$) possessing a significant absorption in the visible and NIR light higher than high pressure and high temperature H_2 reduced black TiO_2. Crystalline core is composed of stoichiometric anatase phase, while amorphous 2 nm thick shell is free of typical defects such as V_{Os} and Ti^{3+} sites and the former are filled by H forming Ti–H bonds.

4.2.2 *Black TiO$_2$ Nanowires and Nanobelts*

The hydrogenation of TiO$_2$ nanowires (NWs) on a Ti mesh scaffold generated a unique structure of a crystalline core coated with an amorphous shell (Figure 4.7). The phase of the single crystal core was still anatase TiO$_2$, as shown by the spacing of the exposed crystallographic plane but a thick amorphous layer (4 nm) covered the NWs surface. The diameter of the nanowire increased from 15 nm to 27 nm, probably because of the deformation and destruction of the crystal lattice and the appearance of the amorphous layer.[30]

One-dimensional hydrogenated TiO$_2$ nanobelts with disordered surface can be obtained by heating pristine nanobelts in H$_2$/Ar (1:1) atmosphere at 1000°C while samples treated at 600°C and

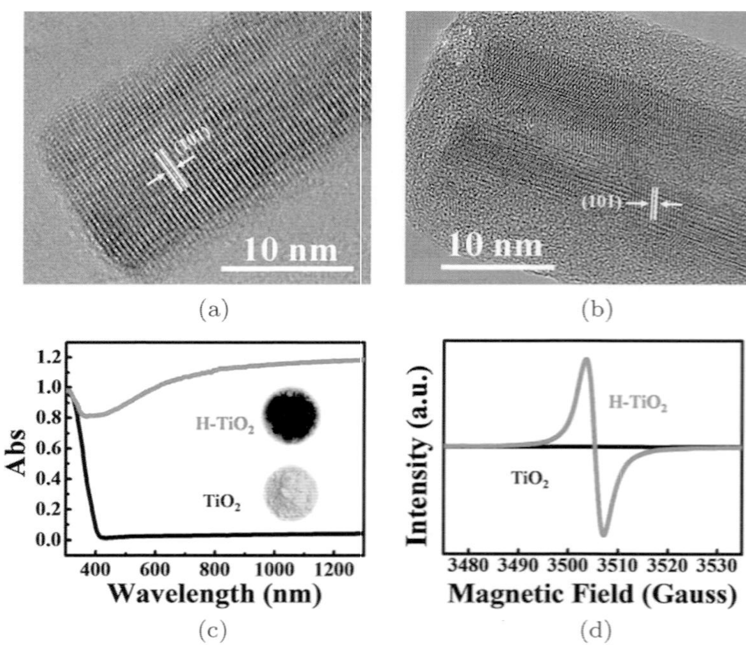

Figure 4.7. High-resolution TEM (HRTEM) images of (a) TiO$_2$ and (b) hydrogenated TiO$_2$ (H–TiO$_2$) NWs. (c) UV–Vis–NIR absorption spectra of TiO$_2$ nanowires and H–TiO$_2$ nanowires. The inset shows the corresponding photographs. (d) Electron paramagnetic resonance (EPR) spectra of TiO$_2$ and H–TiO$_2$ NWs. Reprinted from Ref. [30]. © The Royal Society of Chemistry.

800°C are gray and blue, respectively.[12] Samples show the same crystal-phase composition of nanobelts treated in air atmosphere (100% anatase at 600°C, 800°C, and mainly rutile for 1000°C) but with a small amount of Ti_2O_3 phase. The increase of the amount of V_{Os} in the TiO_2 structure results in disorder-induced lattice strains and reflects that in the progressive intensity decrease of TiO_2 (A and R) peaks with the increase of hydrogenation temperature. Actually, when morphology is concerned, temperature has strong effects: only H-600 sample shows no structural change between parent and hydrogenated nanobelts, but an amorphous shell/crystalline core structure is formed. The amorphous shell has a thickness of 8 nm and results from the formation of high concentrations of Ti^{3+} sites at the surface. Samples show also the presence of around 3% of V_{Os} that do not affect the lattice spacing of the crystalline core. The treatment at 800°C does not alter the morphology but surface becomes smooth and a zigzag contour is observed. At 1000°C, samples sinter to form a dendritic morphology with a smooth surface.

Slightly different results were obtained by Tian *et al.*[31] who reported on the preparation of black TiO_2 nanobelts by annealing the TiO_2 nanobelts in hydrogen and argon gas mixtures (1:1) at temperature at 600°C for 60 min. Hydrogenated nanobelts were mainly composed of highly crystalline anatase phase with low amounts of Ti_2O_3 (revealed by XRD). The Ti^{3+}/Ti^{4+} ratio was calculated to be approximately 88.9:11.1, and the concentration of oxygen defects was calculated to be about 3% (not specified if on surface and/or bulk).

Novel hierarchical TiO_2 nanostructures with wire-in-tube architectures were fabricated by hydrothermal treatment of anodic TiO_2 nanotube arrays.[32] Furthermore, the hydrothermally treated samples were modified into reduced black TiO_2 by annealing under vacuum at 450°C for 2 h. Sintering in vacuum condition led to the growth of nanoparticles decorated on the surface of inner-nanowires and the development of nanoparticles on the tubular outer-shell. Sintering under the vacuum condition leads to the vapor transport playing a dominating role during the sintering, while reducing the driving force for densification due to the coarsening effect. Nanowires and

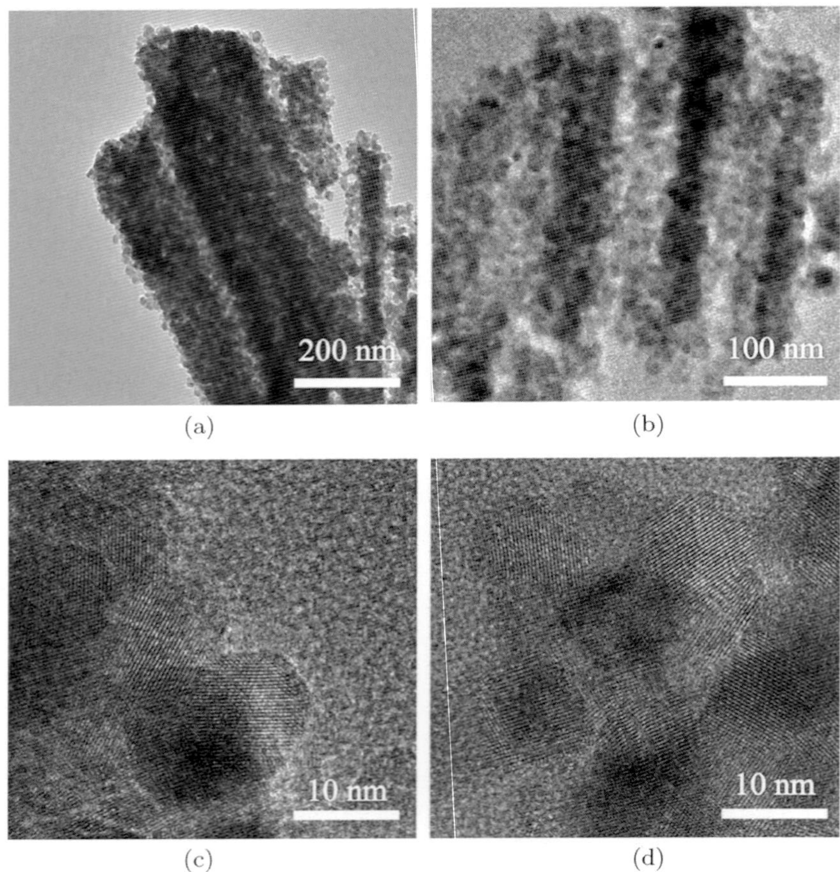

Figure 4.8. TEM images of vacuum-H–TiO$_2$ samples. Reprinted from Ref. [32]. © The Royal Society of Chemistry 2014.

nanotubes were composed of TiO$_2$ nanoparticles with diameter around 10 nm. Similar to the black TiO$_2$ created by hydrogen treatment, the nanoparticles in the vacuum-H–TiO$_2$ samples have a shallow disordered surface layer around 0.5 nm (Figure 4.8).

Phase composition was investigated by means of XRD, Raman, XPS, EPR, and confirmed the presence of anatase with some reduced TiO$_{2-x}$, with presence of Ti^{3+} in the bulk. Conversely, the surface of black vacuum-H–TiO$_2$ sample becomes nearly stoichiometric when exposed to air at room temperature, after vacuum sintering, and this prevents further oxidization of Ti^{3+} in the bulk, and increase stability of black vacuum-H–TiO_2.

4.2.3 *Black TiO_2 Nanotubes Arrays*

Black TiO_2 nanotubes can be obtained by H_2/Ar and Ar treatment of pristine NTs but not by high pressure H_2 treatment, which results in blue-NTs. These black samples do not show any modification of NT morphology nor any variation of phase composition or lattice parameters, i.e. no conversion to suboxide occurred and the amount of Ti^{3+} is below 1%.[3] Namely, only blue sample show so-called noble-metal-free cocatalysis effect. The same cocatalyst effect is further evidenced in TiO_2 nanotubes modified by high-energy proton implantation selectively at their top.[3]

Black TiO_2 NTs[33] obtained by a 450°C 2h H_2 treatment of amorphous as prepared NTs array (prepared by anodization of Ti foil) results in material composed by an oxygen-deficient TiO_{2-x} anatase phase (black TiO_2 phase) with low crystallinity (Figure 4.9).

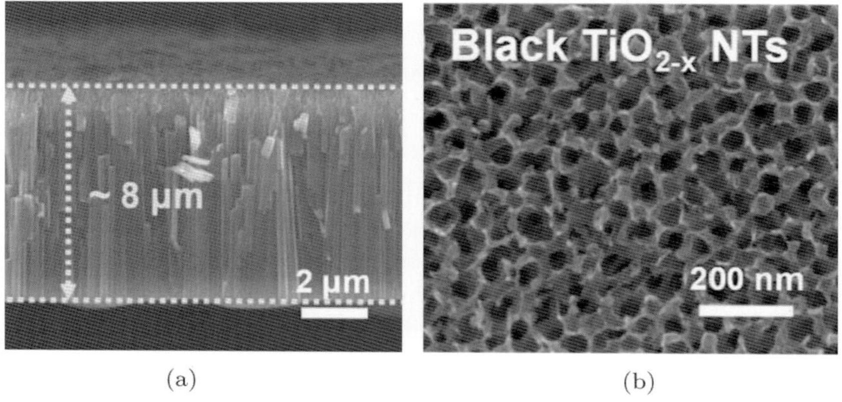

(a) (b)

Figure 4.9. Cross-sectional morphology of the as-prepared TiO_2 NTs and surface morphology of the anatase TiO_2 NTs and black TiO_{2-x} NTs. Reprinted from Ref. [33]. © The Royal Society of Chemistry.

Black TiO$_2$ NTA can be fabricated also by electrochemical self-doping of red TiO$_2$ NTA and annealing at 450°C for 1 h under nitrogen atmosphere. The resulting material kept structure Phase composition (mainly anatase) and morphology of pristine NTs but exhibited a black coloration resulting from the formation of Ti^{3+} sites, or V$_{Os}$ by the cathodic polarization.[34]

Similar results were obtained by Li *et al.*[35] by electrochemical reduction at -1.5 V and the resulting black NTs kept original structure and morphology but the formation of a conductive layer with high concentration of Ti^{3+} defects, up to 22% of total Ti, preferentially confined within shallow internal surface area of the nanotubes. Alternative approaches for electrochemical doping through proton intercalation were reported by the same author.[36]

Lu *et al.*[37] reported the hydrogenation of vertically aligned TiO$_2$ NTs under 5% H$_2$ and 95% Ar flow at 450°C for 1 h. The resulting NTs were fully crystalline, interestingly the amorphous layer present in pristine NTs was completely removed, and only anatase phase was detected. Lattice parameters show no noticeable change along a and b directions, but obvious expansion along the c direction upon hydrogenation treatment. Defects such as Ti^{3+} sites, V$_{Os}$ and hydrogen doping are located only in the bulk, while surface layers are defects-free, in extent as low as lattice parameters are not influenced.

The synthesis of black TiO$_2$ NTs arrays starting from pristine TiO$_2$ NTs arrays obtained via electrochemical anodization, by melted Al reduction treatment, was also reported by Cui *et al.*[38] The resulting material is crystalline and composed only by anatase phase with defects such as Ti^{3+} and V$_{Os}$. Black TiO$_2$ nanotubes codoped with Fe^{3+} and Ti^{3+} were synthesized by annealing Fe-deposited TiO$_2$ NTs under vacuum, resulting in pure anatase phase, without any segregated iron oxide phase.[39]

4.2.4 *Black TiO$_2$ Thin Films*

Thin films are, at the same time, very good model systems and the base of important functional devices, allowing one to have a well-defined material for fundamental studies but in a form suitable for large-scale fabrication. Black TiO_2 thin film obtained by annealing in hydrogen at temperatures 350–500°C a film fabricated by atomic layer deposition (ALD) was reported by V. Gurylev *et al.*[40] Since ALD technique offers several advantages such as atomically smooth surface and precise control of film thickness in the atomic scale, it allows observing even a minor change of the surface. Thermal treatment results in crystallization of amorphous starting phases however, peak intensities of hydrogenated TiO_2 thin films are reduced with respect to the sample annealed in air. This occurrence is usually attributed to the formation of defects in the TiO_2 lattices.[20,27] The shift of a main anatase peak (101) to higher angles is usually attributed to formation of V_{Os}, which are major defects for TiO_2 annealed in hydrogen at such low temperature.[35] The amorphous nature of the as-deposited TiO_2, which is converted to anatase phase after thermal treatment in hydrogen or air was revealed by XRD analysis.

Crystalline core/amorphous external layer mesoporous black TiO_2 nanocrystals were obtained by a Si quantum dot-assisted synthesis by Huang *et al.*[41] and show a crystalline anatase core surrounded by 1 nm thick disordered shell. The presence of Ti^{3+} was reported but no indications about the location were reported.

Nakajima *et al.* reported the formation, by pulsed laser irradiation, of a thin film of black TiO_2, on a $TiO_2/SnO_2{:}Sb/Al_2O_3$ sample, consisting mainly of anatase with some rutile TiO_{2-x} phase with V_{Os} defects.[42] Conversely, Qingli *et al.*[13] obtained a porous black TiO_2 film by a simple hydrothermal treatment of Ti plates using H_2O_2 solution (30 wt.%), the films structure consists in anatase and rutile phases rich of Ti^{3+} sites and V_{Os}, even if the distribution among different phases is not specified, namely, anatase, rutile or an undetected amorphous phase.

Black TiO_2 inverse opals (BTIOs), can be obtained by reduction of WTIOs with H_2 at 500°C for 2 h and show a crystalline anatase scaffold covered by an a disordered layer with a thickness of approximately 1.01–2.07 nm (Figure 4.10).[9] Notably, BTIOs has a specific surface area twice as large as that of the pristine WTIOs,

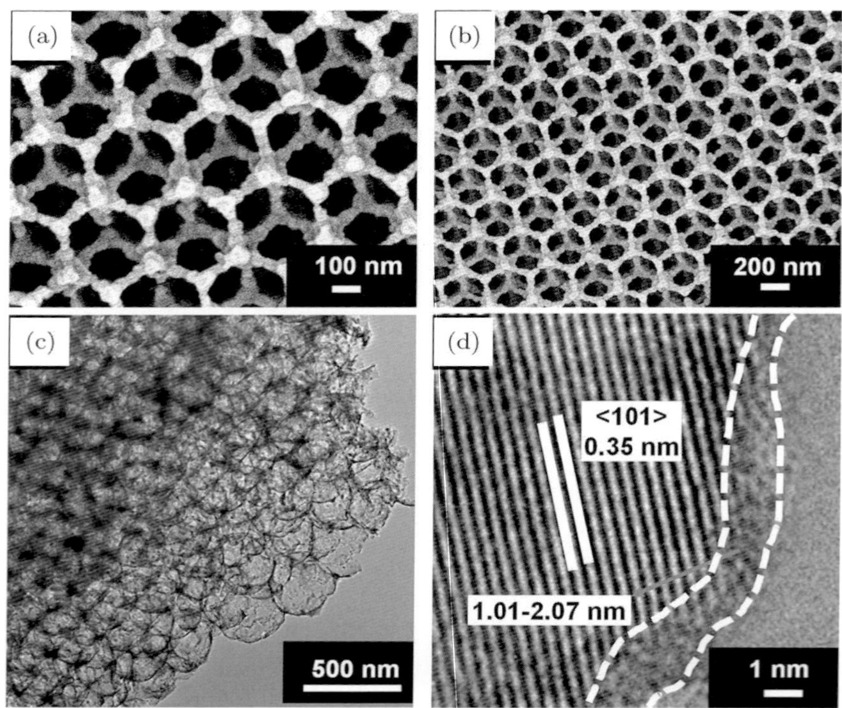

Figure 4.10. FE-SEM (A and B), TEM (C), and HR-TEM (D) image of BTIOs. Reprinted from Ref. [9]. © The Royal Society of Chemistry.

which may be attributed to the presence of more active sites on the hydrogenated surfaces of BTIOs for N_2 adsorption.[9]

Similar inverse opal structures of sulfur-doped black TiO_2 were reported by Liang *et al.*[43] but phase composition is more complex since samples are composed by anatase with lower amount of rutile and brookite, and Ti^{3+}/V_{Os} defects are present. Finally, sulfur is present as polysulfides contained in the three-dimensional matrix both physically trapped and chemically adsorbed on surface.

Hexagonally dimpled of black TiO_{2-x} thin films were obtained by a two-step anodization followed by removing the top oxide layer.[44] The film was constituted by anatase rich in V_{Os} defects but free of reduced Ti^{3+} sites. Such a peculiar structure was attributed to the oxidation process suffering from the insufficient supply of oxygen and to the consumption of Ti^{3+} ions by F^- ion, as they were more active and vulnerable to chemical dissolution.[44]

4.2.5   *Black TiO$_2$ Mesoporous Structures*

Zhou *et al.*[45] developed a facile synthesis of highly ordered mesoporous black TiO_2 materials with a hexagonal mesostructure, a relatively high surface area ($124\,m^2\,g^{-1}$), and a large pore size and pore volume of $\sim 9.6\,nm$ and $0.24\,cm^3\,g^{-1}$, respectively. Obtained material showed typical crystalline-core/disordered layer structure consisting in a highly crystalline anatase core surrounded by a thin disordered reduced surface layer containing Ti^{3+} sites and surface hydroxyl groups resulting from high temperature hydrogenation treatment (Figure 4.11).[45]

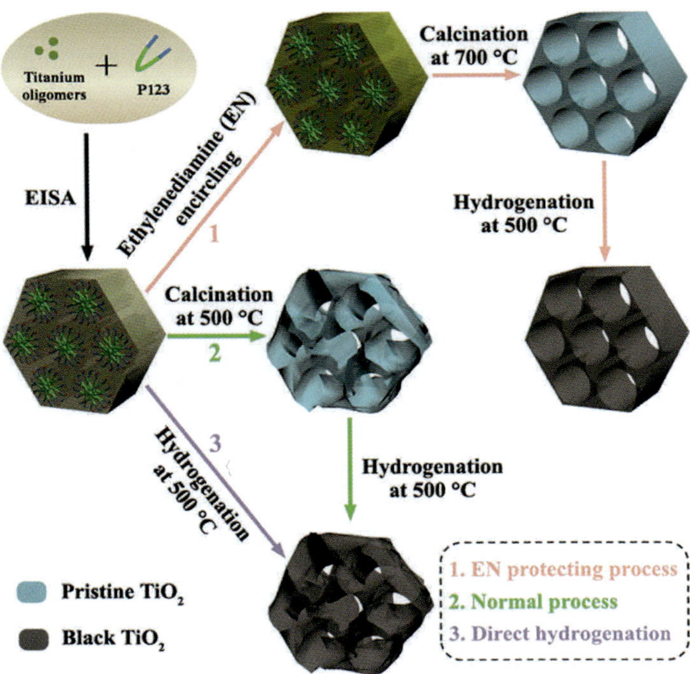

Figure 4.11. Schematic synthesis process for the ordered mesoporous black TiO_2 materials. Reprinted from Ref. [45]. © American Chemical Society.

4.3   Concluding Remarks

Black TiO$_2$ was obtained as a wide number of nanostructures ranging from nanoparticles in powder form to more complex hierarchical one- or two-dimensional nanostructured thin films, such as nanowires — nanotubes arrays. However, this high variability of morphology does not correspond to a similar behavior if structures are concerned: core–shell type structures dominate the scenario. Black TiO$_2$ nanoparticles or nanocrystals usually reveal a crystalline core–amorphous shell structure, and nanowires, nanotubes, and other high aspect ratio systems, show the corresponding crystalline scaffold surrounded by an amorphous layer.

Crystalline phases very often consist in anatase or rutile, which lattice parameter, are strongly influenced by defectivity: V_{Os} induce bonds (i.e. cell parameters) contraction. Obviously, amorphous, disordered shells/layers, irrespectively from the presence or not of defects, show both phenomena contraction and expansion of bonds length. However, some exceptions were reported such as the occurrence of fully crystalline structures.

Point defects are the second feature induced by treatments leading to the formation of black TiO$_2$, and they consist in reduced Ti^{3+} sites, V_{Os}, dangling bonds, and interstitial H, the amount and presence strongly depending on the kind of preparative methodology. It should be stressed that the key role played by these defects inducing the formation of intra bandgap states strongly influences electronic structure (i.e. DOS) and related functional properties (light absorption, photogenerated charge conduction and recombination, etc.).

The location of point defects in crystalline or amorphous disordered structures is a point that, at the same time, is the object of deep studies or is simply not considered, even if it plays an obvious central role in dictating properties of black TiO$_2$ materials.

Another key remark that deserve to be pointed out after a survey of the literature is the difficulty to draw rational correlations between pretreatment parameters (TiO$_2$ starting materials, reducing agent, temperature, heating/cooling cycles) and the resulting structural and electronic features remaining one of the open questions rich of determinant implications not only from a fundamental research point of view but also for the practical applications of black TiO_2.

References

(1) Chen, X.; Liu, L.; Huang, F. *Chem. Soc. Rev.* **2015**, *44*, 1861–1885.

(2) Chen, X.; Liu, L.; Yu, P. Y.; Mao, S. S. *Science* **2011**, *331*, 746–750.

(3) Liu, N.; Häublein, V.; Zhou, X.; Venkatesan, U.; Hartmann, M.; Mačković, M.; Nakajima, T.; Spiecker, E.; Osvet, A.; Frey, L. *et al. Nano Lett.* **2014**, *14*, 3309-3313.

(4) Lin, T.; Yang, C.; Wang, Z.; Yin, H.; Lü, X.; Huang, F.; Lin, J.; Xie, X.; Jiang, M. *Energy Environ. Sci.* **2014**, *7*, 967.

(5) Chen, X.; Shen, S.; Guo, L.; Mao, S. S. *Chem. Rev.* **2010**, *110*, 6503–6570.

(6) Umebayashi, T.; Yamaki, T.; Itoh, H.; Asai, K. *J. Phys. Chem. Solids* **2002**, *63*, 1909–1920.

(7) Chen, X.; Burda, C. *J. Am. Chem. Soc.* **2008**, *130*, 5018–5019.

(8) Naldoni, A.; Allieta, M.; Santangelo, S.; Marelli, M.; Fabbri, F.; Cappelli, S.; Bianchi, C. L.; Psaro, R.; Dal Santo, V. *J. Am. Chem. Soc.* **2012**, *134*, 7600–7603.

(9) Xin, L.; Liu, X.; *RSC Adv.* **2015**, *5*, 71547–71550.

(10) Su, T.; Yang, Y.; Na, Y.; Fan, R.; Li, L.; Wei, L.; Yang, B.; Cao, W. *ACS Appl. Mater. Interfaces* **2015**, *7*, 3754–3763, 150204123703003.

(11) Chen, X.; Zhao, D.; Liu, K.; Wang, C.; Liu, L.; Li, B.; Zhang, Z.; Shen, D. *ACS Appl. Mater. Interfaces* **2015**, *7*, 16070–16077.

(12) Tian, J.; Leng, Y.; Cui, H.; Liu, H. *J. Hazard. Mater.* **2015**, *299*, 165–173.

(13) Qingli, W.; Zhaoguo, Z.; Xudong, C.; Zhengfeng, H.; Peimei, D.; Yi, C.; Xiwen, Z. *J. CO2 Util.* **2015**, *12*, 7–11.

(14) Wang, S.; Zhao, L.; Bai, L.; Yan, J.; Jiang, Q.; Lian, J. *J. Mater. Chem. A* **2014**, *2*, 7439.

(15) Liu, L.; Yu, P. Y.; Chen, X.; Mao, S. S.; Shen, D. Z. *Phys. Rev. Lett.* **2013**, *111*, 065505.

(16) Wang, Z.; Wen, B.; Hao, Q.; Liu, L.-M.; Zhou, C.; Mao, X.; Lang, X.; Yin, W.-J.; Dai, D.; Selloni, A. *et al. J. Am. Chem. Soc.* **2015**, *137*, 9146–9152.

(17) Wei, S.; Wu, R.; Jian, J.; Chen, F.; Sun, Y. *Dalt. Trans.* **2015**, *44*, 1534–1538.

(18) Chen, X.; Liu, L.; Liu, Z.; Marcus, M. A.; Wang, W.-C.; Oyler, N. A.; Grass, M. E.; Mao, B.; Glans, P.-A.; Yu, P. Y. *et al. Sci. Rep.* **2013**, *3*, 1–7.

(19) Naldoni, A.; Fabbri, F.; Altomare, M.; Marelli, M.; Psaro, R.; Selli, E.; Salviati, G.; Dal Santo, V. *Phys. Chem. Chem. Phys.* **2015**, *17*, 4864–4869.

(20) Tian, M.; Mahjouri-Samani, M.; Eres, G.; Sachan, R.; Yoon, M.; Chisholm, M. F.; Wang, K.; Puretzky, A. A.; Rouleau, C. M.; Geohegan, D. B. *et al. ACS Nano* **2015**, *9*, 10482–10488.

(21) Liu, N.; Schneider, C.; Freitag, D.; Venkatesan, U.; Marthala, V. R. R.; Hartmann, M.; Winter, B.; Spiecker, E.; Osvet, A.; Zolnhofer, E. M. *et al. Angew. Chemie Int. Ed.* **2014**, *53*, 14201–14205.

(22) Zhu, G.; Yin, H.; Yang, C.; Cui, H.; Wang, Z.; Xu, J.; Lin, T.; Huang, F. *ChemCatChem* **2015**, *7*, 2614–2619.

(23) Li, M.; Song, W.; Zeng, L.; Zeng, D.; Xie, C. *Mater. Lett.* **2014**, *136*, 258–261.

(24) Wang, S.; Yang, X.; Wang, Y.; Liu, L.; Guo, Y.; Guo, H. *J. Solid State Chem.* **2014**, *213*, 98–103.

(25) Jiang, B.; Tang, Y.; Qu, Y.; Wang, J.-Q.; Xie, Y.; Tian, C.; Zhou, W.; Fu, H. *Nanoscale* **2015**, *7*, 5035–5045.

(26) Wang, H.; Lin, T.; Zhu, G.; Yin, H.; Lü, X.; Li, Y.; Huang, F. *Catal. Commun.* **2015**, *60*, 55–59.

(27) Fujiwara, K.; Deligiannakis, Y.; Skoutelis, C. G.; Pratsinis, S. E. *Appl. Catal. B Environ.* **2014**, *154–155*, 9–15.

(28) Teng, F.; Li, M.; Gao, C.; Zhang, G.; Zhang, P.; Wang, Y.; Chen, L.; Xie, E. *Appl. Catal. B Environ.* **2014**, *148–149*, 339–343.

(29) Wang, Z.; Yang, C.; Lin, T.; Yin, H.; Chen, P.; Wan, D.; Xu, F.; Huang, F.; Lin, J.; Xie, X. *et al. Adv. Funct. Mater.* **2013**, *23*, 5444–5450.

(30) Shan, Y.; Yang, Y.; Cao, Y.; Yin, H.; Long, N. V.; Huang, Z. *RSC Adv.* **2015**, *5*, 34737–34743.

(31) Tian, J.; Leng, Y.; Zhao, Z.; Xia, Y.; Sang, Y.; Hao, P.; Zhan, J.; Li, M.; Liu, H. *Nano Energy* **2015**, *11*, 419–427.

(32) Chen, B.; Beach, J. A.; Maurya, D.; Moore, R. B.; Priya, S. *RSC Adv.* **2014**, *4*, 29443–29449.

(33) Eom, J.-Y.; Lim, S.-J.; Lee, S.-M.; Ryu, W.-H.; Kwon, H.-S. *J. Mater. Chem. A* **2015**, *3*, 11183–11188.

(34) Kim, C.; Kim, S.; Lee, J.; Kim, J.; Yoon, J. *ACS Appl. Mater. Interfaces* **2015**, *7*, 7486–7491.

(35) Li, Z.; Ding, Y.; Kang, W.; Li, C.; Lin, D.; Wang, X.; Chen, Z.; Wu, M.; Pan, D. *Electrochim. Acta* **2015**, *161*, 40–47.

(36) Li, H.; Chen, Z.; Tsang, C. K.; Li, Z.; Ran, X.; Lee, C.; Nie, B.; Zheng, L.; Hung, T.; Lu, J. *et al. J. Mater. Chem. A* **2014**, *2*, 229–236.

(37) Lu, Z.; Yip, C.-T.; Wang, L.; Huang, H.; Zhou, L. *ChemPluschem* **2012**, *77*, 991–1000.

(38) Cui, H.; Zhao, W.; Yang, C.; Yin, H.; Lin, T.; Shan, Y.; Xie, Y.; Gu, H.; Huang, F. *J. Mater. Chem. A* **2014**, *2*, 8612.

(39) Chen, B.; Haring, A. J.; Beach, J. A.; Li, M.; Doucette, G. S.; Morris, A. J.; Moore, R. B.; Priya, S. *RSC Adv.* **2014**, *4*, 18033.

(40) Gurylev, V.; Su, C.-Y.; Perng, T.-P. *J. Catal.* **2015**, *330*, 177–186.

(41) Huang, H.; Zhang, H.; Ma, Z.; Liu, Y.; Zhang, X.; Han, Y.; Kang, Z. *J. Mater. Chem. A* **2013**, *1*, 4162.

(42) Nakajima, T.; Nakamura, T.; Shinoda, K.; Tsuchiya, T. *J. Mater. Chem. A* **2014**, *2*, 6762.

(43) Liang, Z.; Zheng, G.; Li, W.; Seh, Z. W.; Yao, H.; Yan, K.; Kong, D. *ACS Nano* **2014**, *8*, 5249–5256.

(44) Dong, J.; Han, J.; Liu, Y.; Nakajima, A.; Matsushita, S.; Wei, S.; Gao, W. *ACS Appl. Mater. Interfaces* **2014**, *6*, 1385–1388.

(45) Zhou, W.; Li, W.; Wang, J.-Q.; Qu, Y.; Yang, Y.; Xie, Y.; Zhang, K.; Wang, L.; Fu, H.; Zhao, D. *J. Am. Chem. Soc.* **2014**, *136*, 9280–9283.

CHAPTER FIVE

The Black and White Issue of Nanotitania

Guilian Zhu, Tao Xu† and Fuqiang Huang**

**CAS Key Laboratory of Materials for Energy Conversion,
Shanghai Institute of Ceramics, Chinese Academy of Sciences,
Shanghai 200050, China*

*†Department of Chemistry and Biochemistry,
Northern Illinois University, DeKalb, Illinois 60115, USA*

Traditional titanium dioxide (TiO_2) powder, white color, is one of the most extensively studied semiconductors and has a wide bandgap of $\sim$3.0–3.2 eV, corresponding to an absorption edge of $\sim$390–410 nm.[1–5] The white color originates from the highly efficient scattering of the full spectra of visible light by the TiO_2 powders. Therefore, TiO_2 powders become an extensively used pure white pigment, a multibillion industry. In recent two decades, the photocatalytic activity of TiO_2 has been attracting a broad range of research effort. TiO_2 absorbs light with energy larger than its bandgap and produces excited electrons in the conduction band (CB) and excited holes in the valence band (VB).[6–8] Parts of the electrons and holes will transport to the TiO_2 surface and participate in the subsequent electrochemical reaction, accompanying with electron–hole recombination during the charge carrier transport.[9] Consequently, the more light TiO_2 can absorb, the more excited charges are likely to be on the surface and the higher the photoelectrochemical activity is likely to occur. However, the wide bandgap limits the optical absorption of TiO_2 in the ultraviolet (UV) region of the solar spectrum, resulting in both its white color and insufficient utilization of solar energy (less than 5%).[10] Even if TiO$_2$ has a high efficiency in utilizing the UV light, its overall solar activity is thus very limited. This is a bottleneck challenge for TiO$_2$ applications in the areas ranging from photocatalysis and photovoltaics to photo-/electrochromics and sensors. Therefore, it has become a common interest to improve the optical absorption properties of TiO$_2$ in order to enhance its overall photoactivity.

During the past few decades, many efforts have been attempted to enhance solar absorption by creating colored TiO$_2$ through band structure engineering (element doping and oxygen deficiency).[2,4,11–17] All these efforts have brought the absorption of TiO$_2$ into the visible-light region and improved photocatalytic activity has been demonstrated. Nevertheless, the visible-light harvest remains inefficient due to ineffective light absorption and numerous carrier recombination centers.[18] Very recently, a hydrogenated black titania with a bandgap around 1.5 eV is reported to boost the full spectrum of sunlight absorption and the light photocatalytic activity.[19] This discovery has triggered worldwide research interest in black TiO$_2$ nanomaterials.[20–28] As we know, the color of a substance depends on its absorption in visible-light range from red to yellow to blue, while the white color is due to the scattering in full visible spectra and black color is due to full absorption in visible region. Thus, the black coloration of titania suggests its strong absorption in the entire visible-light region.

Black TiO$_2$ can originate from various reasons such as surface disorder, oxygen vacancy (V$_O$) (Ti^{3+}) and hydrogen dopants, etc.[19,21,29,30] In this chapter, we will first describe the possible chemical and structural changes that lead white titania to black titania. Then, we will discuss the properties, synthetic conditions, performance, and applications of black titania.

5.1 Reasons for the Black: Structural and Chemical Changes

The band structure of titania is the most essential factors that determines its various properties, including colors, optical absorption, photocataytic activity, etc. The valence and conduction states of titania are derived mainly from the O 2p orbitals and the Ti 3d orbitals, respectively. Consequently, those factors, which can influence the band structures of titania (introducing mid-gap states/dopant states to O 2p or Ti 3d states, etc.), will result in a coloration change of titania.

The surface disorders were claimed to be a critical factor in changing the band structures and obtaining black titania.[19, 22, 25, 28, 31–36] High-resolution transmission electron microscopy (HRTEM) has been commonly used to differentiate the disordered phase from the crystallaine phase in the black TiO_2 nanoparticles (NPs). A hydrogenated black TiO_2 is successfully prepared by treating pure white TiO_2 NPs under a 20.0-bar pure H_2 atmosphere at about 200°C for 5 days (Figure 5.1(a)). In agreement with

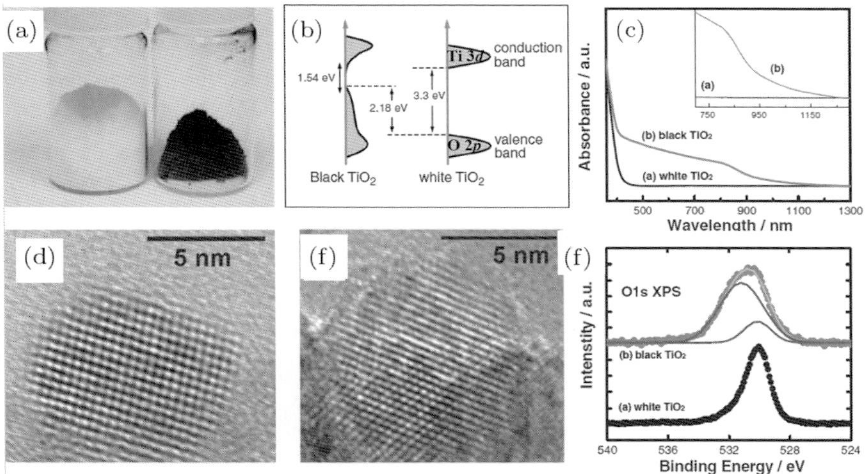

Figure 5.1. (a) Pictures of white and black TiO_2 nanomaterials. (b) Schematic illustration of the DOS of disorder-engineered black TiO_2 nanocrystals, as compared to that of unmodified TiO_2 nanocrystals. (c) Optical absorption spectra of white and black TiO_2 NPs. HRTEM images of (d) white and (e) black TiO_2 NPs. (f) O1s XPS spectra of the white and black TiO_2 nanocrystals. The red and black circles are XPS data. The green curve is the fitting of experimental data for black TiO_2 nanocrystals, which can be decomposed into a superposition of two peaks shown as blue curves.[19] Reproduced from Ref. [19]. © The American Association for the Advancement of Science, 2011.

the color change, the band structures of the two samples are rather different. The density of states (DOS) of disorder-engineered semiconductor nanocrystals, which describes the number of states per interval of energy at each energy level that are available to be occupied in solid-state and condensed matter physics, as compared to those of unmodified nanocrystals, is shown schematically in Figure 5.1(b). The white TiO$_2$ nanocrystals displayed typical DOS characteristics of TiO$_2$, with a bandgap of 3.3 eV. In comparison, many mid-gap states are introduced into the black TiO$_2$. These mid-gap states form a continuum extending to and overlapping with the CB edge, i.e. band tail states. These extended energy states, in combination with the energy levels produced by dopants, greatly reduce the bandgap of black titania to 1.54 eV and can become the dominant centers for optical excitation and relaxation. Correspondingly, the absorption edge shifts from ~375 nm of white TiO$_2$ to ~807 nm of the black one. The onset of optical absorption of the black hydrogenated TiO$_2$ nanocrystals was lowered to about 1.0 eV (~1200 nm) (Figure 5.1(c)). These results demonstrate the substantially narrowed bandgap and enhanced optical absorption of the hydrogenated black titania.

It is noteworthy that whereas the white TiO$_2$ nanocrystals are highly crystallized (Figure 5.1(d)), the black TiO$_2$ NPs are featured with a well-crystallized lattice core surrounded by a lattice disordered shell (Figure 5.1(e)) from the hydrogenation treatment. Many following studies have demonstrated the existence of the disordered surface and its importance in engineering the band structure of synthesized black titania.[21, 22, 26, 28, 31, 33, 37–41] The disordered shell was believed to host the possible hydrogen dopant, and result in the mid-gap states and the black color of the hydrogenated TiO$_2$ NPs. An additional potential advantage of these engineered disorders is that they provide trapping sites for photogenerated carriers and prevent them from rapid recombination, thus promoting electron transfer and photocatalytic reactions.

Additionally, the disordered surface can accommodate abundant hydrogen atoms during the hydrogenation process, as evidenced by the O 1s X-ray photoelectron spectra (XPS) of the white and black

TiO_2 nanocrystals (Figure 5.1(f)). XPS is known as a powerful tool to investigate the chemical binding and VB position on the sample surface. The white TiO_2 displays a single O 1s XPS peak at 530.0 eV, while the O 1s XPS of black TiO_2 can be resolved into two peaks at about 530.0 eV and 530.9 eV. The O 1s peak at 530.0 eV in both the two samples can be assigned to Ti–O bonds in the TiO_2 nanocrystals. The broader peak at 530.9 eV in black TiO_2 is attributed to Ti–OH species, which suggests the hydrogen dopant at the surface. The inserted hydrogens stabilize the lattice disorders by passivating their dangling bonds, but the hydrogen 1s orbital coupling to the Ti atom does not make a substantial contribution to the mid-gap states here.

The surface disorder not only exists in the hydrogenated titania, but also in black titania prepared via other methods. An aluminum reduction method was also reported to successfully prepare black titania,[27] as shown in Figure 5.2(a). Similar crystalline core/amorphous shell structure is observed through the HRTEM images. The pristine white TiO_2 nanocrystals are highly crystallized, as the well-resolved lattice features are shown in Figure 5.2(b). After a 300°C reduction, the TiO_2 nanocrystals have displayed the characteristic core–shell structure with a $\sim$1.5 nm thick disordered surface layer coating on a crystalline core of TiO_2 (Figure 5.2(c)). The thickness of the disordered layer increases with the reduction temperature in the temperature interval of 300–500°C, which can be clearly observed in Figures 5.2(c)–(e). The surface disorder is also a critical factor for the enhancement of the optical absorption and the black coloration of the as synthesized black titania.

In many reported hydrogenated titania, the surface Ti–H and/or Ti–OH bonds are demonstrated to have great influences on their coloration and property variation.[28,29,42–46] The XPS spectra, Fourier transform infrared (FTIR) spectra, ^{1}H nuclear magnetic resonance (NMR) spectra, and X-ray diffraction (XRD) pattern, etc. can all provide information about the Ti–H and Ti–OH bonds in the black titania.[26,29,46–48] In the black titania prepared via a H-plasma assisted method, the Ti–H and Ti–OH bonds are clearly demonstrated to exist.[28] Besides the Ti 2p 3/2 and 2p 1/2 peaks typical for

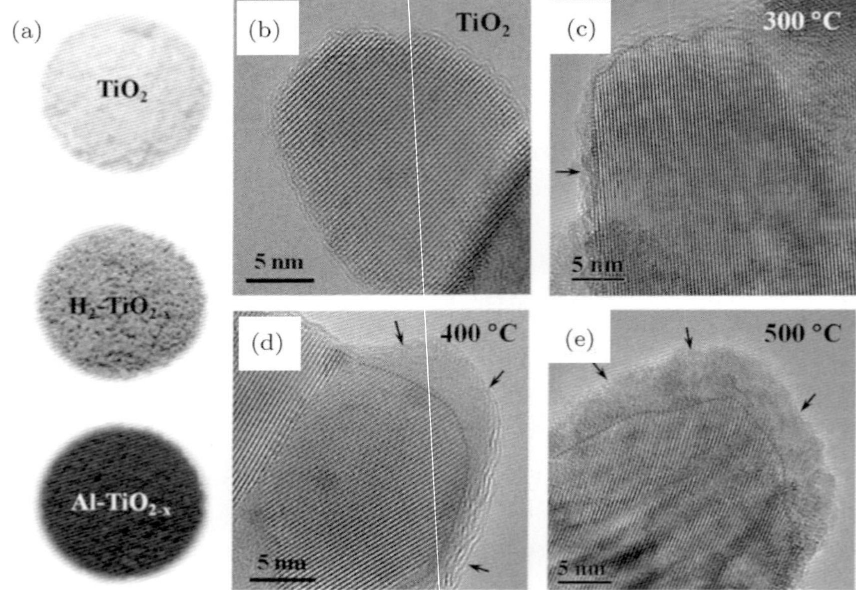

Figure 5.2. (a) Photographs of pristine TiO_2, gray TiO_{2-x} obtained by H_2 anneal (H_2–TiO_{2-x}), and black TiO_{2-x} obtained by Al reduction (Al–TiO_{2-x}); (b–d) HRTEM images of TiO_2 nanocrystals before (a) and after (b–d) the Al reduction at different temperatures for 6 h.[27] Reproduced from Ref. [18]. © The Royal Society of Chemistry.

the Ti^{4+}–O bonds in TiO_2 (centered at binding energies of 458.5 eV and 464.3 eV, Figure 5.3(a)), the hydrogen plasma introduces an additional broad peak centered at 457.1 eV in the Ti 2p XPS spectra, which is attributed to the surface Ti–H bonds. A characteristic small O 1s XPS peak for Ti–OH bonds at about 531.8 eV is also observed in the black titania, as shown in the Figure 5.3(b). Additionally, the four extra absorption peaks at 3645, 3670, 3685, and 3710 cm^{-1} in the FTIR spectra (Figures 5.3(c) and (d)) and the two additional sharp resonances at chemical shifts of 0.4 ppm and 0.01 ppm in the ^1H NMR spectra (Figures 5.3(e) and (f)), are all powerful evidence to verify the existence of Ti–OH bonds in the black titania.

In a hydroxylated TiO_2 with various degrees of blackness derived from amorphous hydrate, synthesized via a one-step aqueous reaction

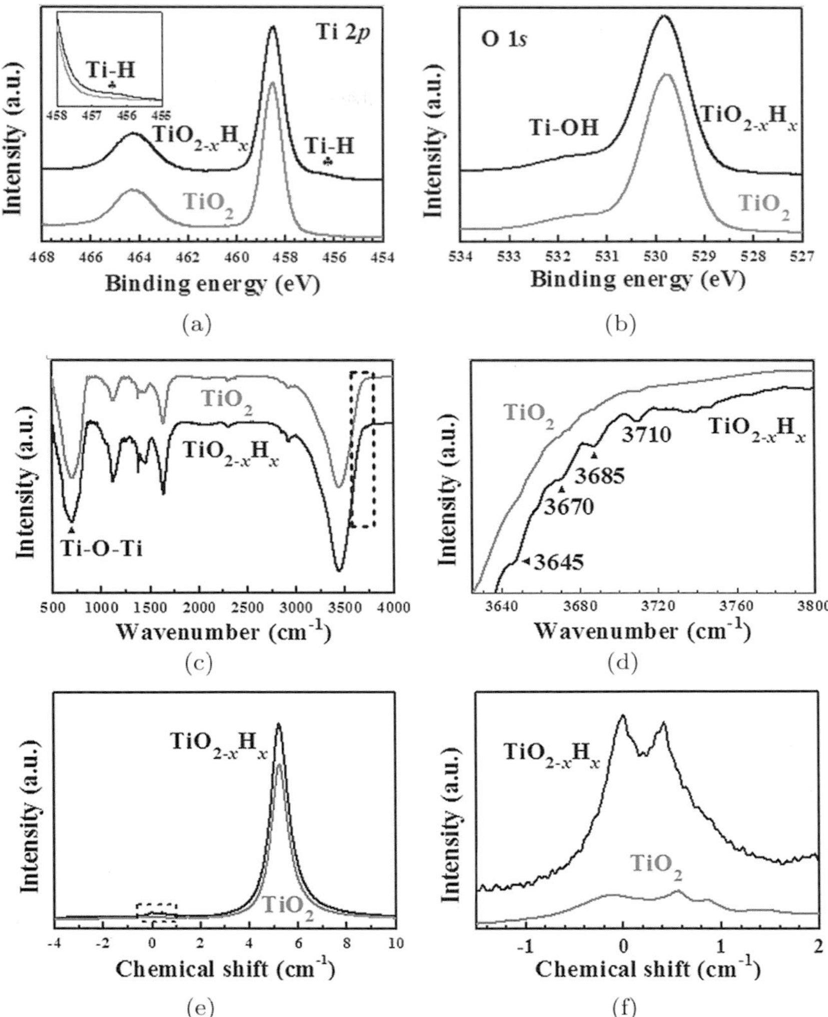

Figure 5.3. Pristine white TiO_2 and black $TiO_{2-x}H_x$: (a) Ti 2p XPS spectra, (b) normalized O 1s XPS spectra (c and d) FTIR and (e and f) ^{1}H NMR spectra.[28] Reprinted from Ref. [28]. © Wiley-VCH.

method assisted by ultrasonic irradiation of high power intensity, the decisive influence of hydroxyls on electronic structure and optical absorption ability of the black titania is demonstrated.[46] The original white TiO_2 was synthesized by simple one-step aqueous reaction of

Ti(SO$_4$)$_2$ and ammonia water and subsequently dried at 80°C. After the sol of original TiO$_2$ was treated under ultrasonic irradiation, the powder of ultrasonic treated TiO$_2$ would turn to black after drying at 80°C, and the extension of ultrasonic time would result in a deeper blackness of ultrasonic treated TiO$_2$. Figure 5.4(a) displays the appearance of ultrasonic treated TiO$_2$ with various degrees of blackness comparing with the original white TiO$_2$. In correspondence, the DOS of the amorphous hydroxylated black TiO$_2$ shows larger blue-shift of valence band maximum (VBM) toward the Fermi energy and further results in the narrower modified bandgap compared with the pristine white TiO$_2$ (marked by blue arrows in Figure 5.4(b)). The easier electronic transitions from tailed VB to CB substantially enhanced the optical absorption of amorphous hydroxylated TiO$_2$ and finally make the deeper blackness in appearance as the ultrasonic time extends.

The underlying mechanism, which results in the VB band tailing and electronic structure change, lies in the ultrasonic induced hydroxyls on the amorphous TiO$_2$, as evidenced by the O 1s XPS spectra in Figures 5.4(c) and (d). The O 1s spectra show similar shapes for the original TiO$_2$ and ultrasonic treated TiO$_2$ for different hours and the single O 1s peak in each spectrum can be divided into two symmetric peaks, one locates at 530 eV typical for the oxygen of Ti–O bonds in TiO$_2$, and the other one locates between 530.9 eV and 532 eV for the oxygen of Ti–OH bonds. As the area of Gauss peaks in each O 1s XPS spectrum represents the amount of Ti–O and Ti–OH bonds in amorphous hydroxylated TiO$_2$ respectively, it is obvious that the hydroxylation degree of amorphous TiO$_2$ became larger with the extension of ultrasonic time, which confirms that the ultrasonication introduced hydroxyls on TiO$_2$. In a word, with the extension of ultrasonic time, there would be more hydroxyls introduced on the amorphous TiO$_2$, which changed the electronic structure and further induced the localized band bending and bandgap narrowing, resulting in the improvement of optical absorption and the deeper blackness of amorphous hydroxylated TiO$_2$.

Besides the surface disorders and hydrogen dopants, V_{Os} and/or Ti^{3+} states are known to be effective in engineering the electronic structures of TiO_2 and synthesizing colored titania.[20, 26, 27, 40, 43, 49–61] It is believed that the defects would introduce localized states in the bandgap of the titania and thus improving the light absorption and resulting in the colorations. The formation of V_{Os} and/or Ti^{3+} species in the black titania is supported by several pieces of experimental evidence: the measured g-tensors typical of the Ti $3d^1$ state or the V_O with a trapped electron in the electron spin resonance spectroscopy (ESR); the shift in the core level binding energies of the reduced Ti atoms in the X-ray photoelectron spectroscopy spectra; the occurrence of a gap state at $\sim 2\,eV$ above the VBM and $\sim 1\,eV$ below the CB minimum, which is observed by photoelectron spectroscopy; the V_O-dependent diethyl ether cataluminescence (CTL) whose intensity in diethyl ether oxidation reaction on the surface of TiO_2 NPs is proportional to the content of V_{Os}. However, the different experimental techniques mentioned above have different ability to detect the Ti^{3+} species: some techniques mainly probe bulk-like Ti^{3+} ions, while others are more sensitive to surface like Ti^{3+} ions. Combined with the difference of synthesis condition, different characterization results will be obtained for different black titania.

In a hydrogenated-titanium dioxide ($H–TiO_2$) nanocrystals prepared via annealing TiO_2 in H_2/N_2 atmosphere,[62] the V_{Os} play decisive roles in the bandgap engineering and synthesizing colored titania. Increasing the annealing temperature result in a continuously deeper coloration and smaller bandgap of the $H–TiO_2$, as shown in Figures 5.5(a) and (b). Correspondingly with this change, the V_{Os}, observed by the symmetrical signals at $g = 2.002$ and $g = 2.001$ in the EPR spectra of $H–TiO_2$ samples (Figure 5.5(c)), displays a regular increase from 300°C to 500°C. It can be concluded that the introduced V_O could downward-shift Fermi level, which leads to the narrowed bandgap and deeper coloration. The density of V_O reaches saturation point on the surface of $H–TiO_2$ at 500°C, resulting in the signal intensity decline of the 600°C-annealed sample. While in the

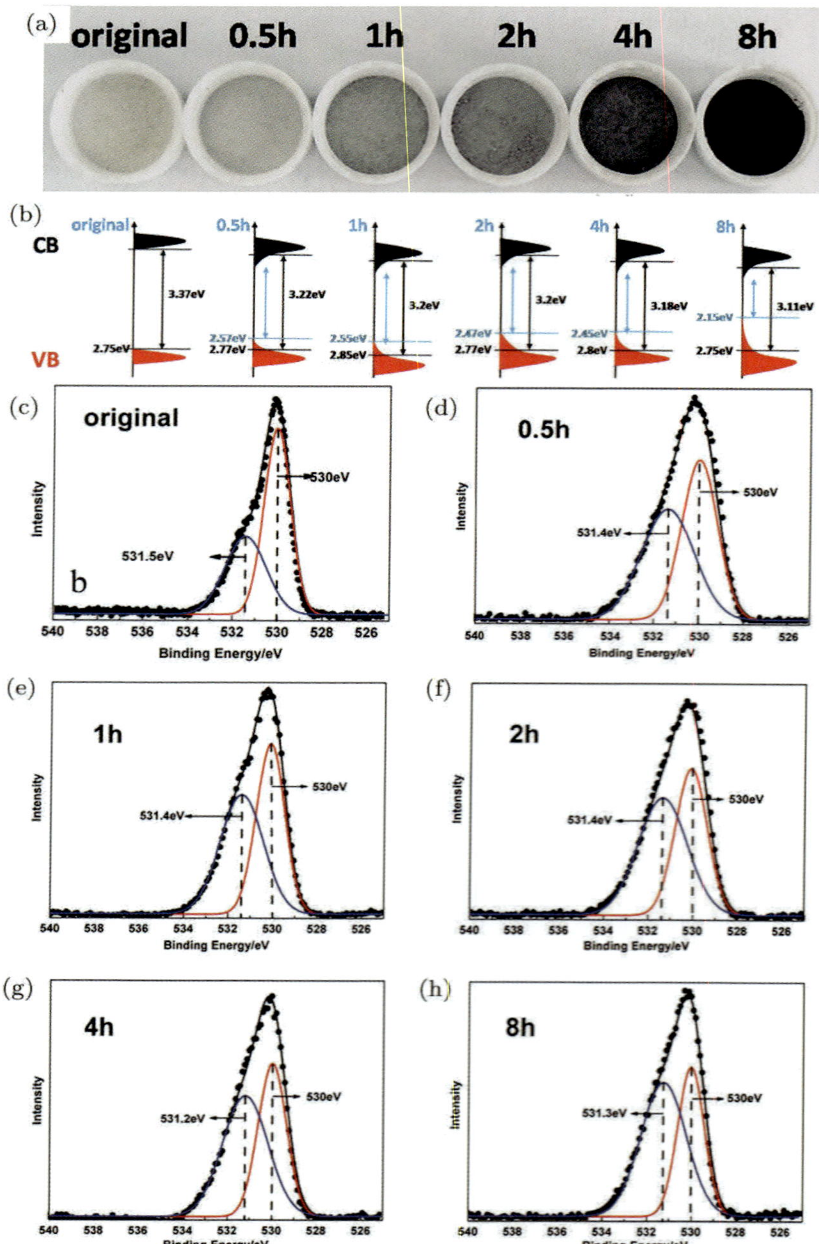

Figure 5.4.

Figure 5.4. (a) A photo comparing the appearance of original TiO$_2$ and ultrasonic treated TiO$_2$ for different hours. (b) The schematic illustrations of DOS of original TiO$_2$ and amorphous hydroxylated TiO$_2$ prepared through ultra-sonication for different hours. The black and blue arrows indicate the bandgaps before and after localized band bending respectively. (c) O 1s XPS spectra of original TiO$_2$ and amorphous hydroxylated TiO$_2$ prepared through ultra-sonication for different hours.[46] Reprinted from Ref. [46]. © Nature Publishing Group.

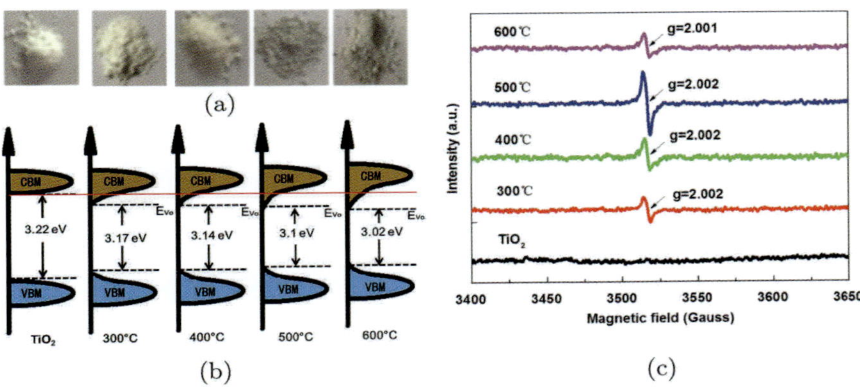

Figure 5.5. (a) Color variation of TiO$_2$ and H–TiO$_2$ samples at different temperature. (b) Schematic diagrams of electronic band structure of TiO$_2$ and H–TiO$_2$. E$_{Vo}$ located below the TiO$_2$ CB represents the energy levels of V$_O$. (c) Electron Paramagnetic Resonance (EPR) spectra recorded at 300 K for TiO$_2$, H–TiO$_2$ samples.[62] Reprinted from Ref. [62]. © American Chemical Society.

black titania prepared through a scalable method of oxidizing TiH$_2$ in H$_2$O$_2$ followed by calcinations in Ar gas (Figure 5.6(a)), the g values of the black rutile titania are consistent with the perpendicular and parallel components of the axially symmetric lattice Ti^{3+} centers in the rutile environment with $g_\perp = 1.975$ and $g_\parallel = 1.943$, respectively. The different g factors observed in the EPR results of these two black titania can be ascribed to the defects location, i.e. defects in the bulk or in the surface. The XPS spectrum of Ti 2p can also shows evidence of the Ti^{3+}. Besides, the typical Ti 2p$_{3/2}$ and 2p$_{1/2}$ XPS peaks for the Ti^{4+}–O bonds in TiO$_2$, peak shifts and/or additional peaks of Ti^{3+} would show in the XPS spectra.[21,63,64]

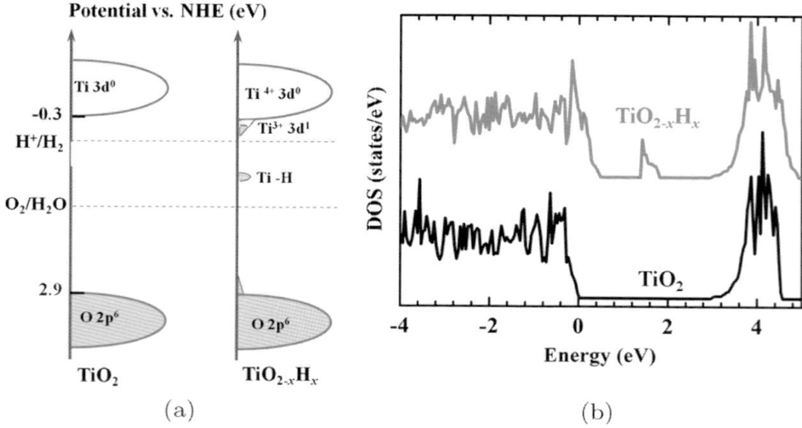

Figure 5.6. (a) Schematic electronic structures for hydrogenated titania (TiO$_{2-x}$H$_x$) and the pristine TiO$_2$. (b) Calculated DOS of pristine TiO$_2$ and black TiO$_{2-x}$H$_x$.[28] Reprinted from Ref. [28]. © Wiley-VCH.

Although the various defects in the black titania (structure disorders, V$_{Os}$, Ti^{3+}, H dopants) can function individually, many reports have demonstrated that the defects would coexist in the black TiO$_2$ and work together to change its band structure and result in the black coloration. For example, in the H-plasma prepared black TiO$_{2-x}$H$_x$, both the Ti^{3+} and H insertion influence its band structure, as shown in Figure 5.6. The surface defects and lattice reconstructions in hydrogenated TiO$_{2-x}$H$_x$ induce shallow energy levels and produce strong band tailing near the CB minimum, which leads to a remarkable bandgap narrowing ($\sim$0.8 eV). Furthermore, the inserted H atoms (Ti–H bondings) introduce localized states at 0.92–1.37 eV below the CB minimum of black TiO$_{2-x}$H$_x$. Therefore, the electronic transitions from the tailed VB and midgap electronic states to CBM combine to result in the more efficient optical absorption in TiO$_{2-x}$H$_x$.[28] In other literatures, the combinations of structure disorder and Ti^{3+},[21,27] the structure disorder and H dopants,[22] and the V$_{Os}$/Ti^{3+}, and H dopants,[65,66] are responsible for the black coloration and band structure change of TiO$_2$.

5.2 Properties of the Black and White TiO$_2$

We can easily draw the conclusion that the properties of the black TiO$_2$ will be rather different from the white one, owing to the dramatic chemical and structural changes. The most apparent difference lies in the optical absorption ability, which can be easily observed by us, the coloration variation from white to black or other colorations (blue, yellow, gray, etc.). The darker coloration reflects an enhanced light absorption ability compared with the white titania, which can be characterized effectively by the optical absorption spectra. The solar absorption ratios can be calculated from the standard solar spectrum and the absorption spectra of the samples. The black titania is reported to absorb more than 80% of the total solar energy by improving visible and infrared absorption, while the pristine TiO$_2$ only absorbs less than 5%.[28,40] For the high-pressure hydrogenated black TiO$_2$,[19] the absorption edge shifts from ~375 nm of the white TiO$_2$ to ~807 nm and the onset of its optical absorption was lowered to about 1.0 eV (~1200 nm), which results in a strong absorption through the whole visible light region in this black TiO$_2$. Similarly, in other hydrogenated P25 TiO$_2$,[36] as the hydrogenation duration time increases up to 15 days, black titania are obtained and the abrupt bandgap absorbance wavelength (absorption edge) shifts to 680 nm from the original ~400 nm, with the onset of the optical absorption lowering to 1.0 eV (Figure 5.7(a)). The huge shifts of absorption edge, which reflects the bandgap changes, are demonstrated by many reports.[37,44,65,67–69] However, in many other reports, the black and white TiO$_2$ show similar absorption edges in spite of the greatly enhanced optical absorption of black TiO$_2$.[27,28,40,43,50,70,71] The enhanced visible and near-infrared (NIR) light absorption are suggested to be ascribed to the isolated states between the forbidden gap in the black TiO$_2$ rather than a shift in the position of either band edge. For example, the Ti^{3+} self-doped blue TiO$_{2-x}$ sample, the considerably large absorption tail in the visible and NIR regions is observed, which is consistent with the change in color of the powders from white to deep blue (Figure 5.7(b)). Thus, the optical spectra of the blue titania show no

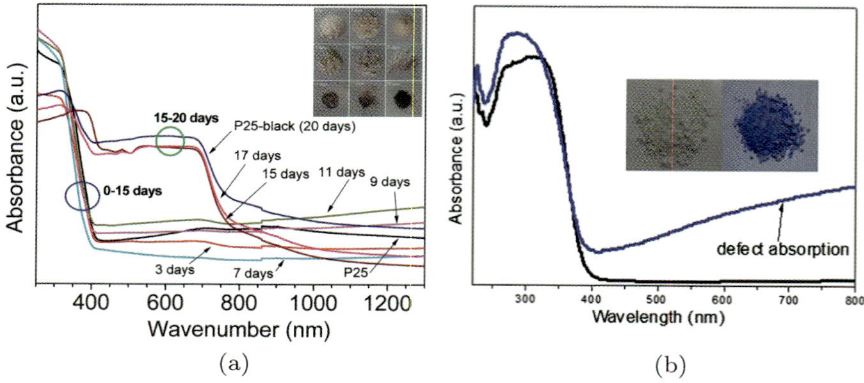

Figure 5.7. (a) UV–Vis spectra of P25 TiO$_2$ treated under hydrogen for 0–20 days at 35 bar hydrogen atmosphere at room temperature.[36] Reprinted from Ref. [36]. © The Royal Society of Chemistry. (b) The UV–Vis diffuse reflectance spectra for the blue TiO$_{2-x}$ (blue line) and the commercial anatase (black line).[50] The insets are the corresponding photographs. Reprinted from Ref. [50]. © The Royal Society of Chemistry.

shifts of the absorption edge and the calculated bandgap of TiO$_{2-x}$ NPs is ∼3.25 eV, which is very close to the typical bandgap of 3.2 eV for anatase.[50]

Additionally, the carrier (electron) concentration and electron transport behavior of the black TiO$_2$ will be greatly changed due to the numerous defects, e.g. Ti^{3+} states or/and V$_{Os}$, which serve as carrier donors. Hydrogen plasma treated n-type TiO$_2$ films obtained a rather high electron concentration (7.3×10^{20} cm^{-3}, 7.8×10^{20} cm^{-3}) resulting in a low sheet resistance ($98.3\,\Omega\,\mathrm{sq}^{-1}$, $69.6\,\Omega\,\mathrm{sq}^{-1}$).[28] In many literatures, the electrochemical impedance spectroscopy (EIS) is used to study charge transfer and recombination properties of the sample. Mott–Schottky plots can provide important information about carrier density and flatband potential of the samples. The Mott–Schottky plots of a hydrogen-treated rutile TiO$_2$ (H:TiO$_2$) nanowire and the corresponding white TiO$_2$ is shown in Figure 5.8.[42] The positive slope in the Mott–Schottky plots for all TiO$_2$ nanowire samples indicates the n-type character of TiO$_2$. The huge difference of the slope of the linear parts in the two plots indicates a tremendous disparity of donor density, as the carrier density is

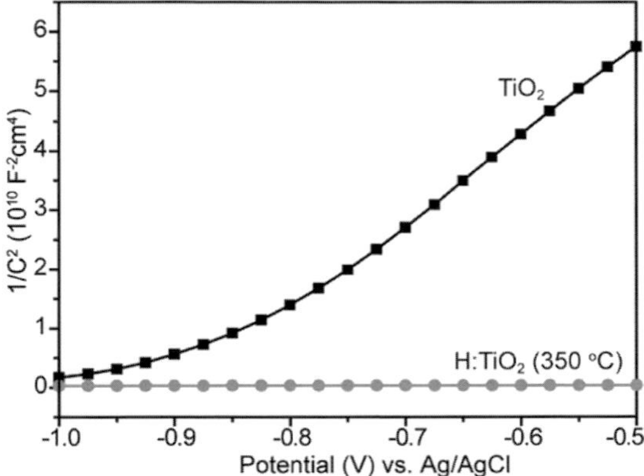

Figure 5.8. Mott–Schottky plots collected at a frequency of 5 kHz in the dark for the pristine TiO$_2$ and the H:TiO$_2$ nanowires annealed at 350°C.[42] Reprinted from Ref. [42]. © American Chemical Society.

inversely proportional to the slope. The calculated electron densities of the pristine TiO$_2$ and black H:TiO$_2$ nanowires were 5.3×10^{18} cm^{-3} and 2.1×10^{22} cm^{-3}, respectively. The enhanced donor density can effectively improve the charge transport in black H:TiO$_2$, which would further reduce the electron–hole recombination. Many black titania or other black nanooxides prepared through various methods show similar results and possess much higher carrier density than the pristine white one.[27,58,72–74]

The photoluminescence (PL) emission spectra is a powerful tool to investigate the efficiency of charge carrier trapping, migration, transfer, and separation, and to understand the fate of photogenerated electrons and holes in semiconductors, since PL emission results from the recombination of free carriers. The PL emission intensity of black TiO$_2$ is widely reported to show apparent decrease compared to the white TiO$_2$, which indicates that the recombination rate of photogenerated electrons and holes has been inhibited in the black TiO$_2$. The suppressed electron–hole recombination is ascribed to formation of V_{Os}/Ti^{3+} or H dopants during the hydrogenation, which can serve as electron capture traps, and hence separates the charge carriers and reduces the recombination.[21,27,38,67,75,76]

5.3 Synthetic Methods to Obtain Black TiO$_2$

Various black TiO$_2$ nanomaterials will display different chemical and physical properties, because their fabrication methods and conditions vary from one another and strongly affect the properties and performance of the obtained TiO$_2$ nanomaterials. High-pressure pure hydrogen treatment was firstly utilized to prepare black TiO$_2$ and triggered the tremendous researches on hydrogenation of black TiO$_2$ thereafter. Additionally, chemical reduction and electrochemical reduction are also demonstrated to be effective in synthesizing black TiO$_2$. All of the above methods can be classed as the reductive type methods, obtaining black titania through inducing defects in the pristine white TiO$_2$. Some oxidative type methods are also reported to synthesize black titania successfully through oxidization of the TiH$_2$.

5.3.1 *Hydrogen Thermal Treatment Methods*

Hydrogen is known as a powerful reductant, it is able to activate facilely by thermal or electromagnetic energy and reduce TiO$_2$ into other chemical species (V$_{Os}$, Ti^{3+}, or other reduction states) or insert into the TiO$_2$ lattice, and thus the lattice structure and other physical/chemical properties of TiO$_2$ will change accordingly. However, the reduction process is rather complex and greatly influenced by many treatment conditions, e.g. the amount of reactants, the chemical properties of the starting TiO$_2$ nanomaterials, the pressure and concentration of the hydrogen gas, and the reaction temperature. Consequently, the variety of hydrogen treatment conditions results in variations of the properties and performance and the accompanying explanations of different black TiO$_2$ prepared by different researchers under different conditions. These variations and complexities, on the other hand, also provide us a large extent of flexibility and many opportunities in tuning the chemical/physical properties and performance of black TiO$_2$ nanomaterials towards our different needs.

Up to now, hydrogen thermal treatments are still the mostly studied methods to prepare black titania, and it does not confine to high-pressure pure hydrogen. Many hydrogenation methods, including ambient or low-pressure pure hydrogen treatment, and ambient hydrogen–argon/nitrogen treatment are developed to prepare black titania. In a typical hydrogen thermal treatment, the pristine white TiO_2 is firstly synthesized or commercially purchased, and then it is treated under a pure H_2 or H_2–Ar/N_2 atmosphere at different temperatures (generally high than 350°C) for a certain time to obtain black TiO_2.[26] High pressure H_2 is used to promote hydrogen insertion or TiO_2 reduction at relatively low temperature. For example, black TiO_2 NPs can be prepared under a 20.0-bar pure H_2 atmosphere at about 200°C for 5 days,[19] or under 35 bar-hydrogen atmosphere at room temperature (about 15°C) after 15 days.[36]

Liu *et al.* compared the color change of anatase TiO_2 nanotubes hydrogenated under various conditions: (i) an anatase TiO_2 nanotube layer (air); (ii) this layer converted with Ar (Ar) or H_2–Ar (H_2–Ar); (iii) a high pressure H_2 treatment (20 bar, 500°C for 1 h) (HP–H_2); and (iv) a high pressure H_2 treatment but mild heating (H_2, 20 bar, 200°C for 5 days) (Sci ref).[65] After the reduction treatments, the color of the TiO_2 nanotubes turns from white to black in H_2–Ar, deep purple in pure Ar (500°C for 1 h), light blue in pressurized H_2 at high temperature (20 bar, 500 1°C for 1 h), and gray in pressurized H_2 at mild temperature (H_2, 20 bar, 200°C for 5 days), as shown in Figure 5.9, which also displays their UV–Vis absorption spectra. Apparently, the hydrogenation condition plays a critical role in the final color and optical properties of the hydrogenated TiO_2 nanotubes.

Considering that hydrogen plasma offers high-energy species such as electrons, atoms, and radicals and has improved activity over the hydrogen atoms, the hydrogen plasma process is utilized as an effective method to prepare black TiO_2 with rather high solar absorption.[28,38,43,77] For example, a black TiO_2 nanotubes sample is successfully prepared through a hydrogen plasma assisted chemical vapor deposition.[78] The hydrogenation was performed in a hot filament chemical vapor deposition apparatus with hydrogen as reaction gas (Figure 5.10(a)). During the treatment process, the TiO_2 was loaded in a corundum boat under the filament. The temperature

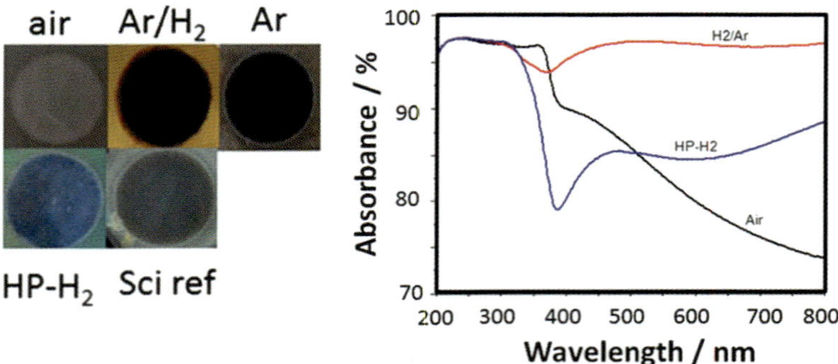

Figure 5.9. Images and the UV–Vis absorption of the TiO$_2$ nanomaterials treated under various conditions.[65] Reprinted from Ref. [65]. © American Chemical Society.

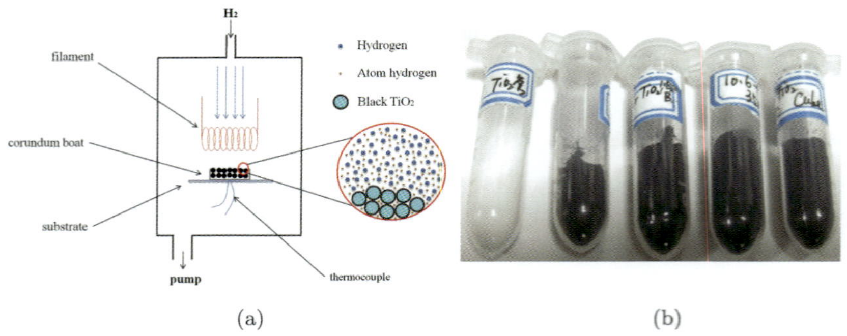

Figure 5.10. (a) Illustration of hydrogen plasma hot-filament chemical vapor deposition setup. (b) Pictures of the pristine and hydrogenated black TiO$_2$ NPs.[78] Reprinted from Ref. [78]. © American Chemical Society.

of the filament was maintained at 2000°C, and the samples were treated for 3 h at 350–500°C. All the hydrogenated-TiO$_2$ NPs had black color, as shown in Figure 5.10(a).

5.3.2 *Chemical Reduction Methods*

In spite of the various benefits of hydrogen thermal treatment, there exist an apparent deficiency that H$_2$ activation lacks uniformity due to the sharply decreased concentration of active hydrogen atoms in the internal TiO$_2$ powder. Additionally, H$_2$ reduction needs high annealing temperature (over 350°C) or high pressure, and H_2 is flammable and explosive. Many researches have been made to develop new reductants instead of the hydrogen, and aluminum,[27,41,56,58,74,79] zinc,[80] CaH_2,[40] $NaBH_4$,[55,81,82] and imidazole[83] are all demonstrated to be effective in preparing black TiO_2.

Aluminum can be used as the reductant for preparing black titania, as illustrated in Figure 5.11(a). TiO_2 samples and aluminum were placed separately in a two-zone tube furnace and then evacuated to a base pressure lower than 0.5 Pa. After that, aluminum was heated at 800°C, and TiO_2 samples were heated at 300–600°C for 6 h. According to Ellingham diagram (Figure 5.11(b)), the reaction

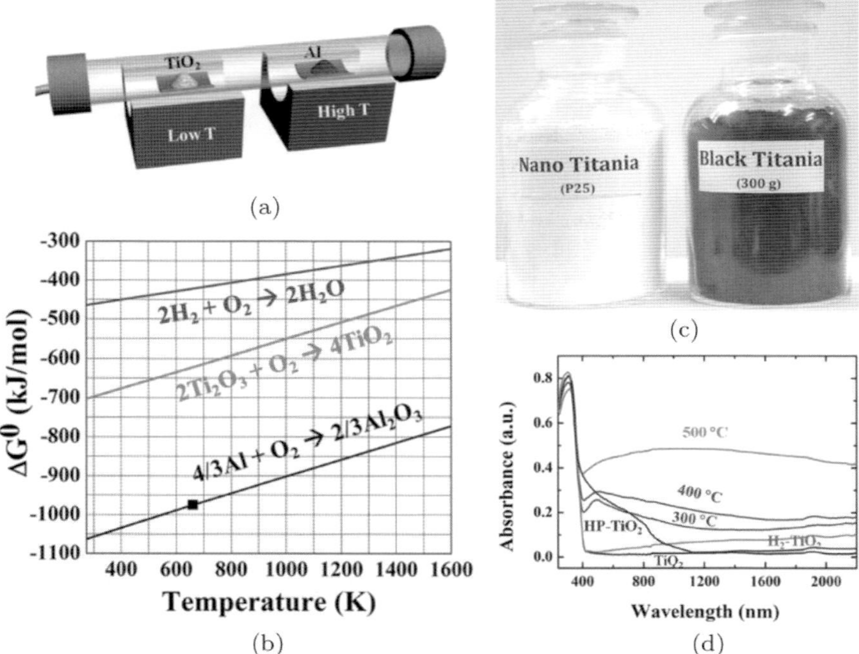

Figure 5.11. (a) Schematic low-temperature reduction of TiO_2 in a two-zone furnace. (b) Ellingham diagram of ΔG vs. temperature. (c) Mass production of black titania (TiO_{2-x}) using our Al-reduction method. (d) Absorption spectra of TiO_{2-x} samples reduced at different temperatures, the H_2 annealed H_2–TiO_{2-x} and high-pressure hydrogenated black titania (HP-TiO_2, from Ref. [19]).[27] Reprinted from Ref. [27]. © The Royal Society of Chemistry.

driving force enables aluminum oxidation and titania reduction thermodynamically. The Gibbs free energy ΔG^0 (T) at different temperature of reaction $2Ti_2O_3 + O_2 \rightarrow 4TiO_2$ lies above that of $4/3Al + O_2 \rightarrow 2/3Al_2O_3$ but below that of $2H_2 + O_2 \rightarrow 2H_2O$ line, so elemental Al can reduce TiO_2 to Ti_2O_3 but H_2 cannot reduce TiO_2 at low temperature. Large amount of black TiO_{2-x} was achieved from one batch by the Al-reduction route (Figure 5.11(c)) and the Al-reduced TiO_{2-x} samples extend the photoresponse from UV light to visible and infrared light region (Figure 5.11(d)). This aluminum reduction method has been applied to prepare other black nano-oxides with high optical absorption, such as black brookite TiO_2, Nb_2O_5, $SrTiO_3$, etc. Additionally, thermal treatment of the mixed aluminum powder and TiO_2 will also produce black titania.

As is known, zinc is also an active metal that has stronger reducibility than many other elements, e.g. Fe, H_2, Ti, etc. It has been reported that stable Ti^{3+} self-doped TiO_2 with tunable phase composition and highly efficient visible-light photoactivity was synthesized via a metallic zinc-assisted method.[76] In this method, Zn not only acts as a reductant leading to the formation of Ti^{3+} ($Zn + Ti^{4+} \rightarrow Zn^{2+} + Ti^{3+}$), but also could stabilize the V_{Os} and hence Ti^{3+} on the surface. $NaBH_4$ can be a promising reducing reagent because of its high reducing ability of reducing Ti (IV) to Ti (III). The reaction is as follows: $NaBH_4 + 8OH^- \rightarrow NaBO_2 + 8e^- + 6H_2O$; $Ti^{4+} + e^- \rightarrow Ti^{3+}$.[81] The $NaBH_4$ treatment introduces V_{Os} on the surface and interior of TiO_2 and extends the absorption of TiO_2 to visible and infrared region. CaH_2, which can decompose and release highly active hydrogen atom at relatively low temperature, can be used to prepare black hydrogenated TiO_2 instead of the hydrogen gas, avoiding the risk of explosion and facilitating the hydrogenation process. Black titania with superior solar absorption is obtained via CaH_2 assisted reduction due to the H-dopants and as-induced defects.[40]

5.3.3   *Electrochemical Reduction Methods*

Through applying a cathodic bias, small ions can be doped into the TiO_2 or Ti^{4+} will be reduced to Ti^{3+}, thus the electrochemical

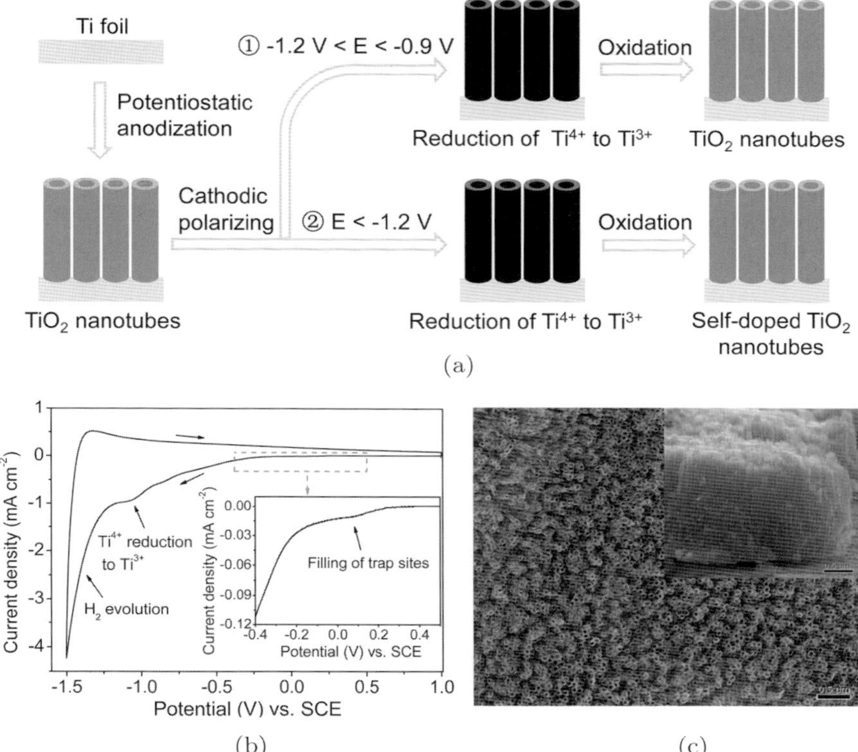

Figure 5.12. (a) Schematic diagram showing the fabrication of electrochemically self-doped TiO$_2$ NTAs where "E" is the potential used in the polarization process. With respect to the cathodic potential applied, two procedures can be distinguished: (1) completely reversible reduction and oxidation of the TiO$_2$ NTAs; (2) some Ti^{3+} remains upon the post-oxidation, leading to the self-doped TiO$_2$ NTAs. (b) CV curve of the TiO$_2$ NTAs in 0.5 M Na$_2$SO$_4$ at a scan rate of $100\,\mathrm{mV\,s^{-1}}$; the inset shows an enlarged view of the region contained within the dashed lined rectangle. (c) FE-SEM image of the $-1.4\,\mathrm{V}$ TiO$_2$ sample with an inset showing a side view.[87] Reprinted from Ref. [87]. © American Chemical Society.

reduction could be very effective in preparing black titania.[84–86] For example, self-doped TiO$_2$ highly ordered nanotube arrays (NTA, SEM images seen in Figure 5.12(c)), were fabricated by a two-step approach, as illustrated in Figure 5.12(a).[87] Initially, the fabrication of TiO$_2$ NTAs was accomplished by anodization of Ti foil at potentiostatic mode in an electrolyte containing F$^-$ ions followed by a thermal treatment at 450°C under ambient atmospheric condition to attain anatase crystal. The as-prepared arrays underwent a cathodic polarization process at suitable negative potential using inert aqueous electrolyte to achieve self-doped TiO$_2$ NTAs. Figure 5.12(b) shows a cyclic voltammetry curve for the TiO$_2$ NTAs recorded in 0.5 M Na$_2$SO$_4$ aqueous solution. The apparent peak at ca. -0.9 V (vs. SCE) actually results from the reduction of Ti^{4+} to Ti^{3+} accompanied by charge compensation via proton (or other small cations such as Li$^+$, Na$^+$, and K$^+$, if present) intercalation (Ti^{4+} + e$^-$ + H$^+$ $\rightarrow$ Ti^{3+}H$^+$), and the further increase at more negative voltage of *ca.* -1.2 V is due to hydrogen evolution (2H$^+$ + 2e$^-$ $\rightarrow$ H$_2$).

5.3.4 *Chemical Oxidation Methods*

All of the above black titania are prepared through a reduction of the pure white TiO$_2$, which initiates from the surface of TiO$_2$. The hydrogen insertion and/or formation of V$_{Os}$/Ti^{3+} can easily occur on the TiO$_2$ surface, while it is hard for the defects form in the bulk of the TiO$_2$ under the same condition, as evidenced by the disordered shell/crystalline core structure of many black TiO$_2$. In another word, the introduced defects are hard to reach in a high-concentration and would be unevenly distributed in the TiO$_2$ matrix, abundant in the surface layer, and gradually decreasing in the bulk. The surface V$_{Os}$ in TiO$_2$ tend to be thermally metastable in air because the Ti^{3+} is easily oxidized, over time even by the dissolved oxygen in water. One solution for this heterogeneous reduction is using the chemical oxidation method, i.e. through oxidation of the materials that contain Ti in low valence (TiH$_2$, Ti, TiO, Ti$_2$O$_3$, etc.) by H$_2$O$_2$ or other oxidants.[44, 49, 57, 88–90] The oxidative reaction e.g. Ti^{2+} $\rightarrow$ Ti^{3+}, is fundamentally different from the reductive conversion of Ti^{4+} $\rightarrow$ Ti^{3+}, high-concentration Ti^{3+} could dope in the matrix of TiO$_2$ during its formation in the oxidative process.

A highly active, yet stable black TiO$_2$ containing a high concentration of Ti^{3+} embedded inside a rutile matrix is prepared through

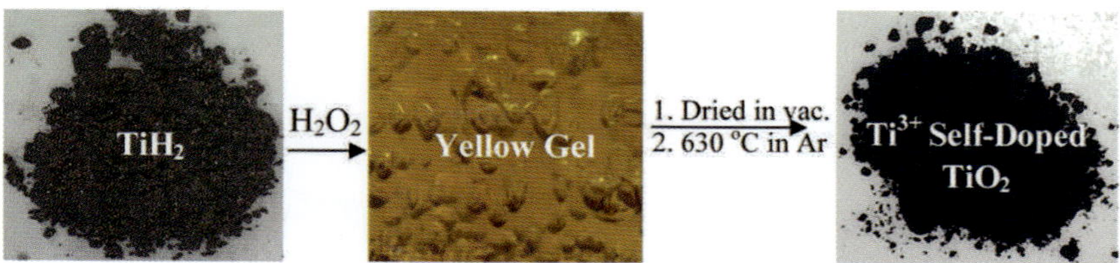

Figure 5.13. (a) Illustration of the synthesis strategy from TiO$_2$ to black TiO$_2$ NPs, along with their images.[88] Reprinted from Ref. [88]. © American Chemical Society.

oxidation of TiH$_2$ by H$_2$O$_2$ (Figure 5.13).[88] In the oxidation process, gray TiH$_2$ powders were firstly reacted with a 30% aqueous solution of H$_2$O$_2$ and gradually converted to a yellow or green gel-like product, presumably forming a cross-linked $[-O-Ti-O-]n-Ti-(OH)x \cdot mH_2O$ matrix, but with substantial V$_{Os}$. Then after calcination in Ar under an optimal temperature, the gel was then desiccated and form a TiO$_2$ lattice with substantial Ti^{3+}. In this way, dopants can be obtained throughout the bulk of the TiO$_2$ matrix. Under hydrothermal conditions, oxidation of TiO and TiH$_2$ is also able to prepare colored titania with abundant Ti^{3+}.[57] Additionally, the annealing in air under certain condition is also competent to prepare defect-rich TiO$_2$.

Water plasma with highly energetic hydroxyl and hydrogen species are demonstrated to be very effective in the oxidation and hydrogenation reactions with the metallic Ti and finally result in the formation of a black H–TiO$_{2-x}$, as shown in Figure 5.14.[44] Firstly, the water plasma was generated in liquid water between a pair of metallic Ti electrodes by applying microsecond bipolar high-voltage pulses with high frequency. Various species, including Ti I (neutral), Ti II (single-charged ions), hydrogen, hydroxyl radicals, and atomic oxygen, were formed, as shown in the optical emission spectrum. Then the surfaces of Ti electrodes were predominantly bombarded with OH$^-$ and O$_2^-$ because the negative ion species were dominantly localized inside the plasma region, resulting in an electrode surface oxidation. Subsequently, the continuous bombardment by highly energetic atomic and molecular species from the water plasma could

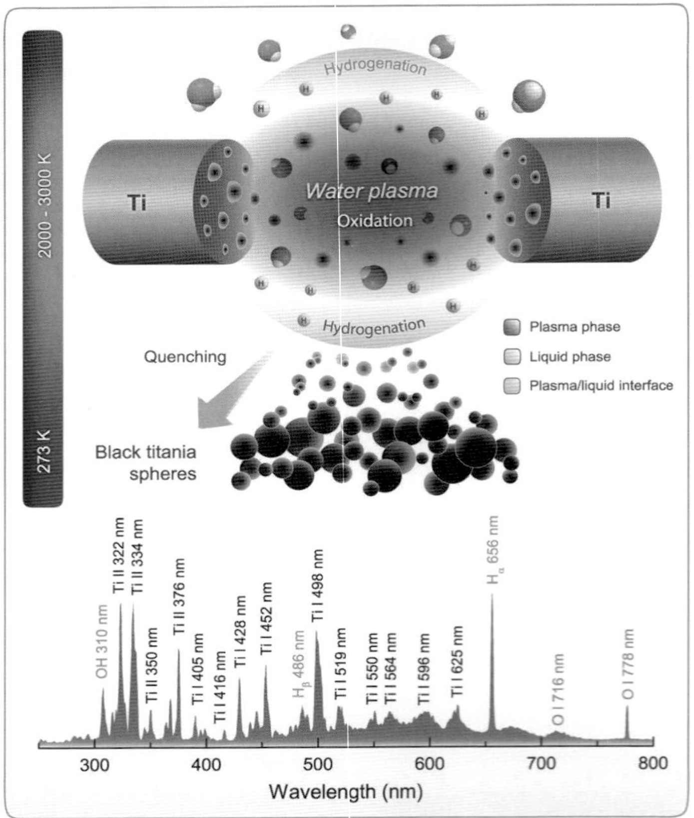

Figure 5.14. The proposed formation mechanism of H–TiO$_{2-x}$ spheres assisted by water plasma and the corresponding optical emission spectrum of the water plasma generated through Ti electrodes.[44] Reprinted from Ref. [44]. © The Royal Society of Chemistry.

create numerous local hot spots on the oxidized electrode surface due to the Joule heating effect. The molten particles disengaged from the electrode surface and then formed a spherical shape due to the effect of surface tension. Due to the abundant hydroxyl radicals with high oxidation ability and the H atoms with high kinetic energy and rapid diffusion, the molten particles were postulated to be readily re-oxidized in the plasma phase and subsequently hydrogenated at/near the plasma–liquid interface before ejecting into the liquid phase. A dramatic quenching of the molten particles in the liquid phase resulted in the black with TiO_2 frozen metastable defects and disordered surfaces.

5.3.5 *Other Methods*

Anodization–annealing is also demonstrated to be a feasible method to prepare defective black TiO_2.[91] TiO_2 nanotubes were first obtained after $10\,h$ anodizations on a Ti foil at $60\,V$ in ethylene glycol containing $0.25\,wt.\%$ NH_4F and $2\,vol\%$ distilled water. This layer was removed before the second anodization under the same condition. The anodized Ti foil was thoroughly washed and dried, and then sintered at $450°C$ for $1\,h$ in ambient atmosphere. A layer of black TiO_2 was obtained on the substrate after removing the top oxide layer.

Self-doped blue TiO_2 can be prepared via a one-step solvothermal method using $TiCl_3$ and TiF_4 as starting materials.[50] The incorporation of Ti^{3+} into the TiO_2 matrix is directly originated from the precursor of $TiCl_3$. The addition of Ti^{4+} in the reaction system inhibit oxidation of Ti^{3+} during the solvothermal treatment:

$$Ti^{3+} + \text{oxygen species} \rightarrow Ti^{4+},$$

which is governed by Le Chatelier's principle. Ti^{3+} will not be completely oxidized to Ti^{4+} during a solvothermal process, and thus the V_O formation is dominantly derived from Ti^{3+}, resulting in a high concentration doping of Ti^{3+} in the bulk TiO_2.

5.4 Performance and Applications of Black TiO_2

Since reported in 2011, not only the development of new preparation methods but also the applications of black TiO_2 have gained wide interest and research. Up to now, black TiO_2 nanomaterials have been demonstrated to show superior performance in photocatalysis and lithium-ion battery over white TiO_2 nanomaterials, and have also opened new applications with promising prospect in supercapacitors, fuel cells, photoelectrochemical sensors, field emission electrodes, microwave absorbers, and cancer photothermal therapy.

5.4.1   *Photocatalysis*

Since the hydrogenated black TiO$_2$ was reported in 2011, the black TiO$_2$ has been widely investigated in photocatalysis, including photocatalytic removal of contaminations, photocatalytic hydrogen production, and photoelectrochemical water splitting. As is known, the photocatalysis process contains three major steps: photon absorption and excitation of electron and holes, charge separation and migration to the photocatalysts' surface, and then photocatalytic reaction between the photogenerated carries with the reactant. Considering the high optical absorption and good electronic transport ability, black/colored TiO$_2$ would tend to have a much better photocatalytic activity than the corresponding white one, which is demonstrated by many reports. The first reported high-pressure hydrogenated black TiO$_2$ had much higher photocatalytic activities in decomposing organic pollutants (methylene blue and phenol) and generating H$_2$ from water–methanol solution than the pristine TiO$_2$.[19] Subsequent studies demonstrate that black titania is very efficient in visible and/or UV driven photodegradation of many organic materials, including methyl blue (MB),[38,44,49,57] methyl orange (MO),[27,40,76,92] rhodamine,[55,93,94] acid fuchsin,[46] 2,4-dichlorophenol,[29] acetaldehyde,[95] etc. Black anatase TiO$_2$ nanosheets with exposed facets achieved through dielectric barrier discharge (DBD) plasma treatment by changing the working gas (Ar, H$_2$, or NH$_3$), and the obtained black TiO$_2$ exhibit enhanced photocatalytic activity in MB degradation due to the induced abundant defects, as shown in Figure 5.15.[38]

Photocatalytic or photoelectrochemical water splitting for hydrogen production is also widely investigated over the black TiO$_2$.[26,29,40,42,53,61,68,92,96–99] The aluminum reduced black titania not only shows much better photocatalytic activity in MO degradation than the pristine TiO$_2$, but also delivers excellent photocatalytic hydrogen production ability under both visible and UV light irradiation and significant photoelectrochemical electrode exhibiting a high solar-to-hydrogen efficiency (1.7%).[27] Wang *et al.* reported that hydrogenated-black TiO$_2$ nanowires treated under

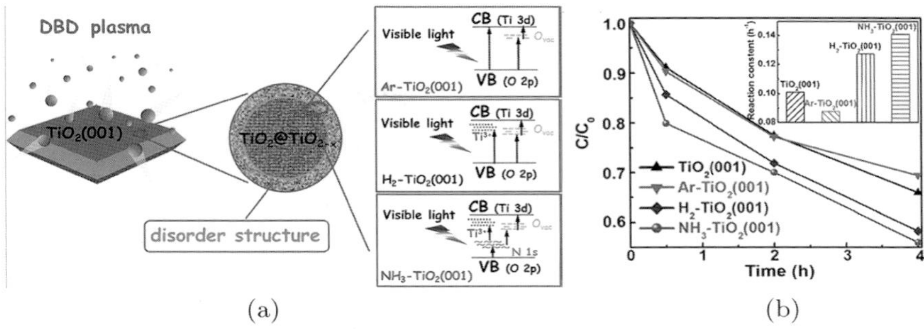

(a)   (b)

Figure 5.15. (a) Proposed mechanisms for the surface reconstruction through DBD plasma discharge with different working gases. (b) Photocatalytic activities of the TiO_2 (001) and DBD-plasma-modified TiO_2 (001) nanosheets suspension in MB, $10\,mg\,L^{-1}$ solution irradiated by visible light.[38] Reprinted from Ref. [38]. © Wiley-VCH.

ultrahigh purity hydrogen atmosphere (ambient pressure) for 3 h at 200–550°C displayed enhanced photoelectrochemical water-splitting performance, the photocurrent of the 350°C annealed black TiO_2 is four times higher than that of the pristine one.[42]

The cocatalyst-free photocatalytic activities were evaluated in generating hydrogen from water of hydrogenated TiO_2 nanotubes and nanorods treated under several conditions: (i) an anatase TiO_2 nanotube layer (air); (ii) this layer converted with Ar or H_2-Ar; (iii) a high pressure H_2 treatment (20 bar, 500°C for 1 h) ($HP-H_2$); and (iv) a high pressure H_2 treatment but mild heating (H_2, 20 bar, 200°C for 5 days) (Sci ref).[65] They found that high pressure H_2-treated black TiO_2 nanotubes had a high open circuit photocatalytic hydrogen production rate without the presence of any cocatalyst, in comparison to the non-activity of those from atmospheric pressure H_2–Ar annealing and the hydrogenated rutile nanorods (Figure 5.16). The high H_2 pressure annealing induced room-temperature stable, isolated Ti^{3+} defect-structure was created in the anatase nanotubes. It was suggested that an adequate H_2 treatment was needed to form these centers and the detailed experimental procedure seems to be crucial. A considerable difference in reactivity existed with respect to the hydrogenation conditions and the phase of the TiO_2 nanomaterials.

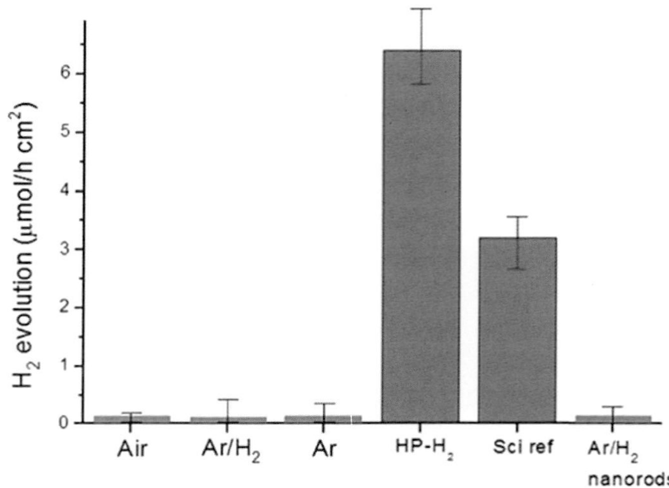

Figure 5.16. Photocatalytic H$_2$ production under open circuit conditions in methanol/water (50/50 vol%) with TiO$_2$ nanotubes and nanorods treated in different atmospheres under AM1.5 (100 mW cm^{-2}) illumination.[65] Reprinted from Ref. [65]. © American Chemical Society.

Black TiO$_2$ NPs obtained from many other method, e.g. oxidation of TiH$_2$, NaBH$_4$, and CaH$_2$ reduction, electrochemical reduction, etc., are all reported to have enhanced photocatalytic activities.[26] However, some results exhibit significantly worse photocatalytic activity of black titania than the pure TiO$_2$. For example, Leshuk *et al.* found that hydrogenated-TiO$_2$ NPs prepared at 200°C under a high-pressure pure H$_2$ gas environment for 5 days or at 400–500°C for 102 h under a 10% H$_2$+ 90% Ar gas flow showed worse photocatalytic activities in decomposing methylene blue because of the propensity to form bulk vacancy defects.[48, 100] Through characterization with positron annihilation and photocurrent measurements of the TiO$_2$ nanocrystals with tunable bulk/surface defects, it was also found that decreasing the relative concentration ratio of bulk defects to surface defects in the TiO$_2$ nanocrystals could significantly improve the electron−hole separation efficiency, thus significantly enhancing the photocatalytic efficiency.[101] In another literature, both the bulk Ti^{3+} and surface V$_{Os}$ are confirmed to contribute to the photocatalysis of TiO$_2$, while a further (dramatic) enhancement in photoactivity can

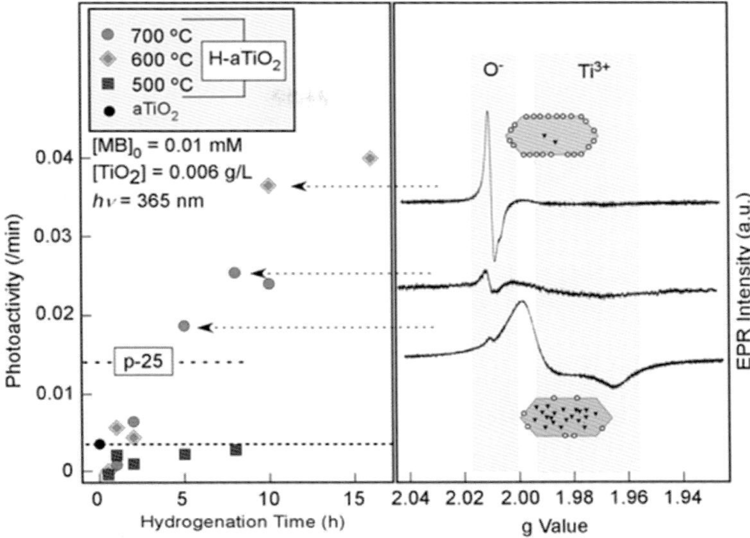

Figure 5.17. Photoactivities (k, min^{-1}) and EPR spectra of H-aTiO$_2$ are shown together against hydrogenation time and temperatures.[101] Reprinted from Ref. [101]. © American Chemical Society.

be obtained for hydrogenated TiO$_2$ nanocrystals treated at 600°C and 700°C with a prolonging of hydrogenation time, which show EPR spectra characterized by a decrease of Ti^{3+} and an increase of surface O$^-$ (Figure 5.17).

Consequently, it can be concluded that the photocatalysis performance of black titania relies heavily on its preparation method. The various defect states introduced during the preparation process account for the variation in their photocatalytic performance.

5.4.2 *Photoelectrochemical Sensor*

The organics in an aqueous phase could be quantified electrochemically by PEC methods by measuring the photocurrents originated from the degradation process. Different oxidation photocurrents at the TiO$_2$ anode can be associated with the concentrations of the organics. Moreover, different organic compounds (C$_y$H$_m$O$_j$N$_k$X$_q$) provide various electron-transfer numbers (n) during the exhaustive oxidization.[53] Based on these principles, the hydrogenated black TiO$_2$ nanorods and nanotubes obtained by treating TiO$_2$ nanorods in H$_2$ or H$_2-$Ar flow under ambient pressure, are applied as a photoelectrochemical sensor for organic compounds under visible light.[53, 68, 102] Taking the hydrogenated-black TiO$_2$ nanorods for example, the hydrogenated TiO$_2$ nanorods showed a highly sensitive and steady photocurrent response (about 100 times larger than the pristine TiO$_2$) in the NaNO$_3$ solution under visible light, due to introduction of V$_O$ and mid-gap energy levels.[68] After addition of malonic acid into the solution, the measured net photocurrents of hydrogenated-TiO$_2$ nanorods photoanode show good linear relationships with the concentrations of malonic acid, and similar results are obtained with glucose and potassium hydrogen phthalate, demonstrating that hydrogenated-TiO$_2$ nanorods can sensitively and steadily detect the concentrations of organic compounds. The increase of the photocurrent was suggested mainly due to the improvement of electronic conductivity that resulted from the creation of mid-energy levels, whose improved electrical conductivity was beneficial to the separation of electron–hole pairs and electron collection from the oxidation of water and organic compounds.

5.4.3 *Photothermal Application*

As we have introduced above, many black titania show great solar absorption across the whole solar spectrum, from the UV to the NIR. This larger solar absorption would tend to result in more electron excitation and relaxation and intensified heat emission, which leads an obvious photothermal effect in the black titania. The photothermic effect of an aluminum reduced black titania was confirmed: the temperature of black TiO$_{2-x}$ disk increases to 37°C, compared to 28°C for pristine TiO$_2$ disk under the irradiation of an AM 1.5G Xe lamp solar simulator for 60 s.[27]

White TiO$_2$ NPs have been widely used for cancer photodynamic therapy based on their UV light-triggered properties. However, biomedical applications using white TiO$_2$ NPs have been limited, since UV light is a mutagen and shallow penetration.

Consequently, there is a great eagerness to develop a new NIR triggered photothermal agent. The dramatic NIR absorption of black TiO_2 NPs make it very promising in this application. After coated by polyethylene glycol (PEG), hydrogenated black TiO_2 (H–TiO_2) NPs are successfully used as photothermal agent for NIR-triggered cancer therapy both *in vitro* and *in vivo*. The photothermal effect of H–TiO_2 NPs can be attributed to their dramatically enhanced non-radiative recombination. After coating, H–TiO_2–PEG NPs exhibit high photothermal conversion efficiency of 40.8%, and stable size distribution in serum solution. The findings herein demonstrate that infrared-irradiated H–TiO_2–PEG NPs exhibit low toxicity, high efficiency as a photothermal agent for cancer therapy, and are promising for further biomedical applications.[103]

5.4.4 *Catalysis*

A hydrogenated TiO_2 NPs was reported to be effective in decomposing gaseous formaldehyde without light irradiation at room temperature, while the pristine white shows very poor activity. The V_{Os} induced by the hydrogen treatment donor have lots of electrons, which would interact with the adsorbed O_2 on the surface of hydrogenated TiO_2 and transform into active oxygen species to degrade formaldehyde without the assistance of the light illumination.[104] Hydrogen reduced TiO_2 was demonstrated to display excellent catalytic activity in the conversion of ethylene to high density polyethylene (HDPE) under mild conditions (room temperature, low pressure, absence of any activator). The Ti^{4-n} defect sites induced during hydrogen treatment serve as ethylene polymerization locations.[105]

5.4.5 *Lithium Rechargeable Battery*

TiO_2 has been suggested as a safer lithium-ion battery anode material with a theoretical capacity value of $335\,mAh\,g^{-1}$, since the insertion/extraction of lithium proceeds within 1.5–1.8 V vs. the Li^+/Li redox couple in TiO_2. However, it is known that the Li-storage performance of TiO_2-using batteries is severely limited by poor chemical diffusion of lithium in all polymorphs resulting from both (i) slow lithium-ion diffusion and (ii) poor electronic conductivity. The disordered structure and/or the Ti^{3+}, which could facilitate the fast charge transfer and donor numerous electrons and increase the electronic conductivity, would make the black TiO$_2$ a promising electrode material for lithium-ion battery.[31,54,78,106–110]

As reported by Xia *et al.*, lithium electrode materials have a great influence on the battery capacities (Figure 5.18).[108] Only a thin surface layer of the host material is available for Li intercalation at high charging–discharging rates for bulk materials due to the low Li$^+$ ion diffusion coefficient and low electronic conductivity. Decreasing the particle size of regulating the morphologies could reach a near-full lithium capacity, but won't improve the limited lithium diffusion. A disordered surface in NPs could lead to an easier charge transport and achieve a full-lithium capacity. The BIEF, across the interface between the amorphized non-stoichiometric outer layer (TiO$_{2-x}$) and the nanocrystalline stoichiometric core (TiO$_2$) within the TiO$_2$ nanocrystals, will facilitate the Li$^+$ ion in both insertion/extraction processes, resulting in a greatly improve lithium-ion rechargeable battery performance: 20 times rate and 340% capacity improvement over crystalline TiO$_2$ nanocrystals.

Hydrogenated mesoporous TiO$_2$ microspheres proceeded under ambient hydrogen at 400°C showed twice the rate capability of mesoporous TiO$_2$ microspheres due to the combination of the short lithium-ion diffusion path and the high electronic conductivity.[106] Shin *et al.* reported that hydrogenated black TiO$_2$ nanocrystals prepared at 450°C under a H$_2$–Ar gas flow displayed excellent rate performance for lithium storage. The high performance was attributed to the well-balanced Li$^+$/e$^-$ diffusion in the hydrogenated-TiO$_2$ nanocrystals reduction reaction.[54] Nanostructured black TiO$_2$, due to its high conductivity and particular structure including a trivalent Ti, resulting from the incorporation of H and N, is demonstrated to be a particularly promising electrode material with a capacity of 127 mA h g^{-1} at 100°C (20 A g^{-1}) and approximately 86% retention after 100 cycles at 25°C.[107]

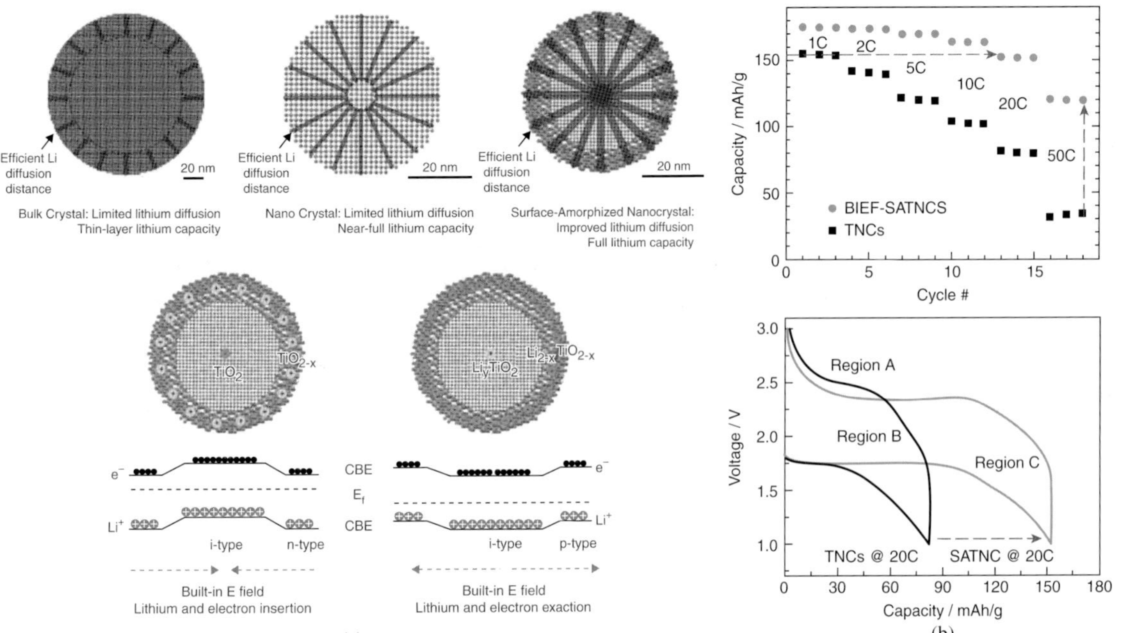

Figure 5.18. (a) Comparison of charge diffusion in electrode materials made of bulk crystals, nanocrystals, surface-amorphized nanocrystals, and illustration of the facilitation of charge transport under the built-in electric field (BIEF) during discharge and charge processes. The length and the width of the arrows illustrate the relative penetration depth and the charge transfer/transport coefficients of the lithium ion in the material, respectively. (b) Up: Rate performance of the TNCs and the SATNCs. The charge rate was changed from 1°C to 50°C while the discharge rate was kept at 1°C. The dashed horizontal line shows the rate improvement at the same capacity, and the dashed vertical line shows the capacity improvement at the same rate; Down: Comparison of the galvanostatic charge/discharge profiles for TNCs and SATNCs at charge rates of 20°C.[108] Reprinted from Ref. [108]. © American Chemical Society.

In the reported literatures, the suggested explanations for the enhanced performance as electrode materials in lithium-ion rechargeable batteries of the black TiO_2 nanomaterials included the creation of V_{Os}, the existence of Ti^{3+} ions, the well-balanced Li^+/e^- diffusion in the hydrogenated TiO_2 nanocrystals, the increased electronic conductivity, short lithium-ion diffusion path, reduced charge diffusion resistance, and pseudo-capacitive lithium storage on the disordered particle surface.

5.4.6 *Supercapacitors*

The black TiO_2 has also been researched as a candidate material for supercapacitors.[84,87,111–113] For example, a self-doped TiO_2 NTAs at $-1.4\,V$ (vs. SCE) exhibited 5 orders of magnitude improvement on carrier density and 39 times enhancement in capacitance compared to those of the pristine TiO_2 NTAs.[87] Another hydrogenated TiO_2 NTAs prepared at $400°C$ in hydrogen gas yields the largest specific capacitance of $3.24\,mF\,cm^{-2}$ at a scan rate of $100\,mV\,s^{-1}$, which is 40 times higher than the capacitance obtained from air-annealed TiO_2 (air$-TiO_2$) NTAs at the same conditions (Figure 5.19(a)). Additionally, the hydrogenated TiO_2 NTAs also show remarkable rate capability with 68% areal capacitance retained when the scan

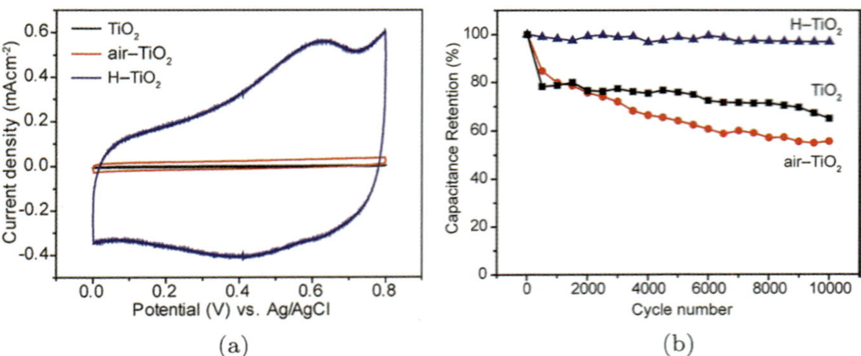

Figure 5.19. (a) CV curves of the untreated TiO_2, air$-TiO_2$, and H$-TiO_2$ NTAs obtained at a scan rate of $100\,mV\,s^{-1}$. (b) Cycle performance of TiO_2 samples measured at a scan rate of $100\,mV\,s^{-1}$ for 10000 cycles.[111] Reprinted from Ref. [111]. © American Chemical Society.

rate increases from $10 \, \mathrm{mV \, s^{-1}}$ to $1000 \, \mathrm{mV \, s^{-1}}$, as well as outstanding long-term cycling stability with only 3.1% reduction of initial specific capacitance after 10000 cycles (Figure 5.19(b)).[111] The increased performance of the black TiO_2 nanomaterials as electrode materials in supercapacitors was suggested due to decreased charge-transport resistance through the solid material, increased densities of charge carrier and hydroxyl groups on the TiO_2 surface, and the higher electrical conductivity.

5.4.7 *Other Applications*

Hydrogenated black titania is also demonstrated to have promising applications in many other fields, including fuel cells,[114,115] field emission,[24,82,116] and microwave absorption,[25,32,71,106] etc. These improved performances are resulted from the defects states induced during the hydrogenation process as well.

References

(1)   Pfaff, G.; Reynders, P. *Chem. Rev.* **1999**, *99*, 1963–1982.

(2)   Kudo, A.; Miseki, Y. *Chem. Soc. Rev.* **2009**, *38*, 253–278.

(3)   Hoffmann, M. R.; Martin, S. T.; Choi, W. Y.; Bahnemann, D. W. *Chem. Rev.* **1995**, *95*, 69–96.

(4)   Schneider, J.; Matsuoka, M.; Takeuchi, M.; Zhang, J.; Horiuchi, Y.; Anpo, M.; Bahnemann, D. W. *Chem. Rev.* **2014**, *114*, 9919–9986.

(5)   Chen, X.; Mao, S. S. *Chem. Rev.* **2007**, *107*, 2891–2959.

(6)   Chen, X.; Li, C.; Gratzel, M.; Kostecki, R.; Mao, S. S. *Chem. Soc. Rev.* **2012**, *41*, 7909–7937.

(7)   Chen, X.; Shen, S.; Guo, L.; Mao, S. S. *Chem. Rev.* **2010**, *110*, 6503–6570.

(8)   Ma, Y.; Wang, X.; Jia, Y.; Chen, X.; Han, H.; Li, C. *Chem. Rev.* **2014**, *114*, 9987–10043.

(9)   Berger, T.; Sterrer, M.; Diwald, O.; Knozinger, E.; Panayotov, D.; Thompson, T. L.; Yates, J. T. *J. Phys. Chem. B* **2005**, *109*, 6061–6068.

(10) Chen, X.; Burda, C. *J. Am. Chem. Soc.* **2008**, *130*, 5018–5019.

(11) Batzill, M.; Morales, E.; Diebold, U. *Phys. Rev. Lett.* **2006**, *96*, 026103.

(12) Subramanian, V.; Wolf, E. E.; Kamat, P. V. *Langmuir* **2003**, *19*, 469–474.

(13) Zhou, H.; Qu, Y.; Zeid, T.; Duan, X. *Energy Environ. Sci.* **2012**, *5*, 6732–6743.

(14) Hong, X. T.; Wang, Z. P.; Cai, W. M.; Lu, F.; Zhang, J.; Yang, Y. Z.; Ma, N.; Liu, Y. J. *Chem. Mater.* **2005**, *17*, 1548–1552.

(15) Asahi, R.; Morikawa, T.; Ohwaki, T.; Aoki, K.; Taga, Y. *Science* **2001**, *293*, 269–271.

(16) Song, S.; Tu, J.; He, Z.; Hong, F.; Liu, W.; Chen, J. *Appl. Catal. A* **2010**, *378*, 169–174.

(17) Suriye, K.; Jongsomjit, B.; Satayaprasert, C.; Praserthdam, P. *Appl. Surf. Sci.* **2008**, *255*, 2759–2766.

(18) Yi, Z.; Ye, J.; Kikugawa, N.; Kako, T.; Ouyang, S.; Stuart-Williams, H.; Yang, H.; Cao, J.; Luo, W.; Li, Z.; Liu, Y.; Withers, R. L. *Nat. Mater.* **2010**, *9*, 559–564.

(19) Chen, X.; Liu, L.; Yu, P. Y.; Mao, S. S. *Science* **2011**, *331*, 746–750.

(20) Hoang, S.; Berglund, S. P.; Hahn, N. T.; Bard, A. J.; Mullins, C. B. *J. Am. Chem. Soc.* **2012**, *134*, 3659–3662.

(21) Jiang, X.; Zhang, Y.; Jiang, J.; Rong, Y.; Wang, Y.; Wu, Y.; Pan, C. *J. Phys. Chem. C* **2012**, *116*, 22619–22624.

(22) Liu, L.; Yu, P. Y.; Chen, X.; Mao, S. S.; Shen, D. Z. *Phys. Rev. Lett.* **2013**, *111*, 065505.

(23) Yang, Y.; Ling, Y. C.; Wang, G. M.; Li, Y. *Eur. J. Inorg. Chem.* **2014**, *4*, 760–766.

(24) Zhu, W. D.; Wang, C. W.; Chen, J. B.; Li, D. S.; Zhou, F.; Zhang, H. L. *Nanotechnology* **2012**, *23*, 455204–455209.

(25) Xia, T.; Zhang, C.; Oyler, N. A.; Chen, X. B. *Adv. Mater.* **2013**, *25*, 6905–6910.

(26) Chen, X.; Liu, L.; Huang, F. *Chem. Soc. Rev.* **2015**, *44*, 1861–1885.

(27) Wang, Z.; Yang, C.; Lin, T.; Yin, H.; Chen, P.; Wan, D.; Xu, F.; Huang, F.; Lin, J.; Xie, X.; Jiang, M. *Energy Environ. Sci.* **2013**, *6*, 3007–3014.

(28) Wang, Z.; Yang, C.; Lin, T.; Yin, H.; Chen, P.; Wan, D.; Xu, F.; Huang, F.; Lin, J.; Xie, X.; Jiang, M. *Adv. Funct. Mater.* **2013**, *23*, 5444–5450.

(29) Zheng, Z.; Huang, B.; Lu, J.; Wang, Z.; Qin, X.; Zhang, X.; Dai, Y.; Whangbo, M. H. *Chem. Commun.* **2012**, *48*, 5733–5735.

(30) Lu, J.; Dai, Y.; Jin, H.; Huang, B. *Phys. Chem. Chem. Phys.* **2011**, *13*, 18063–18068.

(31) Xia, T.; Zhang, W.; Li, W. J.; Oyler, N. A.; Liu, G.; Chen, X. B. *Nano Energy* **2013**, *2*, 826–835.

(32) Dong, J.; Ullal, R.; Han, J.; Wei, S.; Ouyang, X.; Dong, J.; Gao, W. *J. Mater. Chem. A* **2015**, *3*, 5285–5288.

(33) Chen, X.; Liu, L.; Liu, Z.; Marcus, M. A.; Wang, W. C.; Oyler, N. A.; Grass, M. E.; Mao, B.; Glans, P. A.; Yu, P. Y.; Guo, J.; Mao, S. S. *Sci. Rep.* **2013**, *3*, 1510.

(34) Yang, C.; Wang, Z.; Lin, T.; Yin, H.; Lu, X.; Wan, D.; Xu, T.; Zheng, C.; Lin, J.; Huang, F.; Xie, X.; Jiang, M. *J. Am. Chem. Soc.* **2013**, *135*, 17831–17838.

(35) Huo, J.; Hu, Y.; Jiang, H.; Li, C. *Nanoscale* **2014**, *6*, 9078–9084.

(36) Lu, H.; Zhao, B.; Pan, R.; Yao, J.; Qiu, J.; Luo, L.; Liu, Y. *RSC Adv.* **2014**, *4*, 1128–1132.

(37) Plodinec, M.; Gajović, A.; Jakša, G.; Žagar, K.; Čeh, M. *J. Alloys Compd.* **2014**, *591*, 147–155.

(38) Li, B.; Zhao, Z.; Zhou, Q.; Meng, B.; Meng, X.; Qiu, J. *Chem.-Eur. J.* **2014**, *20*, 14763–14770.

(39) Naldoni, A.; Allieta, M.; Santangelo, S.; Marelli, M.; Fabbri, F.; Cappelli, S.; Bianchi, C. L.; Psaro, R.; Dal Santo, V. *J. Am. Chem. Soc.* **2012**, *134*, 7600–7603.

(40) Zhu, G.; Yin, H.; Yang, C.; Cui, H.; Wang, Z.; Xu, J.; Lin, T.; Huang, F. *ChemCatChem* **2015**, *7*, 2614–2619.

(41) Zhu, G.; Lin, T.; Lü, X.; Zhao, W.; Yang, C.; Wang, Z.; Yin, H.; Liu, Z.; Huang, F.; Lin, J. *J. Mater. Chem. A* **2013**, *1*, 9650–9653.

(42) Wang, G.; Wang, H.; Ling, Y.; Tang, Y.; Yang, X.; Fitzmorris, R. C.; Wang, C.; Zhang, J. Z.; Li, Y. *Nano Lett.* **2011**, *11*, 3026–3033.

(43) Teng, F.; Li, M.; Gao, C.; Zhang, G.; Zhang, P.; Wang, Y.; Chen, L.; Xie, E. *Appl. Catal. B* **2014**, *148-149*, 339–343.

(44) Panomsuwan, G.; Watthanaphanit, A.; Ishizaki, T.; Saito, N. *Phys. Chem. Chem. Phys.* **2015**, *17*, 13794–13799.

(45) Fan, C.; Chen, C.; Wang, J.; Fu, X.; Ren, Z.; Qian, G.; Wang, Z. *J. Mater. Chem. A* **2014**, *2*, 16242–16249.

(46) Fan, C.; Chen, C.; Wang, J.; Fu, X.; Ren, Z.; Qian, G.; Wang, Z. *Sci. Rep.* **2015**, *5*, 11712.

(47) Qiu, J.; Dawood, J.; Zhang, S. *Chin. Sci. Bull.* **2014**, *59*, 2144–2161.

(48) Leshuk, T.; Parviz, R.; Everett, P.; Krishnakumar, H.; Varin, R. A.; Gu, F. *ACS Appl. Mater. Interfaces* **2013**, *5*, 1892–1895.

(49) Liu, X.; Xu, H.; Grabstanowicz, L. R.; Gao, S.; Lou, Z.; Wang, W.; Huang, B.; Dai, Y.; Xu, T. *Catal. Today* **2014**, *225*, 80–89.

(50) Zhu, Q.; Peng, Y.; Lin, L.; Fan, C.-M.; Gao, G.-Q.; Wang, R.-X.; Xu, A.-W. *J. Mater. Chem. A* **2014**, *2*, 4429–4437.

(51) Gao, R.; Liu, L.; Hu, Z.; Zhang, P.; Cao, X.; Wang, B.; Liu, X. *J. Mater. Chem. A* **2015**, *3*, 17598–17605.

(52) Wang, X.; Li, Y.; Liu, X.; Gao, S.; Huang, B.; Dai, Y. *Chinese J. Catal.* **2015**, *36*, 389–399.

(53) Li, S.; Qiu, J.; Ling, M.; Peng, F.; Wood, B.; Zhang, S. *ACS Appl. Mater. Interfaces* **2013**, *5*, 11129–11135.

(54) Shin, J.-Y.; Joo, J. H.; Samuelis, D.; Maier, J. *Chem. Mater.* **2012**, *24*, 543–551.

(55) Fang, W.; Xing, M.; Zhang, J. *Appl. Catal. B* **2014**, *160–161*, 240–246.

(56) Yin, H.; Lin, T.; Yang, C.; Wang, Z.; Zhu, G.; Xu, T.; Xie, X.; Huang, F.; Jiang, M. *Chemistry* **2013**, *19*, 13313–13316.

(57) Liu, X.; Gao, S.; Xu, H.; Lou, Z.; Wang, W.; Huang, B.; Dai, Y. *Nanoscale* **2013**, *5*, 1870–1875.

(58) Cui, H.; Zhao, W.; Yang, C.; Yin, H.; Lin, T.; Shan, Y.; Xie, Y.; Gu, H.; Huang, F. *J. Mater. Chem. A* **2014**, *2*, 8612–8616.

(59) Pan, X.; Yang, M.-Q.; Fu, X.; Zhang, N.; Xu, Y.-J.; *Nanoscale* **2013**, *5*, 3601–3614.

(60) Zhang, L.; Wang, S.; Lu, C. *Anal. Chem.* **2015**, *87*, 7313–7320.

(61) Tan, H.; Zhao, Z.; Niu, M.; Mao, C.; Cao, D.; Cheng, D.; Feng, P.; Sun, Z. *Nanoscale* **2014**, *6*, 10216–10223.

(62) Su, T.; Yang, Y.; Na, Y.; Fan, R.; Li, L.; Wei, L.; Yang, B.; Cao, W. *ACS Appl. Mater. Interfaces* **2015**, *7*, 3754–3763.

(63) Yin, H.; Wang, X.; Wang, L.; Zhao, Q. N. H. *J. Alloys Compd.* **2015**, *640*, 68–74.

(64) Zhang, J.; Zhang, L.; Zhang, J.; Zhang, Z.; Wu, Z. *J. Alloys Compd.* **2015**, *642*, 28–33.

(65) Liu, N.; Schneider, C.; Freitag, D.; Hartmann, M.; Venkatesan, U.; Müller, J.; Spiecker, E.; Schmuki, P. *Nano Lett.* **2014**, *14*, 3309–3313.

(66) Wei, W.; Yaru, N.; Chunhua, L.; Zhongzi, X. *RSC Adv.* **2012**, *2*, 8286–8288.

(67) Wei, S.; Wu, R.; Jian, J.; Chen, F.; Sun, Y. *Dalton Trans.* **2015**, *44*, 1534–1538.

(68) Zhang, S.; Zhang, S.; Peng, B.; Wang, H.; Yu, H.; Wang, H.; Peng, F. *Electrochem. Commun.* **2014**, *40*, 24–27.

(69) Mo, L.-B.; Bai, Y.; Xiang, Q.-Y.; Li, Q.; Wang, J.-O.; Ibrahim, K.; Cao, J.-L. *Appl. Phys. Lett.* **2014**, *105*, 202114.

(70) Saputera, W. H.; Mul, G.; Hamdy, M. S. *Catal. Today* **2015**, *246*, 60–66.

(71) Zhang, Z. K.; Bai, M. L.; Guo, D. Z.; Hou, S. M.; Zhang, G. M. *Chem. Commun.* **2011**, *47*, 8439–8441.

(72) Cui, H.; Zhu, G.; Xie, Y.; Zhao, W.; Yang, C.; Lin, T.; Gu, H.; Huang, F. *J. Mater. Chem. A* **2015**, *3*, 11830–11837.

(73) Wang, G.; Ling, Y.; Wang, H.; Xihong, L.; Li, Y. *J. Photoch. Photobio. C* **2014**, *19*, 35–51.

(74) Zhao, W.; Zhao, W.; Zhu, G.; Lin, T.; Xu, F.; Huang, F. *CrystEng-Comm.* **2015**, *17*, 7528–7534.

(75) Wang, H.; Lin, T.; Zhu, G.; Yin, H.; Lü, X.; Li, Y.; Huang, F. *Catal. Commun.* **2015**, *60*, 55–59.

(76) Zheng, Z.; Huang, B.; Meng, X.; Wang, J.; Wang, S.; Lou, Z.; Wang, Z.; Qin, X.; Zhang, X.; Dai, Y. *Chem. Commun.* **2013**, *49*, 868–870.

(77) Mohammadizadeh, M. R.; Bagheri, M.; Aghabagheri, S.; Abdi, Y. *Appl. Surf. Sci.* **2015**, *350*, 43–49.

(78) Yan, Y.; Hao, B.; Wang, D.; Chen, G.; Markweg, E.; Albrecht, A.; Schaaf, P. *J. Mater. Chem. A* **2013**, *1*, 14507–14513.

(79) Lin, T. Q.; Yang, C. Y.; Wang, Z.; Yin, H.; Lu, X. J.; Huang, F. Q.; Lin, J. H.; Xie, X. M.; Jiang, M. H. *Energy Environ. Sci.* **2014**, *7*, 967–972.

(80) Zhao, Z.; Tan, H.; Zhao, H.; Lv, Y.; Zhou, L. J.; Song, Y.; Sun, Z. *Chem. Commun. (Camb)* **2014**, *50*, 2755–2757.

(81) Kang, Q.; Cao, J.; Zhang, Y.; Liu, L.; Xu, H.; Ye, J. *J. Mater. Chem. A* **2013**, *1*, 5766–5774.

(82) Zhang, X.; Wang, C.; Chen, J.; Zhu, W.; Liao, A.; Li, Y.; Wang, J.; Ma, L. *ACS Appl. Mater. Interfaces* **2014**, *6*, 20625–20633.

(83) Zou, X.; Liu, J.; Su, J.; Zuo, F.; Chen, J.; Feng, P. *Chemistry* **2013**, *19*, 2866–2873.

(84) Li, H.; Chen, Z.; Tsang, C. K.; Li, Z.; Ran, X.; Lee, C.; Nie, B.; Zheng, L.; Hung, T.; Lu, J.; Pan, B.; Li, Y. Y. *J. Mater. Chem. A* **2014**, *2*, 229–236.

(85) Zheng, L.; Cheng, H.; Liang, F.; Shu, S.; Tsang, C. K.; Li, H.; Lee, S.-T.; Li, Y. Y. *J. Phys. Chem. C* **2012**, *116*, 5509–5515.

(86) Zhang, Z.; Hedhili, M. N.; Zhu, H.; Wang, P. *Phys. Chem. Chem. Phys.* **2013**, *15*, 15637–15644.

(87) Zhou, H.; Zhang, Y. *J. Phys. Chem. C* **2014**, *118*, 5626–5636.

(88) Grabstanowicz, L. R.; Gao, S.; Li, T.; Rickard, R. M.; Rajh, T.; Liu, D. J.; Xu, T. *Inorg. Chem.* **2013**, *52*, 3884–3890.

(89) Pei, Z.; Ding, L.; Lin, H.; Weng, S.; Zheng, Z.; Hou, Y.; Liu, P. *J. Mater. Chem. A* **2013**, *1*, 10099–10102.

(90) Pei, Z.; Ding, L.; Feng, W.; Weng, S.; Liu, P. *Phys. Chem. Chem. Phys.* **2014**, *16*, 21876–21881.

(91) Dong, J.; Han, J.; Liu, Y.; Nakajima, A.; Matsushita, S.; Wei, S.; Gao, W. *ACS Appl. Mater. Interfaces* **2014**, *6*, 1385–1388.

(92) Tian, J.; Leng, Y.; Zhao, Z.; Xia, Y.; Sang, Y.; Hao, P.; Zhan, J.; Li, M.; Liu, H. *Nano Energy* **2015**, *11*, 419–427.

(93) Li, G.; Lian, Z.; Li, X.; Xu, Y.; Wang, W.; Zhang, D.; Tian, F.; Li, H. *J. Mater. Chem. A* **2015**, *3*, 3748–3756.

(94) Rico-Santacruz, M.; Sepúlveda, Á. E.; Serrano, E.; Lalinde, E.; Berenguer, J. R.; García-Martínez, J. *J. Mater. Chem. C* **2014**, *2*, 9497–9504.

(95) Danon, A.; Bhattacharyya, K.; Vijayan, B. K.; Lu, J.; Sauter, D. J.; Gray, K. A.; Stair, P. C.; Weitz, E. *ACS Catal.* **2012**, *2*, 45–49.

(96) Liu, N.; Schneider, C.; Freitag, D.; Venkatesan, U.; Marthala, V. R.; Hartmann, M.; Winter, B.; Spiecker, E.; Osvet, A.; Zolnhofer, E. M.; Meyer, K.; Nakajima, T.; Zhou, X.; Schmuki, P. *Angew. Chem. Int. Ed. Engl.* **2014**, *53*, 14201–14205.

(97) Xu, C.; Song, Y.; Lu, L.; Cheng, C.; Liu, D.; Fang, X.; Chen, X.; Zhu, X.; Li, D. *Nanoscale Res. Lett.* **2013**, *8*, 391.

(98) Chen, Y.; Tao, Q.; Fu, W.; Yang, H.; Zhou, X.; Su, S.; Ding, D.; Mu, Y.; Li, X.; Li, M. *Chem. Commun. (Camb)* **2014**, *50*, 9509–9512.

(99) Wang, G.; Ling, Y.; Li, Y. *Nanoscale* **2012**, *4*, 6682–6691.

(100) Leshuk, T.; Linley, S.; Gu, F. *Can. J. Chem. Eng.* **2013**, *91*, 799–807.

(101) Kong, M.; Li, Y.; Chen, X.; Tian, T.; Fang, P.; Zheng, F.; Zhao, X. *J. Am. Chem. Soc.* **2011**, *133*, 16414–16417.

(102) Wang, X.; Zhang, S.; Wang, H.; Yu, H.; Wang, H.; Zhang, S.; Peng, F. *RSC Adv.* **2015**, *5*, 76315–76320.

(103) Ren, W.; Yan, Y.; Zeng, L.; Shi, Z.; Gong, A.; Schaaf, P.; Wang, D.; Zhao, J.; Zou, B.; Yu, H.; Chen, G.; Brown, E. M.; Wu, A. *Adv. Healthcare Mater.* **2015**, *4*, 1526–1536.

(104) Zeng, L.; Song, W.; Li, M.; Zeng, D.; Xie, C. *Appl. Catal. B* **2014**, *147*, 490–498.

(105) Barzan, C.; Groppo, E.; Bordiga, S.; Zecchina, A. *ACS Catal.* **2014**, *4*, 986–989.

(106) Li, G.; Zhang, Z.; Peng, H.; Chen, K. *RSC Adv.* **2013**, *3*, 11507–11510.

(107) Myung, S.-T.; Kikuchi, M.; Yoon, C. S.; Yashiro, H.; Kim, S.-J.; Sun, Y.-K.; Scrosati, B. *Energy Environ. Sci.* **2013**, *6*, 2609–2614.

(108) Xia, T.; Zhang, W.; Murowchick, J.; Liu, G.; Chen, X. *Nano Lett.* **2013**, *13*, 5289–5296.

(109) Xia, T.; Zhang, W.; Murowchick, J. B.; Liu, G.; Chen, X. *Adv. Energy. Mater.* **2013**, *3*, 1516–1523.

(110) Qiu, J.; Li, S.; Gray, E.; Liu, H.; Gu, Q.-F.; Sun, C.; Lai, C.; Zhao, H.; Zhang, S. *J. Phys. Chem. C* **2014**, *118*, 8824–8830.

(111) Lu, X.; Wang, G.; Zhai, T.; Yu, M.; Gan, J.; Tong, Y.; Li, Y. *Nano Lett.* **2012**, *12*, 1690–1696.

(112) Cao, X. Y.; Xing, X.; Zhang, N.; Gao, H.; Zhang, M. Y.; Shang, Y. C.; Zhang, X. T. *J. Mater. Chem. A* **2015**, *3*, 3785–3793.

(113) Chen, J.; Xia, Z.; Li, H.; Li, Q.; Zhang, Y. *Electrochim. Acta* **2015**, *166*, 174–182.

(114) Zhang, C.; Yu, H.; Li, Y.; Fu, L.; Gao, Y.; Song, W.; Shao, Z.; Yi, B. *Nanoscale* **2013**, *5*, 6834–6841.

(115) Zhang, C.; Yu, H.; Li, Y.; Gao, Y.; Zhao, Y.; Song, W.; Shao, Z.; Yi, B. *ChemSusChem* **2013**, *6*, 659–666.

(116) Zhang, X.-Q.; Chen, J.-B.; Zhu, W.-D.; Wang, C.-W. *J. Vac. Sci. Technol B* **2014**, *32*, 021808.

CHAPTER SIX

Theoretical Analysis Related to Black Titanium Dioxide Nanomaterials

Lei Liu

State Key Laboratory of Luminescence and Applications,
Changchun Institute of Optics, Fine Mechanics and Physics,
Chinese Academy of Sciences 130033, Changchun, Jilin,
People's Republic of China
liulei@ciomp.ac.cn.

6.1 Introduction

Titanium dioxide (TiO_2) has been extensively investigated for the various photoelectrochemical (PEC) applications,[1] since the discovery of the photo-induced water splitting on TiO_2 electrodes in 1972.[2] For such a photosynthetic process, photons with energy exceeding the bandgap of this semiconductor generate electron–hole pairs, which, after separation and transfer, decompose water into oxygen and hydrogen through oxidation and reduction, respectively. In the similar PEC ways, these electrons and holes can prompt other photocatalytic reactions as well, such as reduction of CO_2, decomposition and synthesis organic chemicals, etc. In particular, its low-cost, inertness, non-toxicity, and abundance make TiO_2 particularly desirable for these photocatalytic applications.

In general, for such photocatalytic processes to be highly efficient upon solar radiation, the semiconductor catalysts have to comply with three requirements. First, the semiconductor should have the right electronic bandgap to absorb the maximum of the solar

"

spectrum. The desirable bandgap of the semiconductor for the optimum utilization of solar energy would be about 1.35 eV.[3] Second, the positions of its conduction band minimum (CBM) and valence band maximum (VBM) should match respectively the reduction and the oxidation potentials, respectively, of the chemicals being catalyzed. Third, the photon-generated electrons and holes should be effectively separated from each other to drive the electrochemistry required for further synthesis. However, although taken as a benchmark standard semiconductor photocatalyst, the intrinsic TiO$_2$ fails to provide enough efficiency for the practical solar energy harvesting. The obvious limitation is its large bandgap of more than 3.0 eV. That means TiO$_2$ can absorb only the UV part of the solar spectrum, which accounts for not more than 5% of the solar radiation on Earth. This inefficiency indeed originates from its intrinsic electronic band structure. In case of solar water splitting, its CBM matches the hydrogen electrode potential quite well, but its VBM stays about 1.6 eV[4] below the oxidation potential. Because of this low-energy VBM, its absorption of the visible and near-infrared solar radiation is suppressed.

In the last few decades, a tremendous amount of research efforts from a variety of angles have been devoted to tuning the electronic properties of TiO$_2$ for achieving the high solar-efficiency in its corresponding photochemistry. Among them, the hydrogen thermal treatment of TiO$_2$ nanocrystals has been demonstrated recently as an effective way in realizing wide-spectrum solar absorption together with excellent photoactivity and stability in photocatalytic hydrogen generation.[5(a)] As the color of these nanoparticles (NPs) turns black after treatment, this TiO$_2$ material has been called black TiO$_2$. In terms of structural changes, the surface layers of these NPs become disordered after treatment, so it is also called "disorder engineering".[1d,1e] While it has been known that normally materials should be highly crystalline to avoid detrimental recombination sites, the high efficiency of this "disorder engineered" material seems interesting and unusual.[5b]

So far, the origin behind the much enhanced solar photocatalytic activity of black TiO$_2$ has not been unequivocally clarified, which has been proposed due to structural disorder, defects of Ti^{3+} or oxygen vacancy (V_O), Ti–H or Ti–OH bonds, etc. In fact, either of these changes has a certain chance to occur after hydrogenation of nanostructured TiO_2. But currently it is still at issue that which factor dominates the PEC behaviors of black TiO_2. In this chapter, with certain selected theoretical works, we try to address the mechanisms behind the enhancement of solar energy conversion of black TiO_2. After a brief introduction on the basic structural and electronic properties of TiO_2 polymorphs, we will discuss the possible ways on engineering the electronic band structures of TiO_2 for better solar absorption and photocatalysis. Then we will summarize the theoretical studies on black TiO_2 so far, and try to provide certain overall understanding on the nature of solar photoelectrochemical conversion of black TiO_2.

6.2 Structural Polymorphs of TiO_2 and their Electronic Band Structures

TiO_2 has been proposed with quite a few structural polymorphs, including the low-pressure forms of rutile, anatase, brookite, TiO_2(B), TiO_2(H), and TiO_2(R), and the high-pressure columbite-, baddeleyite-, cotunnite-, pyrite-, and fluorite-structured forms.[6] As listed in Table 6.1 are their corresponding structural parameters, which hint its versatile bonding styles. In this section, with these polymorphs, we would inspect the change of its elementary electronic characteristics upon structural variation.

Figure 6.1 presents the basic unit-cell structures of three common crystalline phases of TiO_2, i.e. rutile, anatase, and brookite. Among them, the rutile and anatase cells are tetragonal which have six and 12 atoms, respectively, and the brookite is orthorhombic that contains eight formula units. In each lattice, Ti and O atoms bond each other to form TiO_6 octahedrons, with coordination numbers (CN) of 6 and 3, respectively. These three TiO_2 crystal structures had been often described in terms of chains of TiO_6 octahedrons having common edges.[7] Two, three, and four edges are shared in rutile, brookite, and anatase, respectively. The TiO_6 octahedrons

Table 6.1.   The structural parameters of TiO$_2$ phases.[6a]

Phase	Symmetry	a, Å	b, Å	c, Å	β, (°)	Density (g cm^{-3})
Rutile[6b]	*P4$_2$/mnm*	4.5937		2.9587		4.27
Anatase[6c]	*I4$_1$/amd*	3.7848		9.5124		3.89
Brookite[6d]	*Pbca*	9.1840	5.4470	5.1450		4.12
TiO$_2$(B)[6e]	*C2/m*	12.163	3.7350	6.5130	107.29	3.64
Hollandite[6f] TiO$_2$(H)	*I4/m*	10.161		2.9700		3.46
Ramsdellite[6g] TiO$_2$(R)	*Pbnm*	4.9022	9.4590	2.9585		3.80
Columbite[6h] (TiO$_2$-II)	*Pbcn*	4.5318	5.5019	4.9063		4.33
Baddeleyite[6i]	*P2$_1$/c*	4.6400	4.7600	4.8100	99.20	5.08
*Cotunnite[6a]	*Pnma*	5.2460	3.2700	6.1260		5.04
Pyrite[6a]	*Pa-3*	4.8600				4.65
Fluorite[6a]	*Fm-3m*	4.8000				5.15

*Structure at 20.3 GPa. ** Structure at 160 K, 0.1 MPa.

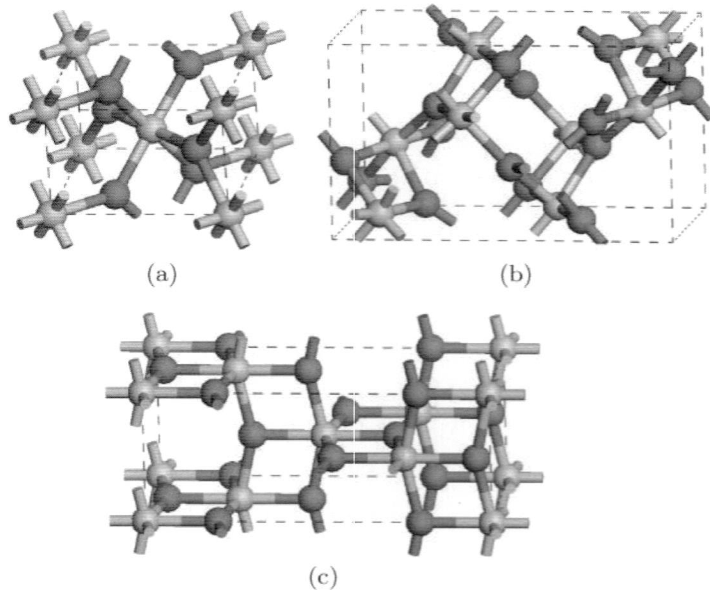

Figure 6.1.   Crystal structures of TiO$_2$ polymorphs. (a) Rutile, (b) Anatase, and (c) Brookite. Red and gray balls represent O and Ti atoms, respectively.

are slightly distorted in rutile and anatase, with two Ti–O bands (1.980 Å for rutile and 1.965 Å for anatase) slightly longer than the other four (1.949 for tutile and 1.937 for anatase), and with some of the O–Ti–O band angles deviating from 90°.[7] In brookite, the TiO$_6$ octahedrons are linked together in a certain twisted way that gives it a more complicated structure. Consequently, brookite presents six different Ti–O bands ranging from 1.87 Å to 2.04 Å and 12 different O–Ti–O band angles from 77° to 105°.[7] In contrast, in rutile and anatase, there are only two kinds of Ti–O bands and O–Ti–O band angles (81.2° and 90.0° for rutile, 77.7° and 92.6° for anatase).[7]

Using first-principles density functional theory (DFT) within the local-density-approximation (LDA) scheme, Mo and Ching calculated the electronic structures of rutile, anatase, and brookite.[7] Shown in Figure 6.2 are the calculated band structures along the symmetry lines of the Brillouin zone for rutile, anatase, and brookite. For rutile, they gave the indirect bandgap at Γ with a rather small value of 1.78 eV.[7] This is expected since the DFT calculations generally underestimate the bandgaps for insulators and semiconductors. These DFT–LDA calculations indicated that the

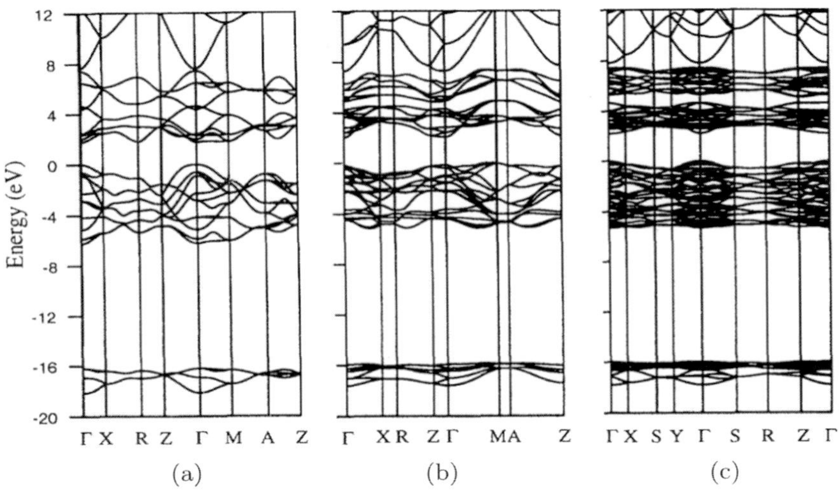

Figure 6.2. Calculated band structures of rutile (a), anatase (b), and brookite (c). Reprinted from Ref. [7]. © 1995 American Physical Society.

upper valence band (VB) of rutile is composed of O 2p orbitals and has a width of 6.22 eV, and its lowest conduction band (CB) consists of two sets of Ti$_{3d}$ bands and has a width of 5.9 eV. The origin of these two sets of Ti$_{3d}$ bands has been ascribed to the hybridized atomic states of t_{2g} and e_g.[7] Unlike rutile, the LDA-calculated minimal bandgap of 2.04 eV for anatase is indirect. Its CBM is at Γ and its VBM is at M. But they argued that as the energy at Γ is only 0.18 eV lower than VBM, anatase can be still considered as a direct-gap insulator.[7] Their calculations also showed the upper VB width of anatase (5.17 eV) is less than that of rutile by about 1 eV. For brookite, their calculations indicated that it has a direct bandgap of 2.20 eV at Γ which is larger than both and anatase, and its upper VB has a total width of about 5.31 eV which is close to that of anatase.

Also with DFT–LDA, Kuo *et al.* studied the structural and electronic properties of a number of high-pressure TiO$_2$ polymorphs,[8] with the aim to find out whether a specific dense polymorph of TiO$_2$ has a narrow bandgap for potential photocatalytic applications under visible light. The selected polymorphs, as plotted in Figure 6.3,[8a] have zero-pressure densities greater than rutile, including α-PbO$_2$ (Columbite), baddeleyite, fluorite, and cotunnite types of structures. Among them, the α-PbO$_2$-type phase is the least dense phase, about 3% denser than rutile. In this phase, atoms still bond similarly as in the ambient rutile, anatase, and brookite phases, with CN of 6 for Ti atoms. But in denser phases of baddeleyite, fluorite, and cotunnite types, Ti atoms will bond to more O neighbors with CN = 7, 8, and 9, respectively.[8a] Their calculated electronic bandgaps show dispersed values, which are listed in Table 6.2 together with the values of rutile and anatase. Among them, the fluorite-type phase had been identified with the lowest bandgap of 1.08 eV, using the calculated lattice parameters of 4.749 Å at ambient pressure.[8a]

Considering the systematic bandgap underestimation of DFT calculations, Kuo *et al.* suggested that the gap value of the fluorite-type TiO$_2$ should be larger than 1.08 eV. They applied the theoretical–experimental difference 1.1 eV for rutile or anatase to this case, and found that the real bandgap of fluorite-type TiO$_2$ is 2.18 eV.[8a] A bandgap of 2.18 eV corresponds to a wavelength 570 nm within the visible light range. Based on the calculated imaginary part of the

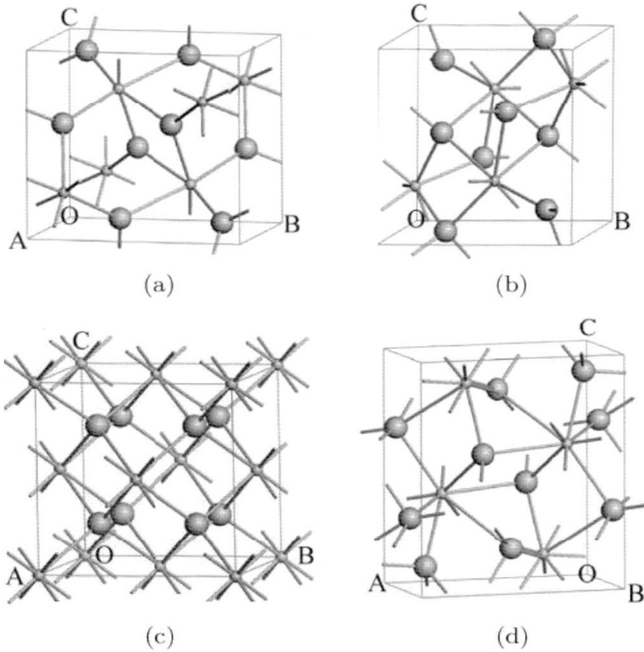

Figure 6.3. Unit cells of TiO_2 polymorphs: (a) R-PbO_2-type, (b) baddeleyite-type, (c) fluorite-type, (d) cotunnite-type. Large spheres represent the O ions, small spheres the Ti ions, O, A, B, and C denote the origin, a-, b-, and c-axes, respectively. Reprinted from Ref. [8a]. © 2005 American Chemical Society.

Table 6.2. The Calculated bandgap for Rutile, Anatase, α-PbO_2-type, Baddeleyite, Fluorite, and Cotunnite types of TiO_2.[8a]

	Band gap (eV)	
Phases	**LDA**	**exptl**
Rutile	1.82(D)[*8d], 1.86(D)[**]	3.0^d
Anatase	2.13(ID)[**]	3.2^e
α-PbO_2-type	2.33(D)[*8e], 2.62(ID)[**]	
Baddeleyite	2.06(D)[*8f], 2.16(ID)[**]	
Fluorite	1.08(D)[**]	
Cotunnite	1.65(D)[*8g], 1.72(ID)[**]	

D is the direct bandgap. ID is the indirect bandgap.
[*]Calculated by using the experimental lattice parameters.
[**]Calculated by using the calculated lattice parameters.

dielectric function, they claimed that the fundamental absorption edge for fluorite-type TiO$_2$ occurs at 1.82 eV, resulting from the transitions between its VBM and CBM.[8a] Therefore, they indicated that fluorite-type TiO$_2$ may be useful in the design of photocatalysts responsive to visible light.

The above DFT simulations have presented the elementary electronic features of TiO$_2$ polymorphs, including the profiles and orbital characters of their CB and VB, and the variation of these bands upon structural modification. Although DFT fails to produce correctly the gap values of semiconductors, that shortcoming can be overcome with those methods beyond DFT, such as the hybrid functionals[9] mixed with a portion of Hartree–Fock (HF) exchange with the traditional density functional, or by GW many-body calculations.[10] With these methods once the electronic band structure is described properly for each TiO$_2$ polymorph, its capability of absorbing solar energy through photoelectric conversion can be easily clarified. But for the overall high PEC efficiency of a semiconductor, the electrochemical conversion thereafter has to be enough high as well, which means the converted electron and hole pairs should be well separated and consumed as much as possible by the chemical reactions on the catalyst interfaces. This dynamic process relies firstly on the alignment of its VBM (for holes) and CBM (for electrons) with respect to other materials or the vacuum energy level. For TiO$_2$, it has been commonly observed that the mixed rutile/anatase systems show more favorable photocatalytic properties than either of their pristine phases.[11] The fundamental band alignment between rutile and anatase had been considered as a dominant factor in understanding carrier dynamics and outperformance of this mixed phase semiconductor.[11a]

To identify the band alignment between anatase and rutile TiO$_2$, Scanlon *et al.* simulated the electronic structure of bulk crystals, electrostatic analysis of the Ti and O environments, and the absolute vacuum alignment from embedded crystals.[11a] In Figure 6.4, they compared the electronic density of states (DOS) for rutile and anatase, which were calculated with the hybrid density functional (HSE06). They found that both phases exhibit similar electronic

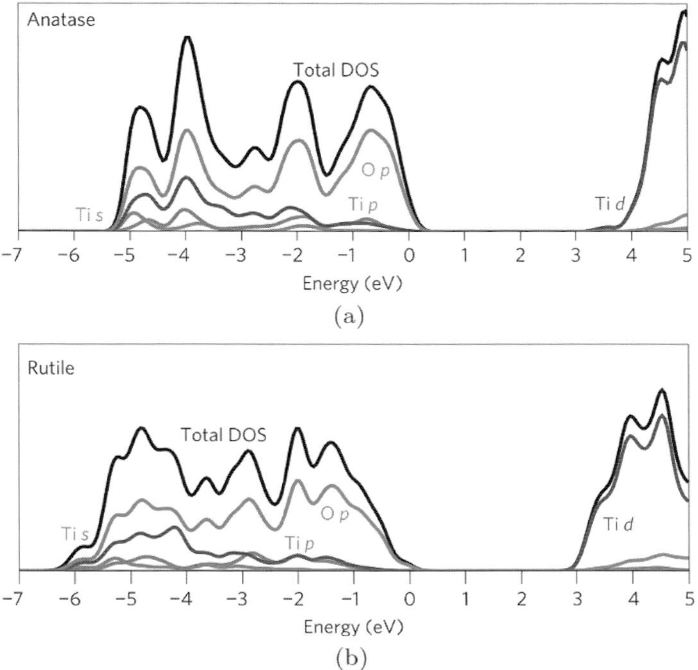

Figure 6.4. Electronic structure of anatase and rutile TiO_2. (a) and (b) Comparison of the total and ion-decomposed electronic DOS of anatase (a) and rutile (b) TiO_2 calculated using the HSE06 hybrid density functional. Reprinted from Ref. [11a]. © 2013 Macmillan Publishers Limited.

band characters and have the similar widths for their upper VBs. They thus suggested that the position of the CB and VBs of ionic TiO_2 be determined by the onsite electrostatic potential and the optical dielectric response. Therefore, they calculated further the local Madelung potentials of Ti and O in anatase and rutile using the polarizable shell model fitted to reproduce the high-frequency dielectric constants of TiO_2. With such analysis, they indicated that the Madelung potential of Ti in anatase (-45.025 V) is 0.17 eV higher than the potential of Ti in rutile (-45.199 V). Similarly, they found that the Madelung potential of O in anatase (26.232 V) is higher than the Madelung potential of O in rutile (25.767 V), and suggested accordingly the VB of rutile 0.47 eV stays above the VB of anatase.[11a] Moreover, they calculated energies of charge carriers propagating at the band edges using the Mott–Littleton defect approach, which includes the high-frequency dielectric response of the material. They found the calculated difference in carrier energies between the two materials support the band alignment based on Madelung potentials. In particular, they obtained a 0.24 eV shift downwards for the electrons at CBM of anatase relative to that of rutile, whereas for holes, they found the VB of rutile to be 0.39 eV higher in energy than that of anatase.

Furthermore, Scanlon *et al.* calculated the ionization potentials of anatase and rutile relative to the vacuum level using a hybrid quantum-mechanical/molecular-mechanical (QM/MM) embedding technique.[11a] In this approach, a part of the crystal was represented by a molecular cluster (treated at a QM level of theory) embedded in an external potential, which represents the system remainder

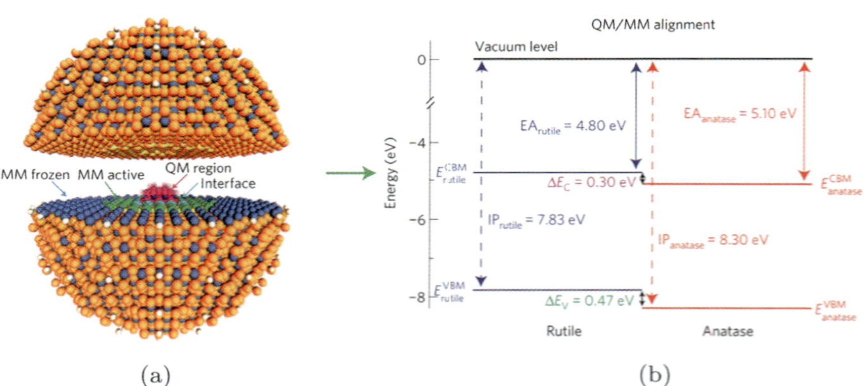

(a) (b)

Figure 6.5. Band alignment between rutile and anatase from QM/MM. (a) Graphic of the hybrid QM/MM cluster used for rutile in the positive charge state. The cluster is divided into hemispheres to highlight the different regions in the model. Hole density iso-surfaces are shown (semi-transparent purple) in the QM region. (b) Schematic of the QM/MM alignment of rutile and anatase TiO$_2$. IP and EA denote ionization potential and electron affinity, respectively. The electron affinity is calculated by adding the experimental bandgaps to the calculated ionization potentials. ΔE_V and ΔE_C are the VB offsets and CB offsets, respectively and E^{CBM} and E^{VBM} denote the positions of the CB and VB respectively. Reprinted from Ref. [11a]. © 2013 Macmillan Publishers Limited.

(treated at an MM level of theory), as illustrated in Figure 6.5(a).[11a] By this MM model, they calculated the system response to ionization, on which all electronic degrees of freedom are fully relaxed while keeping the nuclei frozen. The ionization potential was calculated by taking the energy difference between a (electronically) relaxed system in the neutral and positive charge states. After calculating the ionization potential of both rutile and anatase for a series of QM cluster sizes (from ∼50 atoms to ∼80 atoms), they also found an offset of ∼0.47 eV between rutile and anatase, with the rutile VB higher in energy than anatase, as indicated in Figure 6.5(b).

Together with their X-ray photoelectron spectroscopy (XPS) measurements, Pfeifer *et al.* also examined the band discontinuity at the rutile/anatase interface with DFT and the methods beyond DFT, and they found a staggered type-II energy band alignment between them.[11b] In DOS comparison between rutile and anatase, as shown in Figure 6.6(a), they identified a VB discontinuity of 0.63 eV and a CB discontinuity of 0.39 eV, both of which agree with their XPS values. They found that both rutile and anatase exhibit similar DOS profiles except for the appearance of "tails" at both top and bottom in the case of rutile VBs. As shown in Figure 6.6(b) their calculated band structure of rutile demonstrates that the tails originate from a pronounced splitting of the topmost and bottommost levels in the vicinity of the Γ point, which is entirely absent in the anatase case.[11b] With the help of a Wannier function analysis, they clarified the electronic origin of this feature, i.e. one sp^2- and one p_z-like orbitals for each oxygen atom, as presented in Figures 6.6(c) and (d). The three lobes of the sp^2 orbital are oriented along the O–Ti bonds, whereas the p_z-like orbital is oriented perpendicular to the sp^2 plane. The projection of the band structure on this set of states yielded the relative admixture as illustrated by the color coding in Figure 6.6(b). With this analysis, they revealed that the topmost VB near Γ, where a pronounced separation from the other valence states occurs, is virtually exclusively of p_z character and thus can be interpreted as a lone-pair orbital. As the lone-pair orbital does not participate in the O–Ti bond, the splitting of the energy level can be understood consequently. And they explained the downward dispersion of the

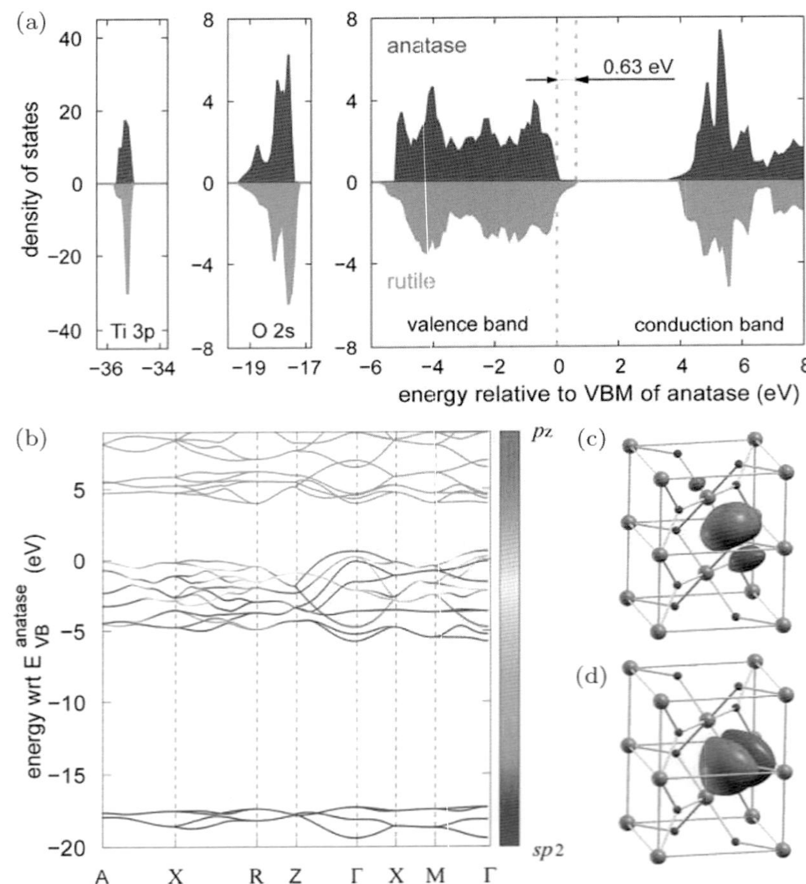

Figure 6.6. (a) Comparison of the DOSs of rutile and anatase. The energy scales have been aligned based on the electrostatic potential at the Ti cores. (b) Band structure of rutile where the color scale indicates the respective admixture of oxygen-centered (c) sp^2- and (d) p$_z$-like orbitals. In (c), only one of the three individual Wannier functions that contribute to the sp^2-like orbital is shown. The remaining lobes are oriented along the other two O–Ti bonds. Reprinted from Ref. [11b]. © 2013 American Chemical Society.

band away from the center of the Brillouin zone Γ with the σ-like overlap between the lone-pair orbitals of neighboring O atoms.

Unlike the case of rutile, as shown in Figure 6.7, the p$_z$-like orbital does not play a prominent role near the VB edges of anatase, and its

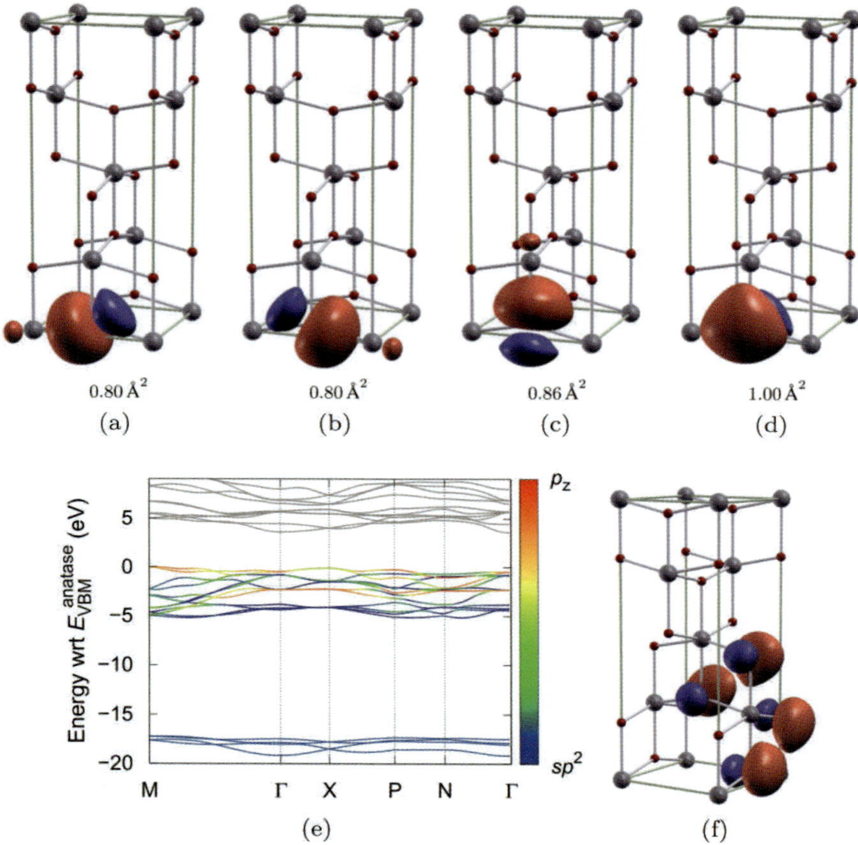

Figure 6.7. Representative ((a)–(c)) sp^2-like and (d) p$_z$-like maximally localized Wannier functions for anatase shown with respect to the conventional unit cell. Orbital spreads are given in the subfigure captions. (e) Band structure of anatase. The color coding indicates the contribution of p$_z$-like Wannier orbitals of the type shown in (d) and (f) to the respective state. (f) Ensemble of p$_z$-like Wannier orbitals. Reprinted from Ref. [11b]. © 2013 American Chemical Society.

band structure also does not exhibit a splitting of states around the Γ point.[11b] Pfeifer *et al.* pointed out that the difference between the two types of behavior originates from the respective orientation of the ensemble of p$_z$-like orbitals in the two different crystal structures. In rutile, the p$_z$-like orbitals are much closer to each other, suggesting stronger interaction and overlap, which results in a larger splitting of the corresponding energy bands and consequently in a higher VBM and the appearance of the tail at the top of the VB. Thus, with such an LCAO picture,[12] they had established a direct connection between the atomic bonding structures of rutile and anatase, their electronic structures, and, their VB offset. In short, their electronic structure analysis shows that the higher VBM of rutile is caused by the stronger overlap between the O $2p_z$ orbitals in rutile compared to those in anatase, leading to a substantial splitting of the resulting energy bands. Their staggered band alignment may explain the enhanced photocatalytic activity of mixed phase TiO$_2$ particles as it provides a driving force for separation of photoexcited charge carriers.[11b]

6.3   Factors Affecting (or "tuning") the Electronic Band Structure of TiO$_2$

In this section, in order to gain insight on understanding the electronic behavior of black TiO$_2$, we will discuss the primary factors in affecting or tuning electronic band structures of TiO$_2$.

6.3.1   *Intrinsic Defects*

The correlation between the defect constitution of solids and their catalytic activity was widely accepted as a formal basis to explain the details of their catalytic process since 1949.[13] For TiO$_2$, it was demonstrated that the very pure stoichiometric condition TiO$_2$ will not be an active catalyst for certain chemical synthesis.[13] In general, six kinds of native-point defects may exist in the TiO$_2$: the vacancies of titanium (V_{Ti}) and oxygen (V_O), the interstitials of titanium (Ti_i) and oxygen (O_i), and two anti-site defects of Ti_O and O_{Ti}. As TiO$_2$ is naturally reduced and n-type, Ti_i and V_O were usually proposed as resulting in the apparent oxygen deficiency (O_d) of TiO$_{2-x}$.[14] Due to the complexity of forming native defects in TiO$_2$, it is challenging to reveal a particular impurity's effect simply by measuring the doping dependent photocatalytic process.[15] With the advantage of building ideal isolated defect models, theoretical analysis — especially those first-principles calculations — were expected to provide some instructive insights into the defect chemistry of TiO$_2$.[15]

Using a semiempirical self-consistent method, Yu *et al.* studied the electronic structures of point defects in reduced rutile.[16] They reported that the donor levels of V_O and Ti_i be around 0.7 eV below the CB edge. With DFT–LDA, Cho *et al.* found that the V_O did not produce a defect level in the bandgap, while Ti_i generated a localized state 0.2 eV below the CB edge.[17] Na-Phattalung *et al.* investigated the atomic structures and electronic properties of point defects in anatase (Figure 6.8).[15] They found that Ti_i was a low-formation energy quadruple donor and caused *n*-type conductivity. V_O had a higher formation energy and a lower kinetic energy to be created from perfect crystal. Post-growth formation of V_O was possible by heating the sample for some time. O_i bonded to lattice oxygen impulsively to form an electrically inactive O_2 dimer, and anti-site defects had high formation energies to automatically break into isolated interstitials

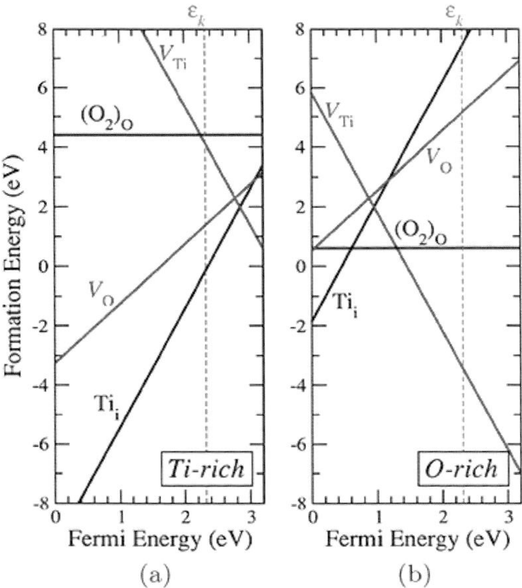

Figure 6.8. DFT-calculated defect formation energies in bulk anatase as a function of the E_F, under the Ti-rich (a) and O-rich (b) growth conditions, respectively. The slope of the line indicates the charge state of the defect. The Fermi energy is referenced to the top of the VB. The vertical dotted line is the calculated bandgap at the special k-point.[15] Reprinted from Ref. [15]. © 2006 American Physical Society.

and vacancies.[15] They indicated further that Ti$_i$, O$_i$, V$_{Ti}$, and V$_O$ produced no defect levels to the bandgap.[15]

Despite the detailed DFT descriptions given on the electronic properties of point defects in TiO$_2$, there were arguments that such approaches based on pure exchange–correlation functionals cannot satisfactorily reproduce the experimental findings.[18] Di Valentin *et al.* pointed out that limited by the insufficient cancelation of the self-interaction energy, pure DFT functionals might fail in describing the localized states, particularly of excess holes and electrons in wide bandgap semiconductors and insulators such as V$_O$ centers in TiO$_2$.[18a,b] They suggested that inclusion of a HF exchange or Hubbard term enhanced the Ti 3d^1 defect state description and induced the polaronic distortion. With the help of spin-polarized hybrid DFT calculations, they investigated the Ti$_i$ atom's charge and spin state in reduced bulk anatase and rutile TiO$_2$.[18b] A Ti$_i$ atom in an interstitial cavity instantly turned into a Ti^{3+} ion, donating three electrons to the lattice Ti ions and keeping one on its 3d shell. Both interstitial and lattice Ti^{3+} ions induced new states near 1.0–1.5 eV below the CBM in the bandgap. These states were spread in an interval of 0.7–0.8 eV as they were exposed to different adjacent environments.[18b] Janotti *et al.* investigated V$_O$ in rutile with the hybrid functionals.[18c] They showed that V$_O$ was a shallow donor in the bandgap and the energy of the +2 charge state (V$_O^{2+}$) was lower than the +1 charge (V$_O^{1+}$) and neutral states (Figure 6.9). With the low formation energy of V$_O^{2+}$, they explained the *n*-type conductivity of TiO$_2$ single crystals heated under O-poor conditions.[18c]

6.3.2 *Doping*

Similar as intrinsic defects, introducing alien atoms into the lattice of a semiconductor may bring electronic states into its bandgap, i.e. doping, has been taken as a typical way in artificially tuning the electronic structures of semiconductors. This is also the case for improving the solar photocatalytic efficiency of TiO$_2$.

Asahi *et al.* had suggested whether visible-light activity could be realized in TiO$_2$ by doping relies on three requirements: (i) doping

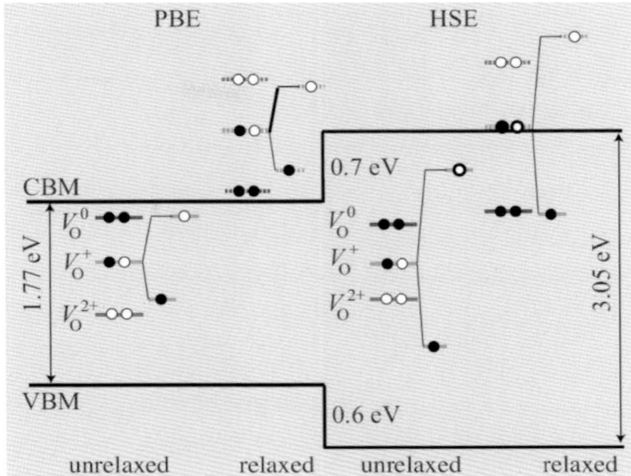

Figure 6.9. Calculated position of the single-particle a_1 state of the V_O in TiO_2. The PBE and HSE band structures were aligned as described in the text. The results for various charge states and for both unrelaxed and relaxed vacancies are shown. For V_O^+, the position of the a_1 state in both spin-up and spin-down channels is also indicated. The positions of states above the CBM were estimated based on projected densities-of-states on the three nearest-neighbor Ti atoms and are represented by dashed lines.[18c] Reprinted from Ref. [18c]. © 2010 American Physical Society.

should produce states in the bandgap of TiO_2 that absorb visible light; (ii) the CBM, including subsequent impurity states, should be as high as that of TiO_2 or higher than the H_2/H_2O level to ensure its photoreduction activity; and (iii) the states in the gap should overlap sufficiently with the band states of TiO_2 to transfer photoexcited carriers to reactive sites at the catalyst surface within their lifetime.[19] With conditions (ii) and (iii), they suggested that the anionic species are desired for the doping rather than cationic metals, which often give quite localized d states deep in the bandgap of TiO_2 and result in recombination centers of carriers. With DFT–LDA, they calculated DOSs of the substitutional doping of C, N, F, P, or S for O in the anatase TiO_2 crystal. Among them, they identified that the substitutional doping of N was the most effective because its p states contribute to the bandgap narrowing by mixing with O 2p states.

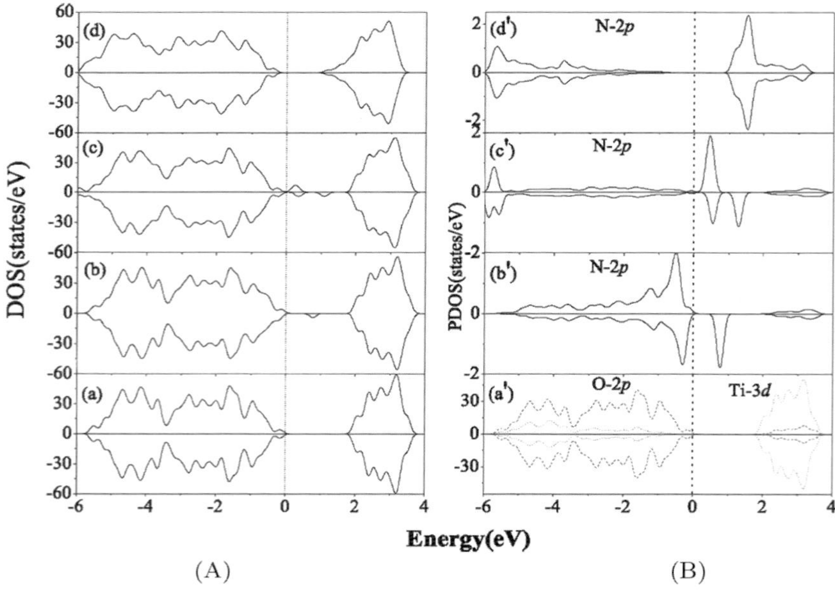

Figure 6.10. (A) TDOS and (B) PDOS for 72-atom rutile supercell. (a) Undoped rutile model, (b) substitutional N–O doped model, (c) interstitial N-doped model, and (d) substitutional N–Ti-doped model. The energy is measured from the top of the VBs of rutile TiO$_2$.[20] Reprinted from Ref. [20]. © 2006 American Chemical Society.

Using DFT with generalized gradient approximation (GGA) functional of PW91, Yang *et al.* studied the N-doping effects on the electronic and optical properties of TiO$_2$ rutile crystal.[20] Figure 6.10 plotted the total density of states (TDOS) and the projected density of states (PDOS) of the different N-doped 72-atom rutile model structures they built.[20] For the model of one substitutional N atom to O atom, they found little shift of CBM and VBM, with respect to the undoped material. They observed that a large part of N 2p states lie in the CB while some N 2p localized states extend into the bandgap about 0.7 eV above VBM, as shown in Figure 6.10(a). For the interstitial N-doped model, their calculated results showed no bandgap narrowing, and the interstitial N atom introduces N 2p states that were mainly located between the O 2p VBM and CBM and that formed two energy bands. These N 2p states led to about

0.7–1.0 eV decreasing of the energy gap, as shown in Figures 6.10(b) and (c). Therefore, they suggested the transition energy of the photon can be reduced because of the transitional energy levels introduced by the N impurities rather than bandgap changes. However, for the model of one substitutional N atom to Ti atom, they found that the N 2p states were mainly located in the CB, and there was a rather larger decline of VBM, as well as CBM. The phenomenon that was explained by the removal of the electrons from the supercell, resulting from the replacement of one Ti with N atom, leads to a reduction of Coulomb repulsion and the shift of the energy band edges. With the decline of both band edges, they found that the energy bandgap was reduced by about 0.25 eV in contrast with the undoped rutile supercell, as shown in Figure 6.10(d).

In principle, both anion and cation dopings can shift the absorption edge of TiO_2 toward the visible-light region for better photocatalytic efficiency. However, such monodopings may also result in partially occupied impurity bands that act as recombination centers and reduce the photogenerated current, or lead to formation of strongly localized d states within the bandgap which reduce carrier mobility.[4] To overcome such issues, Li and Wei *et al.* suggested using a passivated codoping approach to redshift the TiO_2 absorption edge and improve its photocatalytic efficiency for hydrogen production.[4] This is based on the concept that in codoping cases, the electrons on the donor levels passivate the same amount of holes on the acceptor levels, so the systems still keep semiconductor character, leaving much less recombination. After analyzing the band structure of TiO_2 and the chemical potentials of the dopants with DFT–LDA, they proposed that the band edges of TiO_2 can be modified by passivated codopants such as (Mo + C) to shift VBM up significantly, while leaving the CBM almost unchanged, thus satisfying the stringent requirements.

6.3.3 *Surface and Interface Effects*

For TiO_2 in PEC applications, its electrochemical conversion process depends crucially on their surfaces or interfaces of these materials. As truncated from the chemically stoichiometric bulk, the atoms on the semiconductor surfaces become unsaturated and possess many dangling bonds. That will normally cause high surface activities to react with adjacent molecules or particles to form new compounds as addressed in many works.[21] For example, with DFT, Zhao *et al.* investigated the water adsorption and decomposition on the low-index anatase TiO$_2$ surfaces.[22] Their results showed that the surface-structure-sensitive water decomposition reaction is determined not only by the surface energy but also by the surface properties. On these low-index surfaces, they found the chemical activity for the water decomposition reaction decreased in the order $(001) > (103)_f > (100) > (110) > (103)_s > (101)$.

In terms of electronic band structure of semiconductors such as TiO$_2$, the structural discontinuity on surfaces or interfaces can also introduce electronic states into their bandgap, causing long-tail optical absorption and red-shift optical emission. For example, using the linear combination of muffin-tin orbitals energy-band method, Kasowski and Tail calculated the one-electron DOS of thick slabs of rutile TiO$_2$ with (001) and (110) symmetry together with those of bulk TiO$_2$ and Ti$_2$O$_3$.[23] They found that the local Ti–O coordination is the major factor in determining their electronic band structures. In particular, for the unreconstructed (001) face of TiO$_2$, they found that there were a large band of occupied surface states in the bulk bandgap while there were no occupied states for the unreconstructed (but slightly relaxed) (110) surface. In fact, once into the photocatalytic processes, the surfaces of TiO$_2$ may also be bonded with alien atoms and not be clean any more. Obviously, these alien bondings will also bring electronic states more or less into the bandgap and affect the electronic properties of TiO$_2$ consequently.

6.3.4 *Amorphous/Disordered Phases*

If point defects or impurities can be considered as zero-dimensional structural changes, surfaces and interfaces as two-dimensional ones, the amorphous or disordered phases of TiO$_2$ can be taken as three-dimensional structural modifications. That will certainly bring more complicated electronic features than low-dimensional modifications do to TiO$_2$.

Kaur and Singh reported that amorphous TiO_2 nanomaterials had characteristic structural and electronic properties.[24] They studied three amorphous TiO_2 models with supercell dimensions of $2 \times 2 \times 3$ (72 atoms), $2 \times 2 \times 4$ (96 atoms), and $3 \times 3 \times 4$ (216 atoms), based on the MD simulations with Perdew–Burke–Erzerhof (PBE) DFT functionals.[24] They observed that most Ti–O bond lengths were between 1.86 Å and 2.1 Å for these amorphous structures, in contrast that the crystalline rutile Ti–O bond lengths were at 2.017 Å and 2.05 Å. For amorphous TiO_2, they found relatively larger bandgap and band tail states compared to rutile and anatase. These tail states were localized due to the disorders of bond angle and the positions of O and Ti atoms. They indicated further that higher disorder resulted in more tailing bands in the bandgap.[24]

6.4 Lattice Strain Effects

Unlike the factors discussed earlier, lattice strain that happen to the bulk phase with deviated lattice parameters is not an atomic-bond-breaking way in affecting the electronic properties of TiO_2. For semiconductors, lattice strain effects can be generally observed at the interfaces between two phases with mismatched lattice or bonding structures.

Using DFT–GGA band structure calculations, Yin *et al.* showed that unlike the rutile phases, the bandgap of TiO_2 in the anatase phase can be effectively reduced by applying stress along a soft direction.[25] The softening of Ti–O bonds along the c or (001) direction provided an opportunity to tune the bandgap of TiO_2 by applying a stress along this direction. In deformation potential theory, the bandgap change of a semiconductor under stress can be given by

$$E_g(\sigma_a, \sigma_c) = E_g^0 + b_a\sigma_a + b_c\sigma_c, \tag{1}$$

where E_g^0 is bandgap of unstrained bulk crystal; b_a and b_c are bandgap pressure coefficients of epitaxial and uniaxial stress, respectively.[26] The hydrostatic bandgap pressure coefficient is $b_a + b_c$. The b_c and b_a was 35.3 meV/GPa and 39.8 meV/GPa, respectively. As shown in Figure 6.11, their results indicated that the compressive

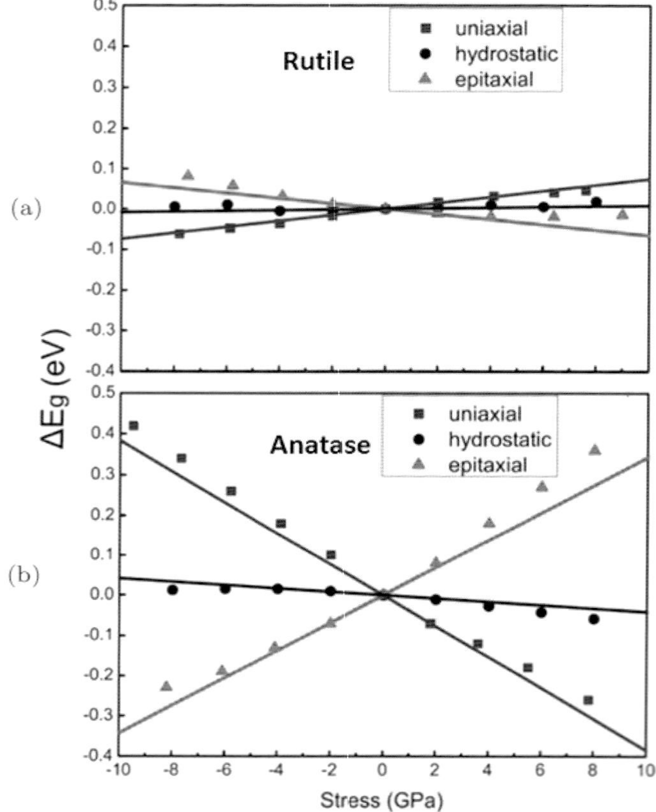

Figure 6.11. The bandgap change in rutile (a) and anatase (b) under stress. Dots: calculated results, lines: fittings.[25] Reprinted from Ref. [25]. © American Institute of Physics.

stress along the c axis was easier than in xy plane to reduce the bandgap as $|b_c| > |b_a|$. In comparison, under a pressure up to 10 GPa, the bandgap change of rutile was smaller than 0.08 eV, as the soft axis in rutile is along both (110) and ($1\bar{1}0$) directions, but all of the soft axes lie along the (001) direction in anatase.[25]

6.5   Nanosize Effects

Like other semiconductors in nanoscale, nanophase TiO$_2$ is also subject to the size effect. When the size of NPs decreases, their electronic bands become discrete and the bandgap increases,[27] and the charge carriers behave quantum mechanically.[21c,28] These changes are more easily observed when the size of these particles approaches the Bohr radius of the first excitation state or the de Broglie wavelength of their electrons and holes. The bandgap energy E_g and the radius R of the NPs have the following relationship:

$$E_g \approx E_g^0 + \frac{h^2}{8\mu' R^2} - \frac{1.8e^2}{\varepsilon' R},\qquad(2)$$

where e, h, E_g^0, ε', and μ' is the electron charge, the Planck constant, the bulk bandgap, the apparent dielectric constant of the semiconductor and the apparent reduced mass of the electron and hole in the quantum region, respectively.[27c,27d] The μ' value of TiO_2 is around 1.63 m_0 (where m_0 is the electron rest mass),[27e] and the ε values are 31 and 173 for anatase and rutile, respectively.[29] A clear bandgap shift was observed as $2R < 2\,\mathrm{nm}$, and the difference in E_g between the rutile and anatase decreased as the $2R$ value was close 1 nm, as shown in Figure 6.12.[30]

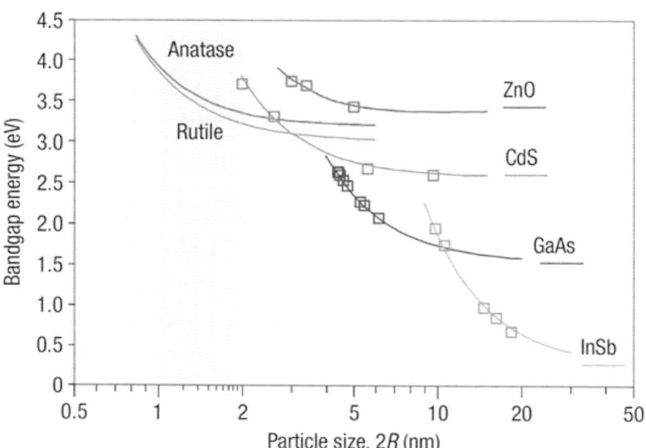

Figure 6.12. Calculated TiO_2 bandgap energy change with size (squares: experimental data).[30] Reprinted from Ref. [30]. © 2008 Nature Publishing Group.

6.6   Theoretical Analysis on Black TiO$_2$

The superior solar-driven photocatalytic activities exhibited by black TiO$_2$ have triggered quite a few theoretical studies to understand the mechanism behind "disorder engineering" and the enhanced solar PEC performance.

To understand the origin of the change in the electronic and optical properties of black TiO$_2$ nanocrystals, Chen *et al.* calculated the energy band structures using DFT with PBE–GGA.[5a] They studied the lattice disorders in TiO$_2$ nanocrystals in the presence of hydrogen and found that, rather than generating levels near the CBM, disorder-induced mid-gap states can up-shift the VB edge of TiO$_2$ nanocrystals. By inspecting the calculated total and projected DOS of a network model structure of TiO$_2$ nanocrystals (with fully relaxed surface dangling bonds), they found that the primary effect of surface reconstruction in TiO$_2$ nanocrystals is to produce strong band tailing near the VB edge. Without the introduction of disorder, they examined the band structures of TiO$_2$ nanocrystals containing four types of intrinsic defects with low formation energies: V_{Ti}, V_O, Ti_i, and O_I, and three types of hydrogen impurities: interstitial H atom (I_H), interstitial H$_2$ molecule (I_{H2}), and H atom forming surface OH bonds with oxygen ($OH_{surface}$). They found, without disorder, among them only the interstitial of Ti_i could yield a gap state in TiO$_2$ nanocrystals, with the value of about $0.5\,eV$ below CBM. After lattice disorder was introduced in the TiO$_2$ nanocrystal models, mid-gap electronic states were created accompanied by a reduced bandgap. A schematic illustration of a supercell of the disordered TiO$_2$ nanocrystal model as compared to that of a bulk anatase TiO$_2$ crystal is shown in Figures 6.13(a) and (b) plots the calculated DOS of the disorder-engineered TiO$_2$ nanocrystals along with those of the bulk and the unmodified TiO$_2$ nanocrystals. Two groups of mid-gap states (centered at about $1.8\,eV$ and $3.0\,eV$) can be observed in the DOS of the disordered TiO$_2$ nanocrystals, for which the Fermi level was found to locate slightly below $2.0\,eV$. The different nature of these two groups of mid-gap states is revealed by the calculated partial DOS (Figure 6.13(c)). While the higher energy mid-gap states ($\sim 3.0\,eV$) are derived from the Ti

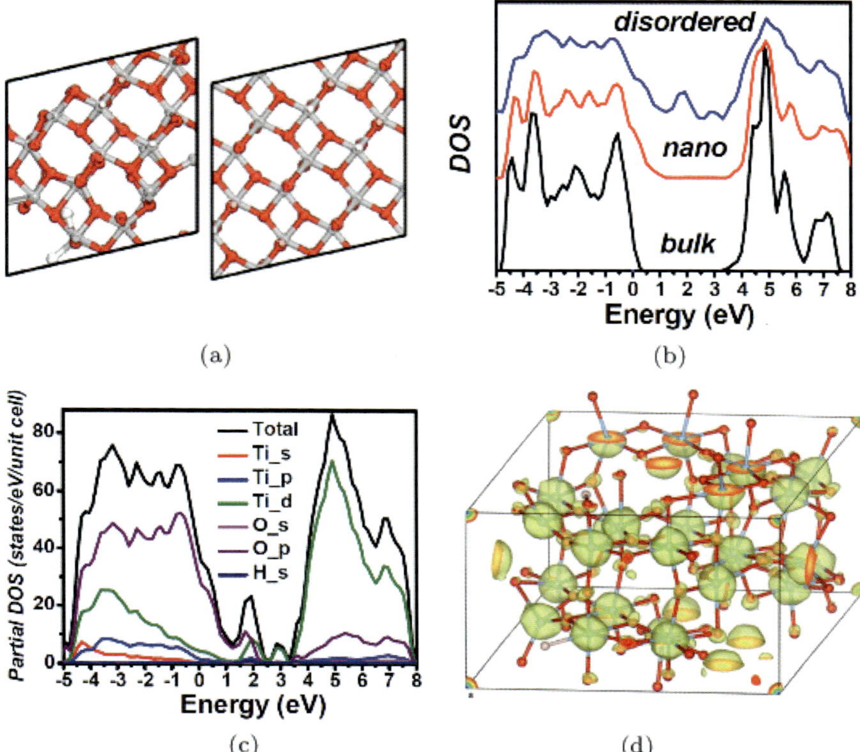

Figure 6.13. (a) Schematic illustration of a supercell for modeling disorder-engineered TiO_2 nanocrystal (red: O atoms, gray: Ti atoms, white: H atoms), as compared to a supercell modeling bulk anatase TiO_2. (b) Calculated DOS of TiO_2 in the form of a disorder-engineered nanocrystal, an un-modified nanocrystal, and a bulk crystal. The energy of the VBM of the bulk phase is taken to be zero. (c) Decomposition of the total DOS of disorder-engineered black TiO_2 nanocrystals into partial DOS of the Ti, O, and H orbitals. (d) three-dimensional plot of calculated charge density distribution of a mid-gap electronic state (at about 1.8 eV) of disorder-engineered TiO_2 nanocrystals. Reprinted from Ref. [5a]. © 2011 by the American Association for the Advancement of Science.

3d orbitals only, the lower energy states ($\sim$1.8 eV) are hybridized from both O 2p orbitals and Ti 3d orbitals, and mainly from the VB states as the result of disorders stabilized by hydrogen. They also examined the three-dimensional charge density distribution of the mid-gap electronic states (Figure 6.13(d)) of disorder-engineered

TiO$_2$ nanocrystals. Their charges associated with the lower energy mid-gap states distribute around every O or Ti atom are indicative of the overall impact of the disorder.

Lu *et al.* investigated the hydrogenation effect on the structural and photocatalytic properties of anatase TiO$_2$ (101) and (001) surfaces with DFT–PBE calculations.[31] They proposed that the high photoactivity of disorder-engineered TiO$_2$ nanocrystals can be ascribed to surface effects (Figure 6.14). They found that hydrogen atoms were chemically absorbed both on Ti$_{5c}$ and O$_{2c}$ atoms for (101), (001), and (100) surfaces by taking into account the synergistic effect of Ti–H and O–H bonds.[304] They suggested that

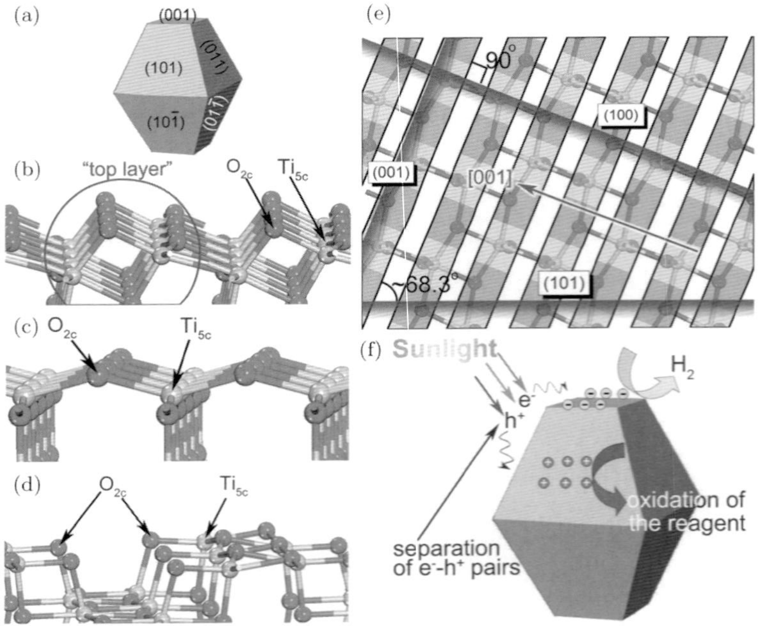

Figure 6.14. (a) Schematic drawings of anatase TiO$_2$ NPs grown in acidic solution. (b–d) Relaxed surfaces of the clean (101), (001), and (100) facets, respectively. (e) Illustration of the layered structure (s-layer) of anatase TiO$_2$ and the (101), (001), and (100) facets sloping, parallel, and perpendicular to the s-layers, respectively; the arrow in (e) denotes the soft axes lying along the (001) direction. (f) Spatial separation of the photogenerated electron–hole pairs as a result of electron–hole flow between the hydrogenated (101) and (001) facets.[31] Reprinted from Ref. [31]. © The Owner Societies 2011.

the hydrogenation-induced lattice distortions on (101) and (100) surfaces of NPs enhanced the intraband coupling within the VB, while the (001) surface was not largely affected. They indicated that the atoms not only induced the lattice disorders but also interacted strongly with the O 2p and Ti 3d states, resulting in a considerable contribution to the mid-gap states. Moreover, they found that the optical absorption was dramatically redshifted due to the mid-gap states and the photogenerated electron–hole separation was substantially promoted as a result of electron–hole flow between different facets of hydrogenated NPs. This accounted for the exceptionally high energy conversion efficiency of black TiO_2 under solar irradiation. In particular, they found that hydrogenation reversed the redox behavior of different surfaces of NPs, and therefore proposed that one could tune the photoexcited electron–hole flow between different surfaces of NPs in accordance to one's request by appropriate chemical surface treatment.[31]

Liu *et al.* pointed out that hydrogenated black TiO_2 required not only hydrogen passivation of Ti and O dangling bonds on the surface.[32] With the DFT–PBE and hybrid functional calculations, they attempted to answer the question of "what exactly is the role of hydrogen in disordering the surface of the anatase nanocrystals?"

Starting with a $2 \times 2 \times 1$ supercell with the formula $Ti_{16}O_{32}$, as shown in Figure 6.15(a), they examined the influence of lattice distortion on the electronic energy bands of anatase TiO_2. By displacing either an O or a Ti atom from its starting equilibrium position with constrained geometry, they analyzed separately the changes of band extrema upon the atomic displacements along the c axis. They found that the lattice distortion effects are quite different for O and Ti sublattices. As shown in Figures 6.15(b) and (c), both O and Ti distortions blueshift VBM effectively, but only the Ti distortion redshifts CBM (the O distortion affects little on CBM). They ascribed the inertia of CBM to the O distortion to the fact that the CBM states arise mainly from the 3d orbitals of the sixfold-coordinated Ti atoms (denoted by Ti_{6c}) and are, therefore, insensitive to changes involving mainly the oxygen sublattice. However, during a displacement of the Ti sublattice, a shift of Ti_{6c} will deform six

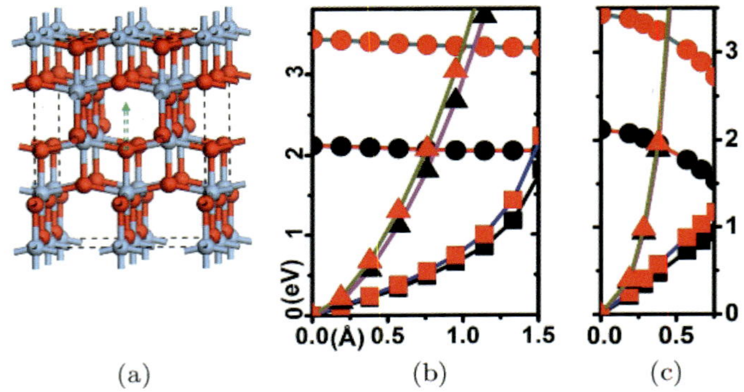

(a)   (b)   (c)

Figure 6.15. (a) $2 \times 2 \times 1$ anatase supercell, where the sky blue and red balls represent Ti and O atoms, respectively. The variations of VBM (represented by squares), CBM (circles), and total energy (triangles) calculated by PBE (colored in black) and by the hybrid functional (red) as a function of the distortion of (b) the O sublattice and (c) the Ti sublattice.[32] Reprinted with permission from Ref. [32]. © 2013 American Physical Society.

Ti–O bonds. This change will affect both the O-2p VBM and the Ti-3d CBM states. Since the Ti atoms are fixed with twice as many Ti–O bonds as the O atoms, it requires much more energy to deform the Ti sublattice than the O one by the same amount. This means that under the relatively low-energy process, the lattice disorder of anatase would be triggered mainly from the O sublattice.

They examined further the hydrogenation effect by comparing the bulk anatase phase, either with an interstitial H$_2$ molecule or with two H atoms bonded to O and Ti separately, as shown in Figure 6.16. For the bulk model of a $2 \times 2 \times 1$ supercell, Figure 6.16(a) showed the charge density distribution of its VBM electronic band, whose "dumbbell shape" distribution surrounding every O atom is characteristic of O 2p orbitals. After an interstitial H$_2$ molecule is introduced into the supercell (labeled as model A), the new highest energy occupied band (HEOB) states appear, resulted from the bonding 1s orbitals of H2 and the 2p orbitals of its neighboring O atoms as plotted in Figure 6.16(b). The hybridization between these orbitals makes the HEOB states behave like VB states. Their energies calculated with the hybrid functional suggested that they are lying about 0.25 eV above the VBM of bulk anatase. In contrast, the

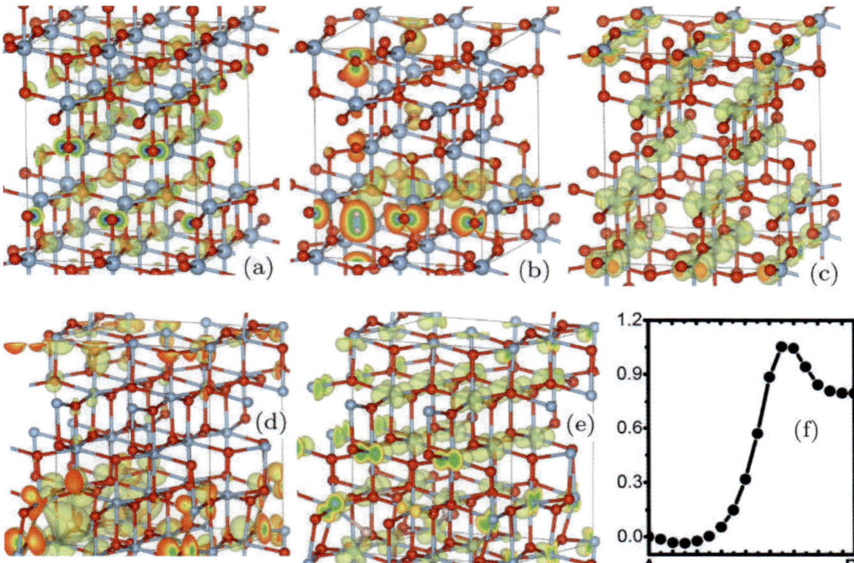

Figure 6.16. Charge density distributions of the VBM band of (a) the $2 \times 2 \times 1$ anatase supercell, (b) the HEOB and (c) CBM bands of model A, (d) the mid-gap band and (e) the CBM band of model B, and (f) the minimum energy path from model A to model B. The locations of H atoms (represented by pink spheres) are highlighted by the black arrows.[32] Reprinted with permission from Ref. [32]. © 2013 American Physical Society.

CBM states in model A are extended over every Ti atom, as shown in Figure 6.16(c). These charge distributions are almost identical to the typical $3d_{xy}$ orbitals. Thus, the itinerant nature of the conduction electrons has not been perturbed by the H_2 dopant. For another model of B (the two H atoms in model A were separated and bonded to Ti and O atoms), they found that in model B a new mid-gap state whose energy is higher than the bulk VBM by 1.14 eV appears. This mid-gap state is also highly localized and composed of a mixture of the H_{1s}–Ti_{3d} bonding states and the 2p orbitals from the nearby O atoms, as shown in Figure 6.16(d). In contrast, the electrons at the CBM band in model B, only 0.12 eV above the bulk CBM, are delocalized and their wave functions (mainly 3d character) are spread over all the Ti atoms, as shown in Figure 6.16(e). In fact, the alien Ti–O and H–O bonds also cause significant local lattice distortion. For example, the formation of Ti–H bonds pushes some neighboring O atom away by as much as 0.44 Å. Since both H bonding and lattice distortion can contribute to the changes of the electronic bands in model B, they studied further the effect of distortion alone by constructing another model: B$'$, where the H atoms are removed from model B while keeping the lattice distortion unchanged. They found that the CBM electrons in model B$'$ have about the same energies as in the bulk anatase, while its mid-gap state is now lowered to 0.55 eV above the bulk VBM. That indicates that both hydrogenation and lattice disorder will play an important role in tuning the electronic bands of hydrogenated TiO$_2$. Therefore, the lattice distortions assisted by H bonding can be taken as a good way to engineer the bandgap of anatase, namely, by blueshifting its VBM while leaving the CBM more or less unchanged. Moreover, the different spatial extent of the mid-gap hole states and the conduction electrons also suggests a spatial separation of the photo-excited electrons and holes, which will contribute to better photocatalytic efficiency as well.

However, such hydrogenated lattice disorder cannot be energetically-favored in bulk anatase. For example, model B is not stable against decay into model A since it is 0.8 eV higher in energy than A. In Figure 6.16(f), Liu *et al.* plotted the minimum energy path from B to A calculated by the elastic band method, which also showed an energy barrier of 0.25 eV from model B to model A. That means the H-assisted lattice distortion in bulk anatase will decay, even at room temperature, following the process ($\sim$Ti–H) + ($\sim$O–H) $\Rightarrow$ ($\sim$Ti–O$\sim$) + H$_2$ $\uparrow$. Considering that surface atoms are less constrained and hence easier to reconstruct than the bulk ones, they examined further the hydrogenation process occurring at surfaces of anatase (001) and nanocrystals. They found that on the surface H, atoms play the role of "scissors," which makes the process, ($\sim$Ti–O$\sim$) + H$_2$ $\Rightarrow$ ($\sim$Ti–H) + ($\sim$O–H), easier to untie the TiO$_2$ lattice, in particular on the surfaces of TiO$_2$ nanocrystals. Combining their bulk and nanocrystal anatase results, they suggested that hydrogenation helps to break up Ti–O bonds near the surface of nanocrystals, while the reverse process allows new Ti–O bonds to reform inside the interior of nanocrystals.

Therefore, they proposed the hydrogenation-induced-disordering mechanism to explain the superior PEC enhancement of the disorder-engineered black TiO_2 nanocrystals.[32] The lattice disorder in black TiO_2 can originate from the hydrogenation that helps to break up Ti–O bonds on the surfaces of anatase nanocrystals by forming Ti–H and O–H bonds. In bulk anatase, the reverse process to release hydrogen molecules is energetically more favorable. The interplay between these two processes stabilizes a disordered shell of anatase NPs. The surface lattice disorder can blueshift the VBM of anatase by introducing mid-gap states while leaving its CBM changed little. The strong localization of the mid-gap holes in the disordered surface layers decreases their spatial overlap with the itinerant electrons. Together with reduced bandgap, this charge separation of the photo-excited electron and hole pairs helps to explain why disorder-engineered black TiO_2 is a highly efficient photo-catalyst for splitting water under solar radiation.

6.7 Summary

There are many factors that can affect the electronic properties of TiO_2 and consequently tune its PEC performance, including defects, dopings, surfaces/interfaces, amorphous/disordered phases, lattice strain, and nanosize effects. For the case of black TiO_2 that is characterized by the nanoscale crystalline-core/disordered-shell structure, all of these factors contribute more or less to tuning its electronic properties. It is the surface disorder that brings the new electronic states into the bandgap of TiO_2 and thus improves its visible and infrared optical absorption. It is the interface between crystalline core and disordered shell nanostructure that promotes the separation between photon-excited electrons and holes and thus improves is electrochemical conversion. Both ordered and disordered parts play important roles in making black TiO_2 unusual and highly efficient in solar PEC applications.

In fact, black TiO$_2$ possesses the multilevel structural modifications that especially count in nanoscale: surfaces with open, reconstructed, or alien terminated bondings, disordered shells with severe distorted lattice, and random point defects even in crystalline cores. While these structural modifications involve the broad variations of bond angle and bond length, it would be necessary to consider the overall changes of the chemical environment rather than to focus locally on the defects, such as Ti^{3+} or O$_v$, to understand the photo-electro-chemical features of black TiO$_2$.

References

(1) (a) Inoue, T.; Fujishima, A.; Konishi, S.; Honda, K. *Nature* **1979**, *277*, 637; (b) Chen, X.; Shen, S.; Guo, L.; Mao, S. S. *Chem. Rev.* **2010**, *110*, 6503; (c) Chen, X.; Li, C.; Graetzel, M.; Kostecki, R.; Mao, S. S. *Chem. Soc. Rev.* **2012**, *41*, 7909; (d) Liu, L.; Chen, X. B. *Chem. Rev.* **2014**, *114*, 9890; (e) Chen, X. B.; Liu, L.; Huang, F. Q. *Chem. Soc. Rev.* **2015**, *44*, 1861.

(2) Fujishima, A.; Honda, K. *Nature* **1972**, *238*, 37.

(3) Hashimoto, K.; Irie, H.; Fujishima, A. *Jap. J. Appl. Phys. 1* **2005**, *44*, 8269.

(4) Gai, Y.; Li, J.; Li, S.-S.; Xia, J.-B.; Wei, S.-H. *Phys. Rev. Lett.* **2009**, *102*, 036402.

(5) (a) Chen, X. B.; Liu, L.; Yu, P. Y.; Mao, S. S. *Science* **2011**, *331*, 746; (b) Diebold, U. *Nat. Chem.* **2011**, *3*, 271; (c) Tachibana, Y.; Vayssieres, L.; Durrant, J. R. *Nat. Photonics* **2012**, *6*, 511.

(6) (a) Swamy, V.; Gale, J. D.; Dubrovinsky, L. S. *J. Phys. Chem. Solids* **2001**, *62*, 887; (b) Abrahams, S. C.; Bernstei, J. L. *J. Chem. Phys.* **1971**, *55*, 3206; (c) Burdett, J. K.; Hughbanks, T.; Miller, G. J.; Richardson, J. W.; Smith, J. V. *J. Am. Chem. Soc.* **1987**, *109*, 3639; (d) Meagher, E. P.; Lager, G. A. *Can. Mineral.* **1979**, *17*, 77; (e) Marchand, R.; Brohan, L.; Tournoux, M. *Mater. Res. Bull.* **1980**, *15*, 1129; (f) Latroche, M.; Brohan, L.; Marchand, R.; Tournoux, M. *J. Solid State Chem.* **1989**, *81*, 78; (g) Akimoto, J.; Gotoh, Y.; Oosawa, Y.; Nonose, N.; Kumagai, T.; Aoki, K.; Takei, H. *J. Solid State Chem.* **1994**, *113*, 27; (h) Grey, I. E.; Li, C.; Madsen, I. C.; Braunshausen, G. *Mater. Res. Bull.* **1988**, *23*, 743; (i) Sato, H.; Endo, S.; Sugiyama, M.; Kikegawa, T.; Shimomura, O.; Kusaba, K. *Science* **1991**, *251*, 786.

(7) Mo, S. D.; Ching, W. Y. *Phys. Rev. B* **1995**, *51*, 13023.

(8) (a) Kuo, M. Y.; Chen, C. L.; Hua, C. Y.; Yang, H. C.; Shen, P. Y. *J. Phys. Chem. B* **2005**, *109*, 8693; (b) Hwang, S. L.; Shen, P. Y.; Chu, H. T.; Yui, T. F. *Science* **2000**, *288*, 321; (c) Simons, P. Y.; Dachille, F. *Acta Crystallogr.* **1967**, *23*, 334; (d) Cromer, D. T.; Herrington, K. *J. Am. Chemi. Soc.* **1955**, *77*, 4708; (e) Chen, S. Y.; Shen, P. Y. *Phys. Rev. Lett.* **2002**, *89*, 096106; (f) Chen, S. Y.; Shen, P. Y. *Jap. J. Appl. Phys. 1* **2004**, *43*, 1519; (g) Dubrovinsky, L. S.; Dubrovinskaia, N. A.; Swamy, V.; Muscat, J.; Harrison, N. M.; Ahuja, R.; Holm, B.; Johansson, B. *Nature* **2001**, *410*, 653.

(9) Zhang, Y. F.; Lin, W.; Li, Y.; Ding, K. N.; Li, J. Q. *J. Phys. Chem. B* **2005**, *109*, 19270.

(10) (a) Chiodo, L.; Garcia-Lastra, J. M.; Iacomino, A.; Ossicini, S.; Zhao, J.; Petek, H.; Rubio, A. *Phys. Rev. B* **2010**, *82*, 045207; (b) Kang, W.; Hybertsen, M. S. *Phys. Rev. B* **2010**, *82*, 085203.

(11) (a) Scanlon, D. O.; Dunnill, C. W.; Buckeridge, J.; Shevlin, S. A.; Logsdail, A. J.; Woodley, S. M.; Catlow, C. R. A.; Powell, M. J.; Palgrave, R. G.; Parkin, I. P.; Watson, G. W.; Keal, T. W.; Sherwood, P.; Walsh, A.; Sokol, A. A. *Nat. Mater.* **2013**, *12*, 798; (b) Pfeifer, V.; Erhart, P.; Li, S. Y.; Rachut, K.; Morasch, J.; Brotz, J.; Reckers, P.; Mayer, T.; Ruhle, S.; Zaban, A.; Sero, I. M.; Bisquert, J.; Jaegermann, W.; Klein, A. *J. Phys. Chem. Lett.* **2013**, *4*, 4182.

(12) Hoffmann, R. *Angew. Chem. Int. Ed. Eng.* **1987**, *26*, 846.

(13) Gray, T. J.; McCain, C. C.; Masse, N. G. *J. Phys. Chem.* **1959**, *63*, 472.

(14) (a) Chester, P. F. *J. Appl. Phys.* **1961**, *32*, 2233; (b) Shannon, R. D. *J. Appl. Phys.* **1964**, *35*, 3414; (c) Purcell, T.; Weeks, R. A. *J. Chem. Phys.* **1971**, *54*, 2800.

(15) Na-Phattalung, S.; Smith, M. F.; Kim, K.; Du, M. H.; Wei, S. H.; Zhang, S. B.; Limpijumnong, S. *Phys. Rev. B* **2006**, *73*, 125205.

(16) Yu, N. C.; Halley, J. W. *Phys. Rev. B* **1995**, *51*, 4768.

(17) Cho, E.; Han, S.; Ahn, H.-S.; Lee, K.-R.; Kim, S. K.; Hwang, C. S. *Phys. Rev. B* **2006**, *73*, 193202.

(18) (a) Di Valentin, C.; Pacchioni, G.; Selloni, A. *Phys. Rev. Lett.* **2006**, *97*, 166803; (b) Finazzi, E.; Di Valentin, C.; Pacchioni, G. *J. Phys. Chem. C* **2009**, *113*, 3382; (c) Janotti, A.; Varley, J. B.; Rinke, P.; Umezawa, N.; Kresse, G.; Van de Walle, C. G. *Phys. Rev. B* **2010**, *81*, 085212.

(19) Asahi, R.; Morikawa, T.; Ohwaki, T.; Aoki, K.; Taga, Y. *Science* **2001**, *293*, 269.

(20) Yang, K.; Dai, Y.; Huang, B.; Han, S. *J. Phys. Chem. B* **2006**, *110*, 24011.

(21) (a) Diebold, U. *Surf. Sci. Rep.* **2003**, *48*, 53; (b) Chen, X.; Mao, S. S. *Chem. Rev.* **2007**, *107*, 2891; (c) Burda, C.; Chen, X. B.; Narayanan, R.; El-Sayed, M. A. *Chem. Rev.* **2005**, *105*, 1025.

(22) Zhao, Z. Y.; Li, Z. S.; Zou, Z. G. *J. Phys. Chem. C* **2012**, *116*, 7430.

(23) Kasowski, R. V.; Tait, R. H. *Phys. Rev. B* **1979**, *20*, 5168.

(24) Kaur, K.; Singh, C. V. *Energy Procedia* **2012**, *29*, 291.

(25) Yin, W.-J.; Chen, S.; Yang, J.-H.; Gong, X.-G.; Yan, Y.; Wei, S.-H. *Appl. Phys. Lett.* **2010**, *96*, 221901.

(26) (a) Wagner, J. M.; Bechstedt, F. *Phys. Rev. B* **2002**, *66*, 115202; (b) Vandewalle, C. G. *Phys. Rev. B* **1989**, *39*, 1871.

(27) (a) Anpo, M.; Shima, T.; Kodama, S.; Kubokawa, Y. *J. Phys. Chem.* **1987**, *91*, 4305; (b) Kavan, L.; Stoto, T.; Gratzel, M.; Fitzmaurice, D.; Shklover, V. *J. Phys. Chem.* **1993**, *97*, 9493; (c) Brus, L. *J. Phys. Chem.* **1986**, *90*, 2555; (d) Nosaka, Y. *J. Phys. Chem.* **1991**, *95*, 5054; (e) Kormann, C.; Bahnemann, D. W.; Hoffmann, M. R. *J. Phys. Chem.* **1988**, *92*, 5196.

(28) (a) Henglein, A. *Chem. Rev.* **1989**, *89*, 1861; (b) Alivisatos, A. P. *J. Phys. Chem.* **1996**, *100*, 13226.

(29) Nakajima, H.; Mori, T.; Shen, Q.; Toyoda, T. *Chem. Phys. Lett.* **2005**, *409*, 81.

(30) Satoh, N.; Nakashima, T.; Kamikura, K.; Yamamoto, K. *Nat. Nanotechnol.* **2008**, *3*, 106.

(31) Lu, J. B.; Dai, Y.; Jin, H.; Huang, B. B. *Phys. Chem. Chem. Phys.* **2011**, *13*, 18063.

(32) Liu, L.; Yu, P. Y.; Chen, X. B.; Mao, S. S.; Shen, D. Z. *Phys. Rev. Lett.* **2013**, *111*, 065505.

CHAPTER SEVEN

Black Titania Coatings: Fabrication Process and Photoelectrochemical/ Photocatalytic Properties

*Tomohiko Nakajima** and *Tetsuo Tsuchiya*

Advanced Coating Technology Research Center,
National Institute of Advanced Industrial Science and Technology,
Tsukuba Central 5, 1-1-1 Higashi, Tsukuba, Ibaraki 305-8565, Japan
**t-nakajima@aist.go.jp*

7.1 Overview of Black Titania Coating Process

Photoelectrochemical (PEC) solar water-splitting has great potential for providing sustainable energy.[1–5] Photoelectrodes containing semiconductor films for solar water-splitting systems have been extensively studied. Improvements in the efficiency of water splitting in the visible light region have been reported.[6–9] The n-type semiconductor, TiO_2, is the most common photocatalyst owing to its high photocatalytic activity, stability, and low cost.[10–12] The wide bandgap ($\sim$3.2 eV) of TiO_2 reduces the photocatalytic response in the visible light region, which covers the majority of the solar spectrum. However, the photocatalytic activity has considerably increased, even under simulated solar light, by the discovery of blackened hydrogen-doped TiO_2 (TiO_2:H) reported by Chen *et al.*[13,14] Other types of black TiO_2, such as anion-doped TiO_2 (TiO_2:S, TiO_2:N, and TiO_2:I)[15] and oxygen-deficient TiO_{2-x},[16,17]

---

*Corresponding author.

were also reported to show high photocatalytic activity, which is very exciting for the reality of H$_2$ production using solar energy. In each case, the blackening largely enhanced incident photon-to-current efficiency (IPCE) in the ultraviolet (UV) range. The defects that emerge around reduced Ti^{3+} ions contribute in increasing the electron donor density in TiO$_2$, resulting in accelerated charge transport and the Fermi level shift to the conduction band, which facilitate charge separation at the reaction interface.[18–20] Meanwhile, black TiO$_2$ still does not sufficiently use visible light[18]; therefore, further improvement of solar-to-hydrogen (STH) is expected by efficient use of the visible light region leading to actual application of solar H$_2$ production. The typical preparation processes are shown in Figure 7.1. In the current stage of development, these black titania powders have shown 5–10 times higher H$_2$ evolution under simulated solar light than the pristine TiO$_2$ powders.[13, 16, 21]

For facile solar H$_2$ production systems, a photocatalyst coating method is also a very effective technology. In general, it is not easy to prepare black titania films by a direct coating process

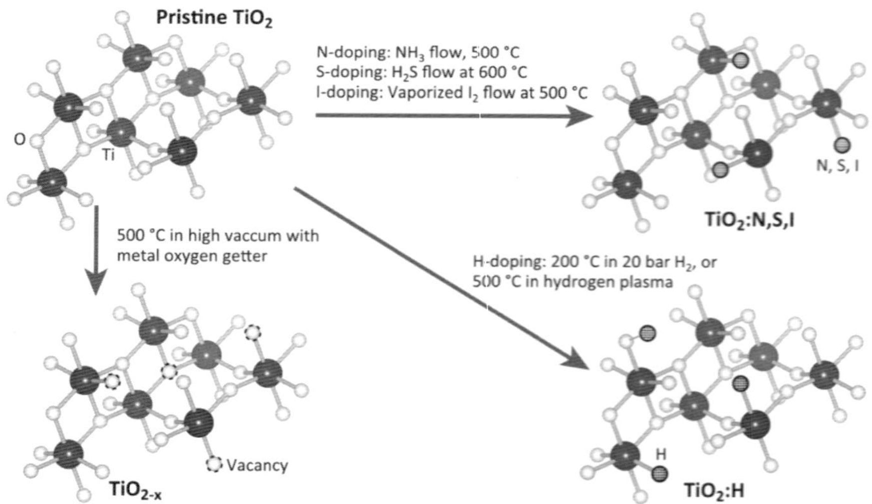

Figure 7.1. Schematic of structures and typical fabrication processes of hydrogenated TiO$_2$:H, anion-doped TiO$_2$:N, S, I, and oxygen-deficient TiO$_{2-x}$.

using black titania powders, since the solidification heating process will damage the blackening source in the samples. Therefore, other post-treatments of pristine TiO_2 films deposited by conventional processes, such as sputtering, chemical solution deposition, and hydrothermal methods, are usually employed. The blackening process of pristine TiO_2 powders requires ingenuity to realize the optimized doping/defect formation for a large enhancement of water-splitting efficiency without excess reduction of Ti^{4+} and destruction of the original crystal structures. For example, TiO_2:H requires annealing at 200°C for 5 days in high-pressure H_2 gas at 20 bar,[13] TiO_2:S requires annealing at 600°C for 4 h in H_2S gas,[15] and TiO_{2-x} requires annealing at 500°C for 6 h under high-vacuum conditions with a metal oxygen getter.[16] Therefore, the fabrication of black titania coating films has also been extensively studied, as listed in Table 7.1. Various techniques are used for the fabrication of black titania films. For preparing precursor pristine TiO_2 films, it is possible to obtain many forms such as compact films, nanoporous layers, and nanotube/rod/wire arrays.[22–26] For the blackening process, annealing in an appropriate gas atmosphere is applicable; pulsed UV laser irradiation,[27] proton implantation,[28] electron beam irradiation,[29] and H_2 plasma treatment[30] are also applicable.

For hydrogenated TiO_2 coatings, Wang *et al.* reported TiO_2:H nanowire arrays with very high STH efficiency of 1.63% under $100\,mW/cm^2$ (1 SUN),[24] and Binetti *et al.* reported almost $1\,mA/cm^2$ of photocurrent for H_2 production by water-splitting,[22] regardless of the dense compact film. H-doping is associated with difficulty in the formation of photoanode coatings since H_2 treatment tends to easily reduce Ti ions leading to destruction of the original TiO_2 lattices. Therefore, it is necessary to optimize the conditions precisely for the formation of hydrogenated TiO_2:H coatings. Compared to the hydrogenation processes, the many fabrication methods are an attractive feature of the oxygen-deficient-type black titania coatings. The fabrication processes of TiO_{2-x} coatings also show difficulty in controlling the appropriate oxygen deficiency, as well as the hydrogenation processes. It is well known that TiO_2 can convert to oxygen-deficient Magnéli Ti_nO_{2n-1} phases,[31,32] which

Table 7.1. Fabrication process and PEC/photocatalytic properties of black titania films.

Types	Fabrication process	Morphology	Property[a]	Measurement conditions	Ref.
TiO$_2$:H	Sputtering TiO$_2$ film $\to$ High pressure H$_2$ annealing	Compact film	Water splitting H$_2$ production: 0.99 mA$\cdot$cm^{-2} at 1.23 V$_{\mathrm{RHE}}$	Xe lamp at 100 mW$\cdot$cm^{-2}, 1 M KOH	22
	Atomic layer deposition of TiO$_2$ film $\to$ High pressure H$_2$ annealing	Compact film	MB decomposition: $k = 5.77 \times 10^{-3}$ min^{-1} ($k = 2.12 \times 10^{-3}$ in reference TiO$_2$ film)	1 SUN, 5×10^{-5} M MB	35
	TiO$_2$ nanowires obtained by hydrothermal process $\to$ H$_2$ annealing	Nanowire array	Water splitting H$_2$ production: 1.63% (STH) at 0.40 V$_{\mathrm{RHE}}$	1 SUN, 1 M NaOH	24
	Anodized TiO$_2$ nanotubes $\to$ High pressure H$_2$ annealing	Nanotube array	Water splitting H$_2$ production: 7 μmol$\cdot$h$^{-1}\cdot$cm^{-2} (at open circuit potential)	1 SUN, 50 vol% methanol, co-catalyst-free	25
	Anodized TiO$_2$ nanotubes $\to$ High energy proton implantation	Nanotube array	Water splitting H$_2$ production: 15 μL$\cdot$h$^{-1}\cdot$cm^{-2} (at open circuit potential)	1 SUN, 50 vol% methanol, co-catalyst-free	28
TiO$_{2-x}$	Sputtering TiO$_2$ film $\to$ Al reduction	Compact film	Water splitting H$_2$ production: 1.7% (STH) at 0.65 V$_{\mathrm{RHE}}$	1 SUN, 1 M NaOH	33
	Spin coating of TiO$_2$ nanoparticles (NPs) $\to$ Pulsed UV laser irradiation	Compact film	Water splitting H$_2$ production: 0.52% (STH) at 0.56 V$_{\mathrm{RHE}}$	1 SUN, 1 M KOH	27

	Preparation	Morphology	Photocatalytic performance	Light source/electrolyte	Ref.
	Spin coating of TiO_{2-x} particles that was obtained by annealing with CaH_2	Compact film	Water splitting H_2 production: $1.99\,mA\cdot cm^{-2}$ at $1.23\,V_{RHE}$	1 SUN, 1 M NaOH	36
	Anodized TiO_2 nanotubes $\rightarrow$ Al reduction	Nanotube array	Water splitting H_2 production: 1.2% (STH) at $0.68\,V_{RHE}$	1 SUN, 1 M NaOH	23
	Anodized TiO_2 nanotubes $\rightarrow$ Microwave treatment	Nanotube array	Water splitting H_2 production: 1.66% (STH) at $0.78\,V_{RHE}$	1 SUN, 1 M KOH	37
	Anodized TiO_2 nanotubes $\rightarrow$ Electrochemical treatment	Nanotube array	Water splitting H_2 production: 0.30% (STH) at $0.48\,V_{RHE}$	1 SUN, 1 M KOH and 1% ethylene glycol	38
TiO_2:N	Sputtering TiO_2 film in N_2	Compact film	Water splitting H_2 production: $601\,\mu mol\cdot h^{-1}\cdot g^{-1}$ ($12\,\mu mol\cdot h^{-1}\cdot g^{-1}$ in reference P25 TiO_2 film)	300 W Xe lamp (2.6 W at $\lambda < 390\,nm$)	39
	TiO_2 nanowires obtained by hydrothermal process $\rightarrow$ Nitrogen implantation	Nanowire array	Water splitting H_2 production: $1.92\,mA\cdot cm^{-2}$ at $0.50\,V$ vs. Ag/AgCl	Xe lamp at $100\,mW\cdot cm^{-2}$, 1 M NaOH	40
TiO_2:S	Spin coating of TiO_2 NPs $\rightarrow$ Al reduction $\rightarrow$ H_2S post-annealing	Compact film	Water splitting H_2 production: 1.67% (STH) at $0.57\,V_{RHE}$	1 SUN, 1 M NaOH	34

[a]MB: methylene blue. STH efficiency $= J(1.23 - E)/I_{AM1.5}$, where J is the photocurrent density (mA/cm^2), E is the applied bias voltage, and $I_{AM1.5}$ is the irradiance of AM1.5 simulated sunlight at $100\,mW/cm^2$ (1 SUN).

have a metallic property without photocatalytic activity. Therefore, the fabrication of TiO$_{2-x}$ coatings also requires precisely controlled moderate reduction conditions. Among the various studies, Wang *et al.* reported a very high STH efficiency (1.7%), which was comparable to the H-doped efficiency.[33] In addition, we developed a rapid (only a few minutes) conversion process from pristine TiO$_2$ to oxygen-deficient TiO$_{2-x}$ by means of pulsed UV laser irradiation.[27] The variety of fabrication processes will be very vital for actual application. The anion-doped-type TiO$_2$, such as TiO$_2$:S, has also shown very high STH (1.67%), as reported by Yang *et al.*[34] In this chapter, we introduce some black titania coating processes with a focus on hydrogenated and oxygen-deficient types of black titania.

7.2   Hydrogenated TiO$_2$ Photoelectrode Coatings

Hydrogenation of TiO$_2$ films has generally been carried out by high-pressure H$_2$ annealing of precursor pristine TiO$_2$ films. Binetti *et al.* accomplished hydrogenation of sputtered TiO$_2$ thin films by means of high-pressure H$_2$ annealing that was fundamentally based on Chen's process for NPs.[13, 22] They revealed that pure hydrogenated anatase TiO$_2$ films could be obtained using amorphous precursor films at 300°C for only 10 min at 1.0 bar H$_2$, while the phase turned rutile when the crystalline anatase precursor films. The reported hydrogenation conditions differ in annealing temperature, time, and pressure. It is important not to form the Magnéli phase (excessively reduced form) or excessive rutile reconstruction. The best hydrogenation conditions depend on the morphology, crystal form, and thickness of the precursor films. These methods will require further systematic investigation for a complete understanding of the hydrogenation process.

Hydrogenation processes have succeeded in preparing nanostructured materials, such as nanowires and nanotube arrays that show very good PEC water-splitting properties. Wang *et al.* reported H$_2$-treated TiO$_2$ nanowire arrays.[24] Figure 7.2(a) shows the SEM image of nanowire arrays with 10–20 nm diameters and 2–3 μm lengths, grown via hydrothermal treatment. Figure 7.2(b) shows the

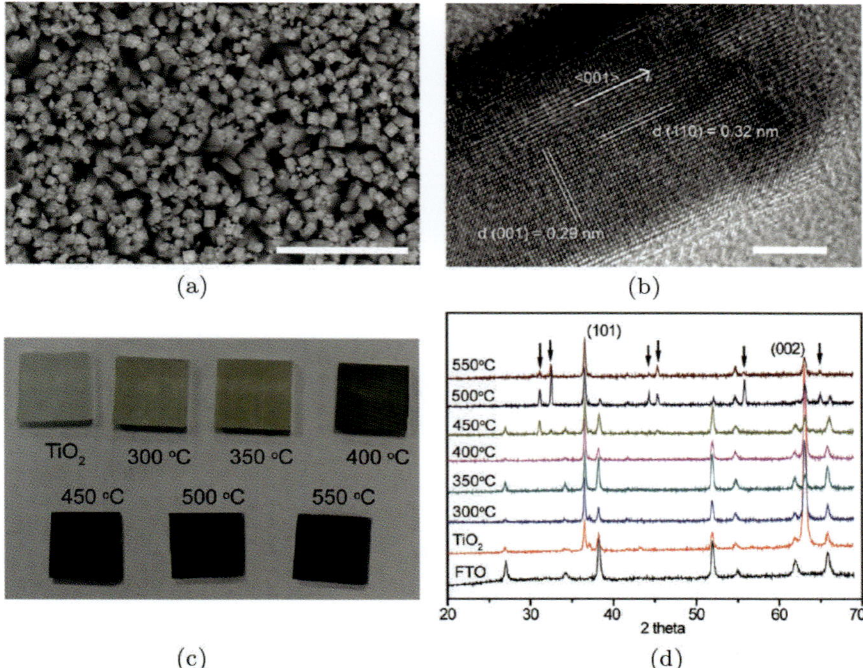

Figure 7.2. (a) Scanning electron microscopy (SEM) image of vertically aligned TiO$_2$ nanowire arrays prepared on a fluorine tin oxide (FTO) substrate (scale bar is $4\,\mu$m). (b) Lattice-resolved transmission electron microscopy (TEM) image of a single TiO$_2$ nanowire (scale bar is $5\,$nm). (c) Photographic images, and (d) X-ray diffraction (XRD) patterns of pristine and hydrogenated TiO$_2$ nanowires annealed under H$_2$ at various temperatures.[24] Adapted with permission from the American Chemical Society.

HRTEM TEM image of the nanowire having a high crystallinity. The obtained nanowires in the study were annealed in H$_2$ without high pressure. With the increased annealing temperature, the sample color changed from yellowish green to black (Figure 7.2(c)). From the XRD data (Figure 7.2(d)), it was confirmed that the long-range ordering of the crystal lattice gradually collapsed, indicating an increase in defect density in the original TiO$_2$ lattice, which is consistent with the case of hydrogenated NPs.[13] Figure 7.3 shows the PEC water-splitting properties of hydrogenated TiO$_2$ nanowire arrays.[24] The photocurrent density under 1 SUN illumination exhibited

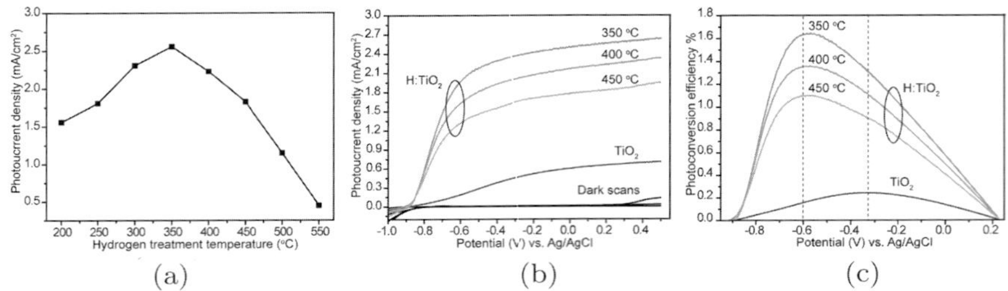

Figure 7.3. Photocurrent density for pristine and hydrogenated TiO$_2$ nanowire arrays as functions of (a) H$_2$ treatment temperature and (b) applied potential in a 1 M NaOH solution under 1 SUN illumination. (c) Photoconversion efficiency of the samples.[24] Adapted with permission from the American Chemical Society.

over 2.5 mA/cm^2 in the sample annealed in H$_2$ at 350°C. The STH efficiency of this sample was considerably high at 1.63%. However, it should be noted that the valence band edge of the H$_2$-treated nanowires by Wang *et al.* was not shifted, whereas it was substantially shifted in the hydrogenated NPs by Chen *et al.*[13] This would be a result of the difference between the H$_2$ treatment conditions; low annealing temperature (200°C for 5 days) at high H$_2$ pressure (20 bar) vs. high annealing temperature (350–450°C for 30 min) at ambient pressure. Wang *et al.* reported that the dark color of the obtained samples was a result of the defects in the bandgap of TiO$_2$ caused by the H$_2$ treatment, which is related to the formation of oxygen vacancies and surface hydroxyl groups on TiO$_2$.[24]

Liu *et al.* reported the hydrogenation for anodic TiO$_2$ nanotube arrays using high-pressure H$_2$ annealing conditions (20 bar H$_2$ at 500°C for 1 h).[25] Figure 7.4 shows the photocatalytic H$_2$ production under an open-circuit condition and the SEM images of pristine, Ar/H$_2$-annealed (Ar/H$_2$), and high-pressure H$_2$-annealed (HP–H$_2$) TiO$_2$ nanotube arrays.[25] Both Ar/H$_2$ and HP–H$_2$ nanotubes maintained the original array structure even after the H$_2$ treatments. The obtained hydrogenated TiO$_2$ nanotube arrays showed very high H$_2$ production at 7 μmol/h/cm^2. This value exceeded the nanotubes treated at a longer annealing time of 5 days (Sci ref.). In this study, the H$_2$ annealing of nanotube and nanorod arrays under

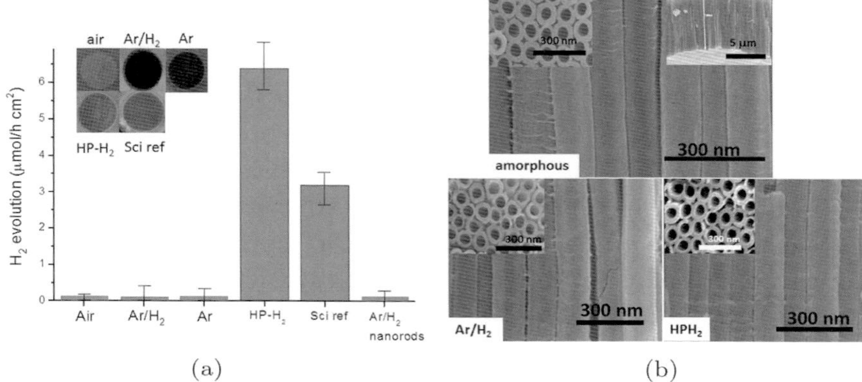

Figure 7.4. (a) Photocatalytic H_2 production under open-circuit conditions with TiO_2 nanotubes and nanorods treated in different atmospheres under 1 SUN illumination. Air, heat treatment in air at $450°C$; Ar, heat treatment in pure argon at $500°C$; Ar/H_2, heat treatment in H_2/Ar (5 vol%) at $500°C$; HP–H_2, heat treatment in H_2 at 20 bar and $500°C$; Sci ref., heat treatment in H_2 at 20 bar at $200°C$ for 5 days. (b) SEM images of the TiO_2 nanotube layer as-formed and after hydrogenation treatments in Ar/H_2 and high-pressure H_2.[25] Adapted with permission from the American Chemical Society.

atmospheric pressure (Ar/H_2) did not enhance the H_2 production under open-circuit conditions. However, the water-splitting experiments under applied external bias potential (PEC water-splitting condition) interestingly exhibited a higher photocurrent density in the Ar/H_2-treated nanorod arrays than the nanotube arrays treated with high-pressure H_2. They concluded that this phenomenon might be strongly related to the electron conductivity of the sample between the surface and the bottom electrode. Actually, higher conductivity was observed in the H_2/Ar sample than the HP–H_2 sample. Surface modification of TiO_2 nanomaterials is very sensitive to H_2 treatment; therefore, the crystal quality, including the amorphous/crystalline variation and formation of defects, was easily changed by the slight difference in annealing conditions, resulting in the large variation in photocatalytic/PEC properties. In other words, TiO_2 surface modifications still have large possibilities for increasing the photocatalytic/PEC activity.

Liu *et al.* reported another intriguing approach for hydrogenation of TiO$_2$ nanostructured coating.[28] They carried out high-energy proton implantation to anodic TiO$_2$ nanotube arrays. The TiO$_2$ nanotube arrays prepared by the self-organized anodization technique were ion-implanted with protons at an energy of 30 keV and a dose of 10^{16} ions/cm^2. The H-implanted TiO$_2$ nanotube arrays with 12 μm tube lengths showed considerably high photocatalytic H$_2$ production at 15 μL $\cdot$ h^{-1} $\cdot$ cm^{-2} (at open-circuit potential under 1 SUN illumination), regardless of no noble metal (e.g. Pt) cocatalyst additions (Figure 7.5(a)). Figure 7.5(b) shows the calculated ion and damage depth distribution in TiO$_2$ nanotubes by a Monte Carlo simulation. The implantation range was estimated to be 1.2 μm tube length with a maximum of H implanted at a depth of ~900 nm and a maximum of lattice defects at a depth of ~800 nm. The surface morphology of the original nanotubes was unchanged after H-implantation, as shown in Figure 7.5(c).

They characterized the structural and morphological changes in H-implanted samples by XRD and TEM. XRD (Figures 7.6(a) and (b)) and selected area electron diffraction (SAED) (Figures 7.6(c)–(h)) were measured for a 1-μm-long tube layer.[28] The Rietveld refinement of XRD patterns for the non-implanted reference sample show the typical anatase pattern of conventionally annealed TiO$_2$ nanotubes. After implantation, they observed a significant decrease in the anatase peak intensities and broadening of peaks, indicating reduction of the structural coherence length due to amorphization and/or discretization of crystallites. In the reference pristine TiO$_2$ nanotubes, the diffraction pattern showed weak preferential (001) orientation. Although the Lotgering factor $F(001)$ was calculated as 49.2%, it reduced to 2.2% after p^+-implantation. The peak broadening and large reduction in the degree of orientation means that the H-implantation efficiently interacts with the TiO$_2$ lattice. Figures 2.5(c) and (d) show high-resolution TEM (HRTEM) images of the reference and H-implanted nanotubes. From these images, the lattice constant was evaluated to be 0.35 nm, corresponding to the (101) lattice plane of anatase. In addition, randomly oriented nanocrystallites with various shapes and sizes were confirmed (Figure 7.6(d)). HRTEM exhibited

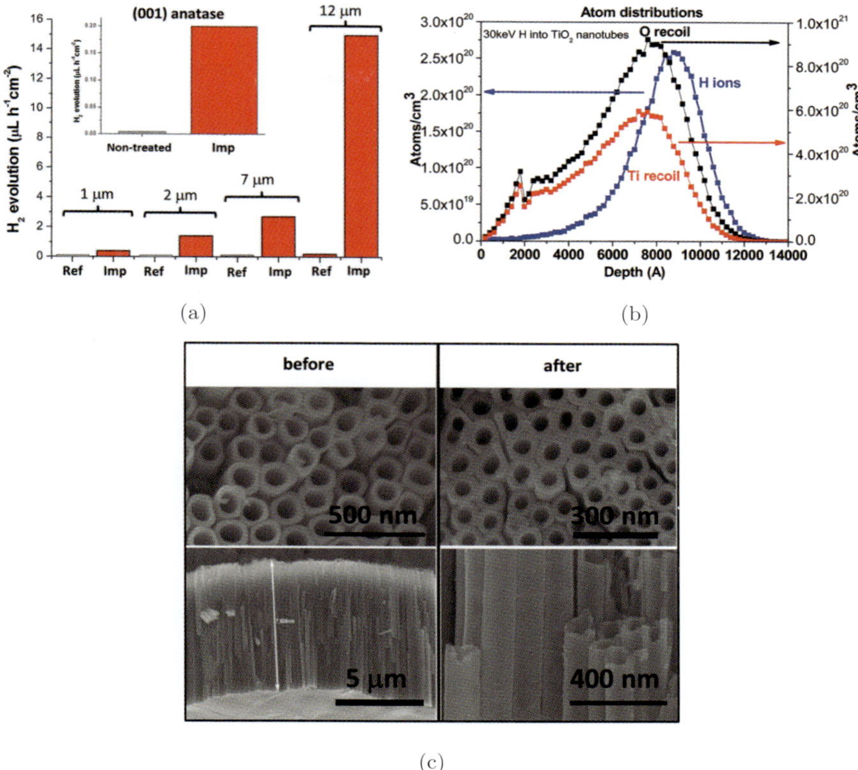

(a)   (b)

(c)

Figure 7.5. (a) Photocatalytic H_2 production under open-circuit conditions in methanol/water (50/50 vol%) of TiO_2 nanotube layers of different thickness illumination before and after H-implantation under 1 SUN (inset: photocatalytic H_2 production of (001) single crystal anatase before and after H-implantation), (gray box represents no detected H_2 evolution; red box represents detected amount of H_2 evolution). (b) Calculation depth distribution of implanted ions (H ions) and crystal damage (Ti-, Orecoil-) for TiO_2 nanotubes. (c) SEM images of TiO_2 nanotube layer before and after H-implantation.[28] Adapted with permission from the American Chemical Society.

thin amorphous rims around the samples (this feature was confirmed in both non-implanted and implanted samples). However, the amount of the amorphous region was apparently higher in the proton-implanted nanotubes. This is not only in line with the XRD results, but also apparent from the SAED patterns (Figures 7.6(e) and (f)). However,

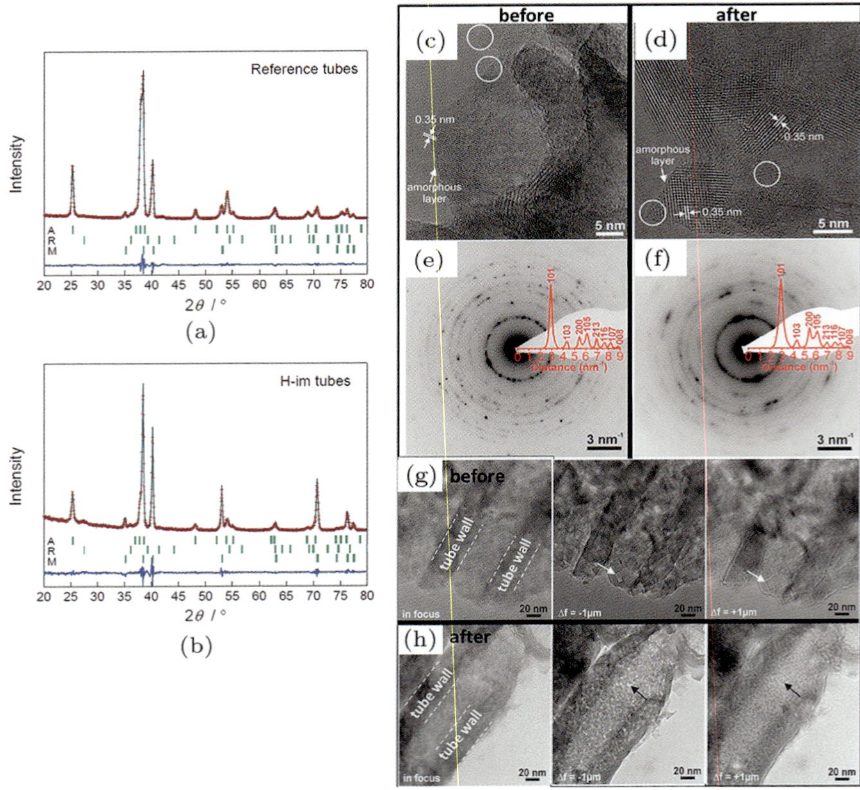

Figure 7.6. XRD spectra and Rietveld refinement of TiO_2 nanotubes before (a) and after (b) implantation; high-resolution TEM (HRTEM) images (c), (d) and inverted SAED patterns (e), (f) of TiO_2 nanotubes before and after H-implantation, respectively. Insets in the SAED patterns are intensity profiles, which are obtained by radial-averaging the respective diffraction patterns; bright-field (BF) TEM images taken in underfocus (left), underfocus (center), and overfocus (right) conditions for TiO_2 nanotubes before (g) and after (h) implantation, showing characteristic Fresnel contrast indicating the presence of voids. The white and black arrows indicate exemplary voids, which become visible in under- and overfocus.[28] Adapted with permission from the American Chemical Society.

the SAED pattern of the nanotubes after p^+-implantation exhibited a halo reflection in addition to the regular diffraction rings. This indicates the presence of a higher amount of amorphous species near the nanocrystallites. Furthermore, a broadening of the diffraction rings is observed, and is particularly visible for the (101) reflections in the SAED patterns, which further confirms a reduced average crystallite size in the nanotubes after p^+-implantation. The changes induced by implantation are particularly apparent from the BF TEM images taken under defocus conditions (Figures 7.6(g) and (h)). They confirmed that H-implantation increased the amount of voids and reduced the crystallite size, as shown in Figure 2.5(f). Voids that emerge after proton implantation are commonly observed,[41] and can be ascribed to point defects diffusion, and agglomeration. In this case, the voids in the implanted material discretize the tube wall on a length scale of 10–20 nm. While square facets were observed before implantation, they disappeared after implantation. This behavior is consistent with the XRD spectrum that shows a clear reduction of the orientation degree. They concluded the effect of H-implantation as follows[28]: a synergistic interaction between the implanted and intact tube segments may be attributed to a coupling between the intact lower tube part as the light absorber and the catalytically active center in the upper tube part. This H-implantation technique is an interesting example of enabling local modifications to provide different functions for each specific part in the nanomaterial coating by the post-treatment.

7.3 Oxygen-Deficient TiO_{2-x} Photoelectrode Coatings

Blackening pristine TiO_2 NPs takes a minimum of several hours because of the low process temperature and hazardous conditions necessary for some methods (e.g. high-pressure H_2 annealing for TiO_2:H and H_2S gas annealing for TiO_2:S). Among the several types of black titania, oxygen-deficient-type TiO_{2-x} is relatively facile to be obtained by high-vacuum annealing with a metal oxygen getter,[33] pulsed UV laser irradiation,[27] electron beam irradiation,[38] RF sputtering deposition,[42] or microwave treatment processes.[37] In particular, pulsed UV laser processing has great potential for industrial manufacturing of black titania coatings.[27]

Recently, UV laser-induced crystal nucleation and the growth of oxide thin films from solutions have been studied.[43–45] Efficient polycrystalline growth of TiO$_2$ thin films from a precursor was reported.[46–48] Oxygen deficiency induced in a rutile TiO$_2$ single crystal surface by pulsed UV laser irradiation has also been studied.[49] This treatment produced an unconventional metallic state in the oxygen-deficient TiO$_{2-x}$ that was not converted to Ti$_n$O$_{2n-1}$ Magnéli phases,[31, 32] and maintained its original rutile crystal structure with stacking faults. This oxygen deficiency was created by instantaneous photothermal heating under pulsed UV laser irradiation. Considering this background, an efficient fabrication of polycrystalline black titania photoanodes was attempted via UV laser irradiation of TiO$_2$ films. Because of this study, very rapid conversion of a pristine TiO$_2$ film to an oxygen-deficient TiO$_{2-x}$ film was achieved. The conversion process was finished within a few minutes. The blackened film showed a 2.6-fold higher STH efficiency than the pristine film. We introduce the details of this fabrication process.

The schematic for the preparation of TiO$_{2-x}$ photoelectrodes is shown in Figure 7.7. The bottom electrode for the photoanode consisted of 10 mol% antimony-doped tin oxide (SnO$_2$:Sb) films with a thickness of 1.2 μm (We employed SnO$_2$:Sb as the bottom electrode in this study since the conducting property of SnO$_2$:Sb is stable for thermal treatments). A polycrystalline Al$_2$O$_3$ substrate (96%, Kyocera) was spin-coated with a dispersion of SnO$_2$:Sb NPs (99.5%, Alfa Aesar) at 2000 rpm for 5 s. The films were pre-heated at 400°C for 10 min and fired at 700°C for 1 h in air. TiO$_2$ films were prepared by metal-organic deposition. The SnO$_2$:Sb/Al$_2$O$_3$ substrates were spin-coated with the starting metal-organic solution (0.5 M Ti, SYM-TI05; Symetrix) at 3000 rpm for 10 s. The spin-coated films were preheated at 300°C for 10 min to convert it to precursor amorphous oxide films by decomposing the organic components. The spin coating and preheating were repeated five times to increase the film thickness to 250 nm. The precursor TiO$_2$ films were heated at 500°C for 1 h in air. The TiO$_2$/SnO$_2$:Sb/Al$_2$O$_3$ specimen was treated as a pristine reference sample. Two post-treatments were carried on the reference TiO$_2$ films: (1) pulsed KrF laser (Compex110, Lambda Physik)

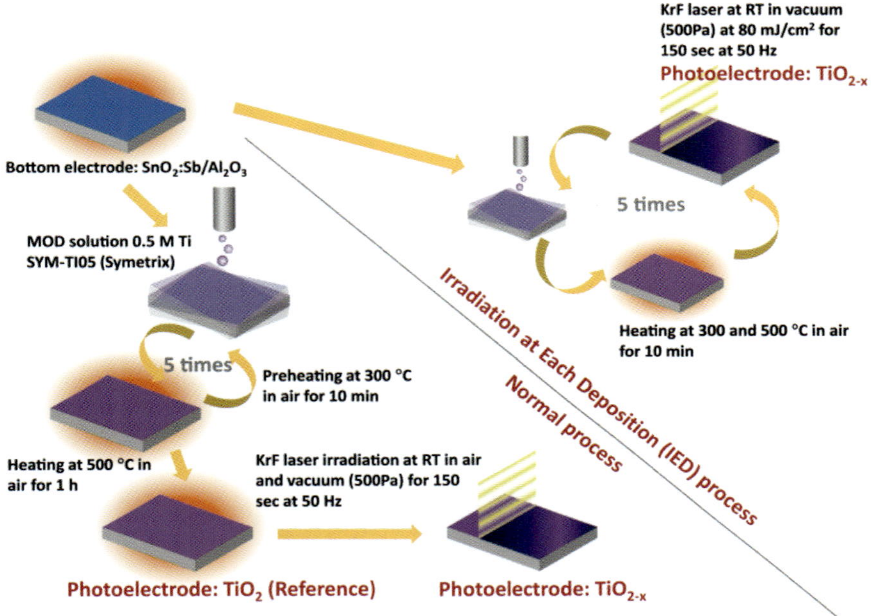

Figure 7.7. Fabrication flowchart for preparation of TiO$_{2-x}$ photoelectrodes by pulsed UV laser irradiation.

irradiation at a fluence of 50–170 mJ · cm^{-2} for 150 s at 50 Hz with a pulse duration of 26 ns at room temperature under a vacuum of 500 Pa using a rotary pump or in air, and (2) post-annealing at 700°C for 6 h in air. For optical absorption and electrical resistivity measurements, silica glass substrates were used without the bottom electrode films. We also prepared a titania film using KrF laser irradiation through the following procedure, called the "irradiation at each deposition" (IED) process. The SnO$_2$:Sb/Al$_2$O$_3$ substrate was spin-coated at 3000 rpm for 10 s from the same starting solution as the TiO$_2$ film. The film was preheated at 300°C for 10 min and then heated at 500°C for 10 min in air. The precursor film was irradiated using the KrF laser at 80 mJ · cm^{-2} for 100 s at 50 Hz under a vacuum of 500 Pa at room temperature. This process was repeated five times to increase the film thickness to 250 nm.

Figure 7.8. SEM images of surface morphology for (a) pristine TiO_2/SnO_2:Sb film, and TiO_{2-x}/SnO_2:Sb films laser irradiated at (b) $50\,\mathrm{mJ\cdot cm^{-2}}$, (c) $80\,\mathrm{mJ\cdot cm^{-2}}$, (d) $110\,\mathrm{mJ\cdot cm^{-2}}$, (e) $140\,\mathrm{mJ\cdot cm^{-2}}$, and (f) $80\,\mathrm{mJ\cdot cm^{-2}}$ (IED).

Figure 7.8 shows the surface morphology of the pristine and laser-irradiated samples. As the irradiation fluence increased, a mud-crack pattern associated with volume shrinkage caused by rapid crystallite growth was observed. The in-plane discrete island size was 0.5–$1\,\mu$m at fluences below $110\,\mathrm{mJ\cdot cm^{-2}}$ and 1–$2\,\mu$m above $140\,\mathrm{mJ\cdot cm^{-2}}$. Above $110\,\mathrm{mJ\cdot cm^{-2}}$, the top TiO_2 surface may have also partially melted, as indicated by the simulated pulsed photothermal heating profiles.

The crystallinity and structure of the samples were analyzed using grazing incidence X-ray diffraction (GIXRD) (Figure 7.9(a)). The pristine TiO_2 thin film crystallized at $500°C$ showed a pure anatase phase. Post-annealing at $700°C$ and pulsed UV laser irradiation caused a small amount of phase conversion from anatase to rutile TiO_2. The GIXRD peak intensity of the TiO_2 film increased to a fluence of $110\,\mathrm{mJ\cdot cm^{-2}}$ after laser irradiation in

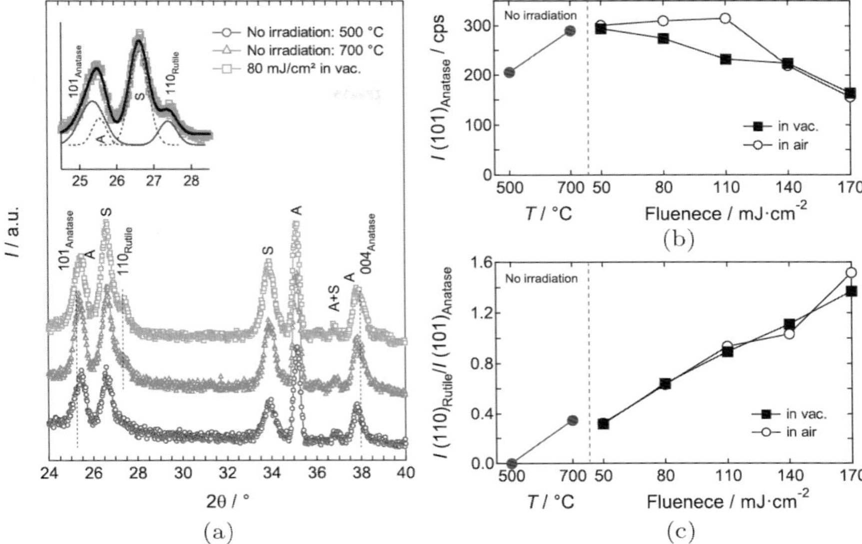

Figure 7.9. (a) GIXRD patterns of the pristine TiO_2/SnO_2:Sb film, $TiO_2/$ SnO_2:Sb film post-annealed at $700°C$, and laser-irradiated TiO_{2-x}/SnO_2:Sb films at $80\,mJ\cdot cm^{-2}$. The subscripts A and S indicate Al_2O_3 and SnO_2:Sb, respectively. (b) Anatase 101 peak intensity of TiO_2 (TiO_{2-x}) films and (c) the peak intensity ratio of TiO_2 rutile 110 and anatase 101 (TiO_{2-x}) films as a function of annealing temperature or laser fluence.

air, and then decreased above $140\,mJ\cdot cm^{-2}$. For vacuum irradiation, the diffraction intensity decreased at fluence values above $50\,mJ\cdot cm^{-2}$, in contrast to the pristine sample wherein crystallite growth occurred from $50\,mJ\cdot cm^{-2}$ to $110\,mJ\cdot cm^{-2}$ (Figure 7.9(b)). Larger laser-induced crystallites should grow closer to the surface than the interface, as reported for typical laser-irradiated oxide thin films.[43, 45, 50] The reduction in the peak intensity indicates that there was no sufficiently coherent periodic structure for stoichiometric TiO_2 near the irradiated surface. The rutile TiO_2 peaks also appeared after laser irradiation in both air and under vacuum. The quantitative ratio of the rutile phase increased with the increase in laser fluence (Figure 7.9(c)).

Figure 7.10 shows photographs of the sample shades. The pristine TiO_2/SnO_2:Sb film was light gray because of the bottom electrode

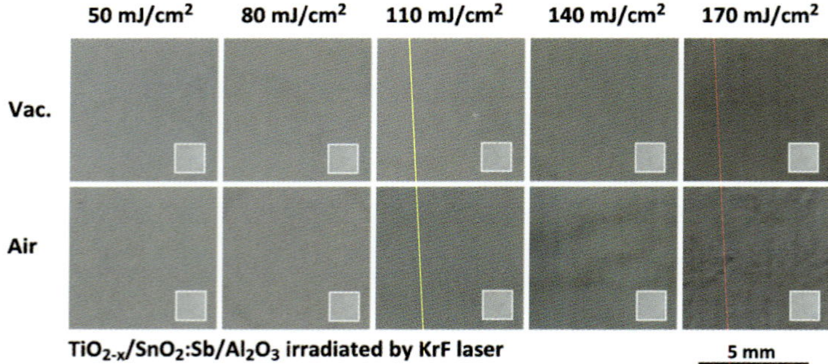

Figure 7.10. Photographic images of the TiO$_2$(TiO$_{2-x}$)/SnO$_2$:Sb/Al$_2$O$_3$ samples with pulsed laser irradiation. The insets are the reference shade of pristine TiO$_2$/SnO$_2$:Sb/Al$_2$O$_3$.

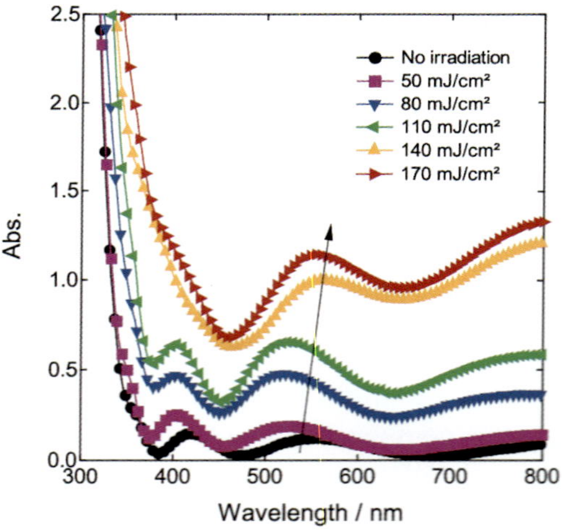

Figure 7.11. Optical absorption spectra of TiO$_2$(TiO$_{2-x}$)/silica glass with and without pulsed laser irradiation.

SnO$_2$:Sb film. After pulsed laser irradiation in air and under vacuum, the film surface darkened. However, the samples irradiated under vacuum clearly exhibited a homogeneous shade, in contrast to the uneven darkening exhibited by the samples irradiated in air. The blackening effect of laser irradiation was also confirmed by the optical absorbance measurements of the TiO$_2$ thin films on silica glass, for laser irradiation under vacuum at 500 Pa (Figure 7.11). The intrinsic optical absorption in the visible range ($<400\,\mathrm{nm}$) was clear in the samples irradiated at fluences $>80\,\mathrm{mJ\cdot cm^{-2}}$. The increasing optical absorption of visible light also indicates the presence of an oxygen deficiency in TiO_2.[33]

Figure 7.12(a) shows cross-sectional SEM images of the reference sample. A compact TiO_2 film on the bottom SnO_2:Sb electrode is visible. The laser irradiation at $80\,\mathrm{mJ\cdot cm^{-2}}$ produced a mud-cracked TiO_{2-x} film with curled edges, which also caused the upper portion of the SnO_2:Sb bottom electrode to crack. The compact TiO_{2-x} film only covered the top of the bottom electrode, and did not coat the surface in the cracks (Figure 7.12(b)). Therefore, the

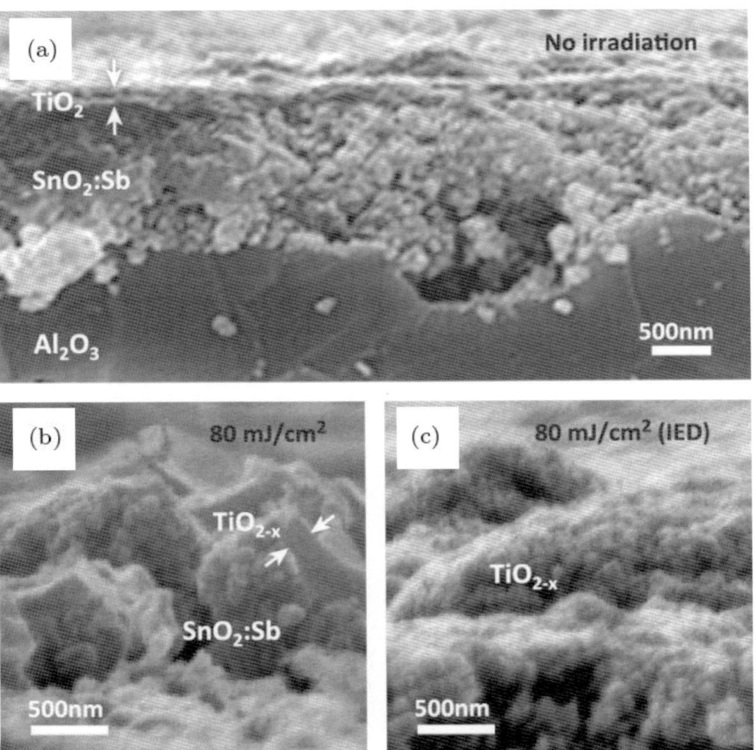

Figure 7.12. Cross-sectional SEM images of (a) reference TiO_2/SnO_2:Sb/Al_2O_3 and TiO_{2-x}/SnO_2:Sb/Al_2O_3 prepared by (b) conventional and (c) IED irradiation at $80\,\mathrm{mJ\cdot cm^{-2}}$.

roughened surface did not increase the surface area of the TiO$_{2-x}$ film. In contrast, the TiO$_{2-x}$ layer covered the entire surface of the mud-cracked bottom electrode in the sample prepared by the IED process at 80 mJ $\cdot$ cm^{-2} (Figure 7.12(c)) through repeated coating and irradiation. In this case, the coated cracks increased the surface area of the TiO$_{2-x}$ film; thus, the IED process was expected to enhance PEC activity.

Figure 7.13(a) shows the AFM images for these samples. The surface areas of the samples for the observed region $(4 \times 4\,\mu\text{m}^2)$ prepared by conventional and IED laser irradiation were actually

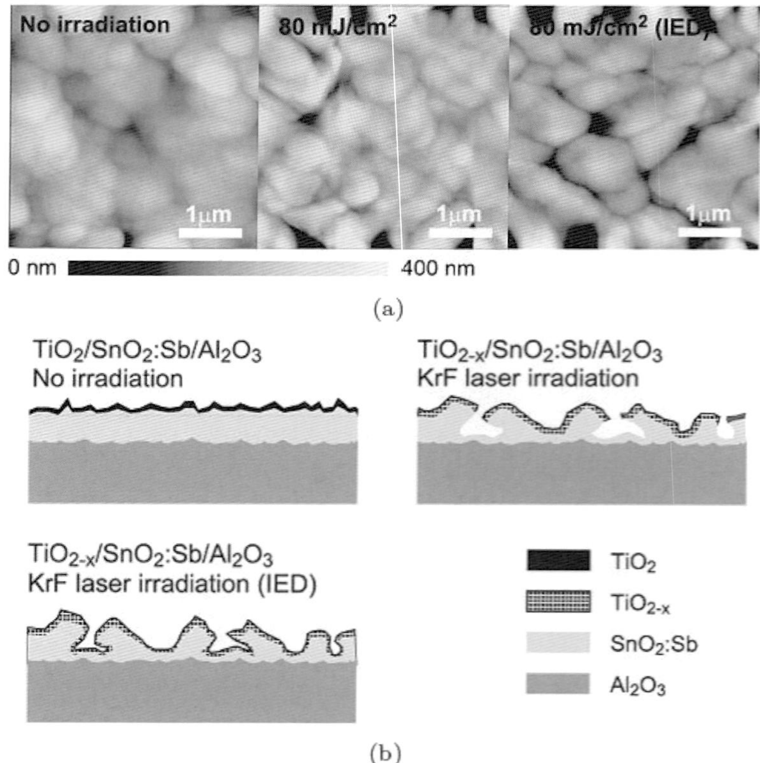

Figure 7.13. (a) AFM images for reference TiO$_2$/SnO$_2$:Sb/Al$_2$O$_3$ and TiO$_{2-x}$/SnO$_2$:Sb/Al$_2$O$_3$ prepared by conventional and IED irradiation at 80 mJ $\cdot$ cm^{-2}. (b) Schematic illustrations of the cross-sectional view of the TiO$_2$(TiO$_{2-x}$)/SnO$_2$:Sb/Al$_2$O$_3$ layers.

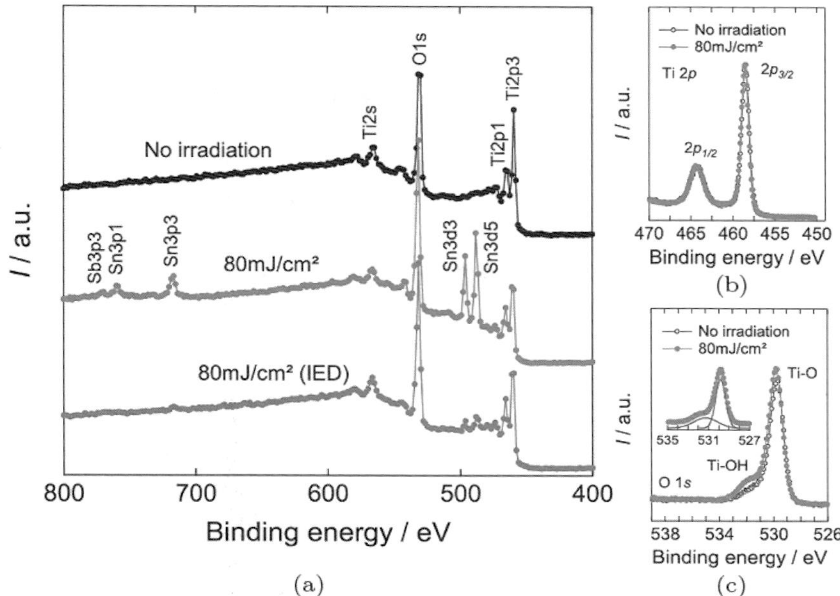

Figure 7.14. (a) XPS survey scan for reference TiO_2/SnO_2:Sb/Al_2O_3 and for TiO_{2-x}/SnO_2:Sb/Al_2O_3 prepared by conventional and IED irradiation at $80\,mJ\cdot cm^{-2}$. (b) Ti 2p, and (c) O 1s spectra for the reference and laser-irradiated ($80\,mJ\cdot cm^{-2}$) samples.

calculated to be 123% ($20.3\,\mu m^2$) and 119% ($19.6\,\mu m^2$) of the reference sample ($16.5\,\mu m^2$). The surface morphology of each sample is summarized in Figure 7.13(b).

Figure 7.14(a) shows the X-ray photoelectron spectra (XPS) survey scans of the samples. The peaks for the reference sample were assigned to Ti ($2s$, $2p_{1/2}$, and $2p_{3/2}$) and O ($1s$) from the TiO_2 layer. In the TiO_{2-x} film that was irradiated at $80\,mJ\cdot cm^{-2}$, additional peaks assigned to Sn ($3d_{3/2}$, $3d_{5/2}$, $3p_{1/2}$, and $3p_{3/2}$) and Sb ($3p_{3/2}$) derived from the bottom electrode were observed because of the surface microcracks (Figures 7.8(d) and 7.12(b)). However, the Sn and Sb peaks disappeared in the TiO_{2-x} film prepared by IED. The top surface of the cracked SnO_2:Sb bottom electrode in this sample was almost completely covered by TiO_{2-x}. This also suggests that the surface morphology proposed in Figure 7.13(b) is correct.

Figure 7.14(b) shows the Ti $2p_{1/2}$ and $2p_{3/2}$ XPS peaks of the titania films before and after irradiation at $80\,\text{mJ}\cdot\text{cm}^{-2}$. The binding energies of Ti $2p_{1/2}$ and $2p_{3/2}$ were $464.3\,\text{eV}$ and $458.5\,\text{eV}$, respectively, in both samples. These peaks were assigned as Ti^{4+} ions in TiO$_2$. There were two signals for the O 1s XPS peak, assigned to the Ti–O bond at $529.8\,\text{eV}$ and Ti–OH bond at $531.2\,\text{eV}$ (Figure 7.14(c)).[51] These results suggest that the top surface of the oxygen-deficient TiO$_{2-x}$ was reoxidized under the ambient atmosphere after laser irradiation. The lack of a shift in the Ti and O XPS peaks, indicating reoxidization of the top surface, has been reported previously for oxygen-deficient black titania.[24,33] A slight increase in the XPS peak corresponding to the Ti–OH bond in the laser-irradiated sample originated from the adsorption of atmospheric humidity at the oxygen-deficient sites created during laser irradiation.

Reoxidation of the top surface after the pulsed laser irradiation was confirmed by XPS studies; nevertheless, the oxygen deficiency in the laser-irradiated samples was clearly proved by Raman spectroscopy. Figure 7.15 shows the irradiated laser fluence dependence of Raman spectra. In the pristine TiO$_2$ film, the observed peaks were assigned to the E_g modes at $144\,\text{cm}^{-1}$, $202\,\text{cm}^{-1}$, and $645\,\text{cm}^{-1}$,

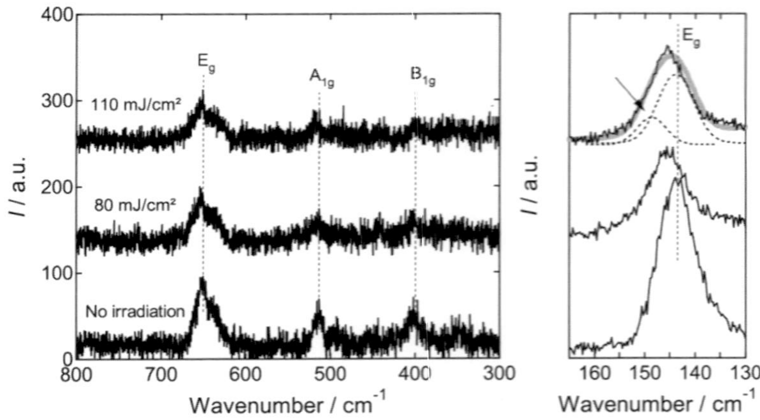

Figure 7.15. Raman spectra of pristine TiO$_2$/silica glass and KrF-laser-irradiated TiO$_{2-x}$/silica glass at $80\,\text{mJ}\cdot\text{cm}^{-2}$ and $110\,\text{mJ}\cdot\text{cm}^{-2}$.

the B_{1g} mode at $397\,\mathrm{cm}^{-1}$, and the A_{1g} mode at $515\,\mathrm{cm}^{-1}$. After the pulsed laser irradiation, these peaks were clearly weakened and broadened with an increase in laser fluence. Additionally, a slight blue shift of the peak at $144\,\mathrm{cm}^{-1}$ was observed due to an emergence of a new peak at around $150\ \mathrm{cm}^{-1}$. This result strongly supports the oxygen deficiency of laser-irradiated TiO_2 films, as reported in previous works on the oxygen-deficient TiO_{2-x}.[33,52,53]

Figure 7.16(a) shows the electrical resistivity (ρ) during laser irradiation. Before irradiation, the pristine TiO_2 film had a very high resistivity of $\rho > 1.5 \times 10^5\,\Omega\cdot\mathrm{cm}$, which was above the measurement limit of the instrument. However, ρ dropped suddenly once laser irradiation at $80\,\mathrm{mJ\cdot cm^{-2}}$ under vacuum began, and reached $7.9 \times 10^2\,\Omega\cdot\mathrm{cm}$ at $180\,\mathrm{s}$ after the first laser pulse. This indicates that the conductivity was increased by the conduction electrons of Ti^{3+}, which were generated by the oxygen deficiency. After laser irradiation ceased, ρ gradually increased in air and was saturated at $7.8 \times 10^3\,\Omega\cdot\mathrm{cm}$. This result strongly suggests that oxygen was released by laser irradiation and the top surface was reoxidized after the laser pulse ceased. The laser on/off dependence of ρ for the TiO_{2-x} films is shown in Figure 7.16(b). ρ cyclically

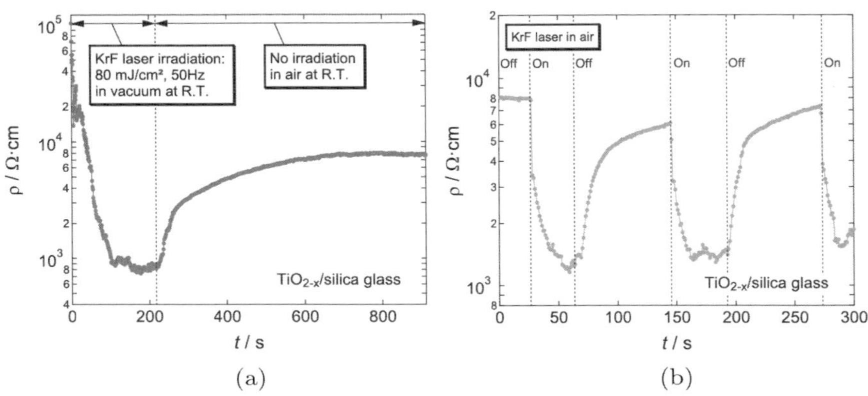

Figure 7.16. (a) Electrical resistivity of TiO_2/silica glass under KrF laser irradiation at $80\,\mathrm{mJ\cdot cm^{-2}}$ with a repetition of $50\,\mathrm{Hz}$ under vacuum. The laser was switched off at $210\,\mathrm{s}$, and air was introduced in the chamber. (b) The KrF laser's on/off dependence of ρ for TiO_{2-x}/silica glass.

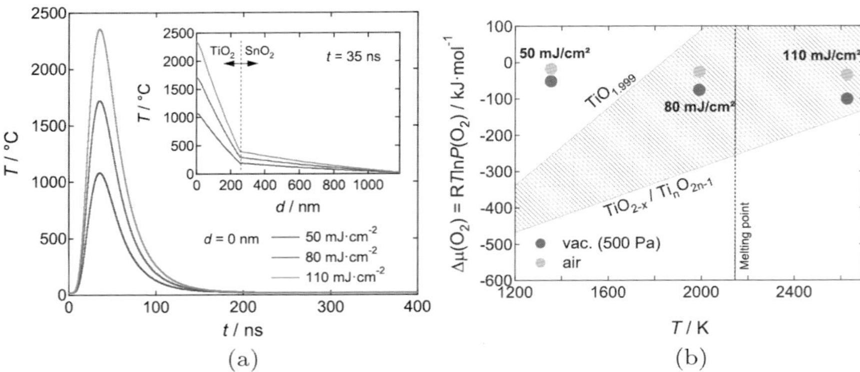

Figure 7.17. (a) Simulated temperature variations of TiO$_2$/SnO$_2$ under KrF laser irradiation at $50\,\text{mJ}\cdot\text{cm}^{-2}$, $80\,\text{mJ}\cdot\text{cm}^{-2}$, and $110\,\text{mJ}\cdot\text{cm}^{-2}$ as a function of time at the film surface and as a function of depth 35 ns after the incident laser pulse (inset). (b) Ellingham diagram for TiO$_2$/TiO$_{2-x}$, and the positions of the samples processed at $50\,\text{mJ}\cdot\text{cm}^{-2}$, $80\,\text{mJ}\cdot\text{cm}^{-2}$, and $110\,\text{mJ}\cdot\text{cm}^{-2}$ under vacuum at 500 Pa and in air are plotted from the simulated maximum temperature.

decreased and increased when the laser was alternately switched on and off, respectively. This reversible behavior is also consistent with laser-induced oxygen release and subsequent reoxidation.

Next, we discuss the laser irradiation process for the TiO$_2$ films using the temperature simulations shown in Figure 7.17(a). The calculation for the bottom SnO$_2$:Sb electrode was approximated by undoped SnO$_2$. The irradiated KrF laser pulse causes instantaneous heating, and the calculated temperature reaches a maximum of 1081°C, 1718°C, and 2354°C at 35 ns after the incident pulse at fluences of $50\,\text{mJ}\cdot\text{cm}^{-2}$, $80\,\text{mJ}\cdot\text{cm}^{-2}$, and $110\,\text{mJ}\cdot\text{cm}^{-2}$, respectively. Thereafter, the temperature is quenched to 20°C. The depth profile reveals that the elevated temperature was highest at the film surface and drops toward the substrate interface. In the Ellingham diagram,[54] the calculated maximum temperature at 80–$110\,\text{mJ}\cdot\text{cm}^{-2}$ at the TiO$_2$ film surface in air and under vacuum at 500 Pa was in the ideal region for oxygen-deficient TiO$_{2-x}$, whereas TiO$_2$ appeared to remain stable at $50\,\text{mJ}\cdot\text{cm}^{-2}$ (Figure 7.17(b)). The photothermal heating and rapid quenching achieved by pulsed laser irradiation successfully captured the high-temperature phase. The oxygen-deficient content (x) may have had a gradient profile because of the temperature gradient from the surface (Figure 7.17(a)) if sufficient oxygen diffusion did not occur after the laser pulse ceased (quenching temperature). Irradiation at $110\,\mathrm{mJ \cdot cm^{-2}}$ would be suitable for creating oxygen deficiencies in the TiO_2 film; however, the film reaches temperatures above the melting point of TiO_2 ($1843^\circ C$).[55] This produced the large cracks observed in the samples irradiated at $140\,\mathrm{mJ \cdot cm^{-2}}$ and $170\,\mathrm{mJ \cdot cm^{-2}}$. Thus, based on the optical absorption, electrical resistivity, and temperature simulations under laser irradiation, we conclude that the pulsed UV laser irradiation successfully induced oxygen deficiencies in TiO_2 films, and this process was effective for the rapid formation of black titania photoanodes.

Linear-sweep photovoltammetry was carried out to evaluate the PEC property of pristine TiO_2 and oxygen-deficient TiO_{2-x} photoanodes for solar water-splitting (Figure 7.18(a)). The photocurrent density of the pristine TiO_2 film at $1.23\,\mathrm{V_{RHE}}$ was $0.503\,\mathrm{mA \cdot cm^{-2}}$, and the onset potential was $0.200\,\mathrm{V_{RHE}}$. The maximum efficiency (η) was 0.20% at $0.645\,\mathrm{V_{CE}}$ (Figure 7.18(b)). In the TiO_{2-x} films prepared by laser irradiation under vacuum, the photocurrent density at $1.23\,\mathrm{V_{RHE}}$ increased substantially with the irradiated laser fluence up to $110\,\mathrm{mJ \cdot cm^{-2}}$, and then decreased above $140\,\mathrm{mJ \cdot cm^{-2}}$.

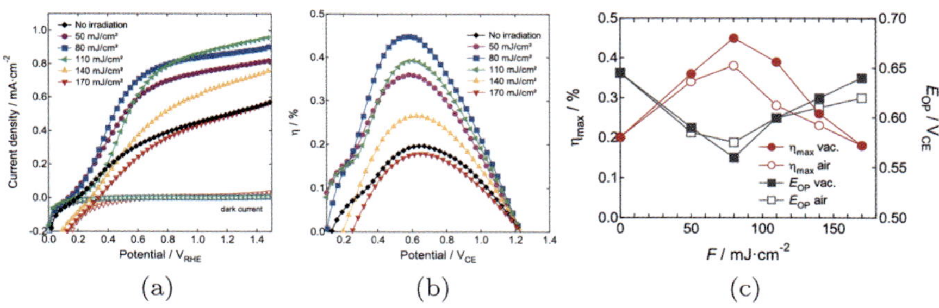

Figure 7.18. (a) J–V curves and (b) η of pristine TiO_2 and TiO_{2-x} photoanodes laser irradiated under vacuum, measured under simulated solar light at $100\,\mathrm{mW \cdot cm^{-2}}$ in 1 M KOH electrolyte (pH = 13.6). (c) η_{max} and E_{OP} for the TiO_{2-x} photoanodes irradiated under vacuum and in air as a function of laser fluence.

Table 7.2. PEC properties of TiO$_2$ and TiO$_{2-x}$ photoanodes.

Preparation condition[a]	$J_{1.23V}^{b}$ (mA·cm^{-2})	J_{OP}^{c} (mA·cm^{-2})	E_{OP} (V$_{CE}$)	η_{max} (%)	E_{onset} (V$_{RHE}$)
HT: 500°C	0.503	0.310	0.645	0.20	0.200
HT: 700°C	0.381	0.290	0.575	0.19	0.100
LV: 50 mJ·cm^{-2}	0.779	0.585	0.590	0.36	0.135
LV: 80 mJ·cm^{-2}	0.863	0.667	0.560	0.45	0.145
LV: 110 mJ·cm^{-2}	0.908	0.660	0.600	0.39	0.185
LV: 140 mJ·cm^{-2}	0.685	0.443	0.620	0.26	0.290
LV: 170 mJ·cm^{-2}	0.491	0.307	0.640	0.18	0.340
LV2: 80 mJ·cm^{-2}	1.029	0.768	0.560	0.52	0.175
LA: 50 mJ·cm^{-2}	0.761	0.523	0.585	0.34	0.190
LA: 80 mJ·cm^{-2}	0.786	0.590	0.575	0.38	0.115
LA: 110 mJ·cm^{-2}	0.670	0.455	0.605	0.28	0.225
LA: 140 mJ·cm^{-2}	0.578	0.396	0.610	0.23	0.235
LA: 170 mJ·cm^{-2}	0.509	0.308	0.620	0.18	0.300

[a]HT, heat treatment; LV, laser treatment under vacuum; LV2, laser treatment under vacuum using the IED process; and LA, laser treatment in air. [b]Current density at 1.23 V$_{RHE}$. [c]Current density at E_{OP} V$_{CE}$.

Moreover, the onset potential (E_{onset}) was lower in samples that were irradiated at laser fluences of 50–110 mJ·cm^{-2} compared with the pristine TiO$_2$ photoanode.

The sample irradiated at 80 mJ·cm^{-2} showed the highest maximum η (η_{max}) of 0.45%, because the optimum potential for η_{max} (E_{OP}) reached a minimum (0.560 V$_{CE}$) at this laser fluence (Table 7.2 and Figure 7.18(c)). High laser fluences ($>$110 mJ·cm^{-2}) that resulted in excessive photothermal heating should negatively influence extrinsic transport in TiO$_{2-x}$/SnO$_2$:Sb. The variation in E_{OP} and η_{max} as a function of the laser fluence for irradiation in air was the same as that under vacuum (Figure 7.18(c)). However, the vacuum process produced a higher η_{max} and reduced the effective bias voltage. This is probably because of the homogeneous oxygen release at the film surface.

A concern generally arises for the origin of photocurrent in the oxygen-deficient photoanodes, whether the observed photocurrent is derived from the water-splitting or a self-oxidation of photoanodes. However, the possibility of the self-oxidation effect was eliminated

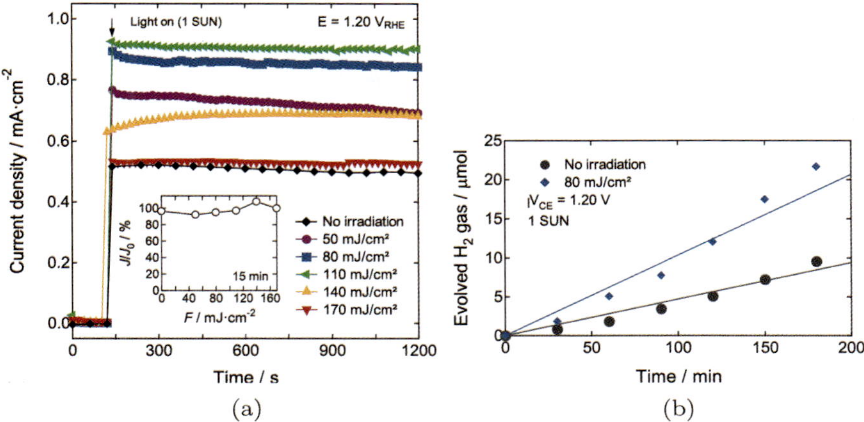

Figure 7.19. Time dependences of (a) photocurrent and (b) H_2 evolution of pristine TiO_2 and TiO_{2-x} photoanodes laser irradiated under vacuum, measured under simulated solar light at $100\,\mathrm{mW\cdot cm^{-2}}$ in $1\,M$ KOH electrolyte (pH = 13.6). The inset of (a) shows the ratio of photocurrent at 15 min from the initial excitation as a function of irradiated laser fluence.

in our samples by the following results. Figure 7.19(a) shows the photocurrent stability of pristine TiO_2 and oxygen-deficient TiO_{2-x} photoanodes at $1.20\,\mathrm{V_{RHE}}$. The photocurrent of pristine TiO_2 after 15 min from the starting excitation remained 96% from the initial value. The laser-irradiated samples also showed comparably high stability of photocurrent of 92–107% at 15 min, indicating the continuous PEC reaction. Moreover, we successfully observed H_2 evolution from both samples with and without the laser irradiation under $100\,\mathrm{mW\cdot cm^{-2}}$ excitation at $1.20\,\mathrm{V_{CE}}$. The H_2 gas evolution rates for pristine and laser-irradiated ($80\,\mathrm{mJ\cdot cm^{-2}}$) samples were calculated as $8.4\,\mu\mathrm{mol\cdot h^{-1}\cdot cm^{-2}}$ and $18.7\,\mu\mathrm{mol\cdot h^{-1}\cdot cm^{-2}}$, respectively, as shown in Figure 7.19(b). The enhanced H_2 evolution rate in the laser-irradiated sample almost corresponded with the increase in photocurrent. These results provide evidence that the laser-irradiated TiO_{2-x} photoanode demonstrated intrinsic water splitting and enhancement of the PEC property.

Figure 7.20(a) shows the photocurrent density and η of the TiO_2 photoanode post-annealed at $700°$C. Post-annealing reduced E_{onset},

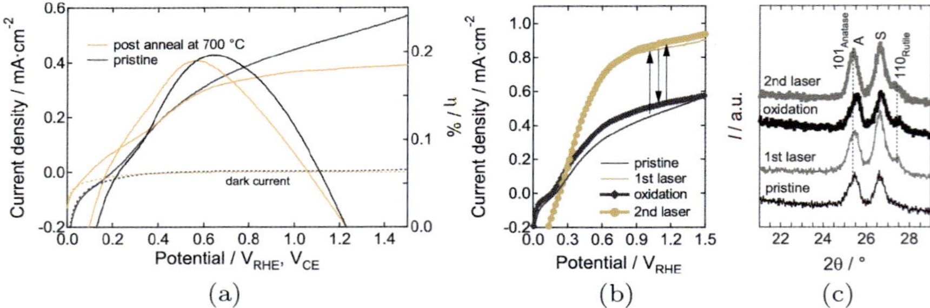

(a)   (b)   (c)

Figure 7.20. (a) J–V curves and η of the pristine and post-annealed (700°C) TiO₂ photoanodes under simulated solar light at $100\,\mathrm{mW\cdot cm^{-2}}$ in 1 M KOH electrolyte (pH = 13.6). (b) The J–V curves of $\mathrm{TiO_{2-x}}$ (TiO₂) photoanodes for first laser irradiation, oxidation, and second laser irradiation. (c) GIXRD spectra of the $\mathrm{TiO_{2-x}}$ (TiO₂) photoanodes corresponding to (b). The subscripts A and S represent Al₂O₃ and SnO₂:Sb, respectively.

although it did not enhance the η value, indicating that the formation of the interface between the anatase and rutile phases did not increase the photocurrent in this system. In addition, we examined the continuous post-treatment dependence of the photocurrent density of the $\mathrm{TiO_{2-x}}$ (TiO₂) photoanodes. Figure 7.20(a) shows that the laser irradiation at $80\,\mathrm{mJ\cdot cm^{-2}}$ under vacuum produced a large increase in the photocurrent. This sample was post-annealed at 500°C for 1 h, and the photocurrent of the sample was reduced to a value similar to that for the pristine TiO₂ photoanode (Figure 7.20(b)). After post-annealing, the sample was irradiated again at $80\,\mathrm{mJ\cdot cm^{-2}}$ under vacuum, and the photocurrent recovered the original value for the $\mathrm{TiO_{2-x}}$ photoanode. During these treatments, the peak intensity of the GIXRD patterns and the rutile content were almost unchanged (Figure 7.20(c)). This means that the enhancement of the PEC property in the laser-irradiated samples did not originate from the heterointerface, but arose from the oxygen deficiency of the TiO₂ films that improved the absorption of the excitatory light over a wide range of wavelengths.

Figure 7.21(a) shows the photocurrent density of the laser-irradiated $\mathrm{TiO_{2-x}}$ photoanode prepared by IED with an optimum fluence at $80\,\mathrm{mJ\cdot cm^{-2}}$ under vacuum. E_{onset} was reduced to 0.175

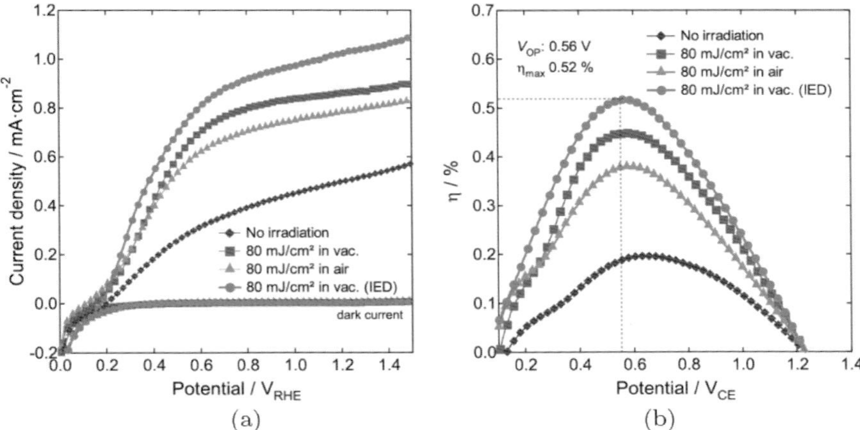

Figure 7.21. (a) J–V curves and (b) η for the pristine TiO$_2$ and laser-irradiated TiO$_{2-x}$ photoanodes under simulated solar light at $100\,\mathrm{mW \cdot cm^{-2}}$ in 1 M KOH electrolyte (pH = 13.6). The laser irradiation was carried out at $80\,\mathrm{mJ \cdot cm^{-2}}$ under vacuum (conventional and IED) and in air.

V_{RHE} and the photocurrent at 1.23 V_{RHE} reached $1.029\,\mathrm{mA \cdot cm^{-2}}$. η_{max} was 0.52% at 0.560 V_{RHE} (Figure 7.21(b)). This high η_{max} was 2.6-fold higher than that of the pristine TiO$_2$ photoanode, which was probably caused by the increase in the surface area by IED.

Figure 7.22 shows the IPCE spectra of pristine TiO$_2$ and laser-irradiated ($80\,\mathrm{mJ \cdot cm^{-2}}$ under vacuum) TiO$_{2-x}$ photoanodes at 0.55 V_{RHE}. The IPCE of the TiO$_2$ photoanode was 6.9% at 340 nm, and decreased to almost zero at around 400 nm. However, in the laser-irradiated TiO$_{2-x}$ photoanodes, the photoresponse was maintained at 480 nm because of the large absorbance of visible light wavelengths (Figure 7.22), and the IPCE in the UV range also increased to 12.5% at 340 nm. The IPCE reached 14.3% in the TiO$_{2-x}$ photoanode prepared by IED and its visible range response was 20-fold greater than that of the pristine TiO$_2$ photoanode at 450 nm.

The main origin of PEC property enhancement in the oxygen-deficient TiO$_{2-x}$ photoanodes would be the high IPCE in the UV wavelength range. The oxygen vacancy in TiO$_2$ can increase the electron donor density, resulting in the improvement of charge transport and the shifting Fermi level of TiO$_2$ toward the conduction band.[18–20]

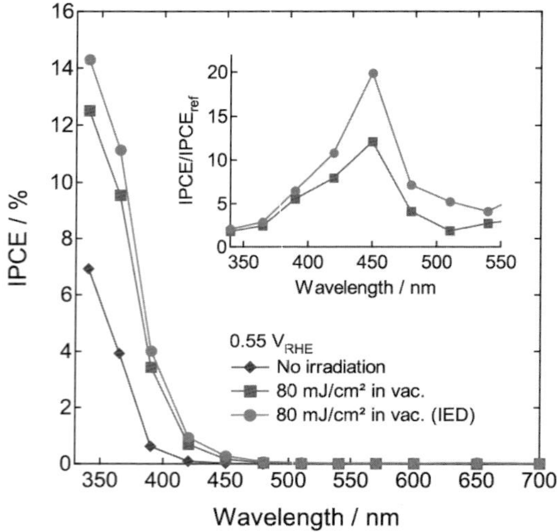

Figure 7.22. IPCE spectra of the photoanodes as a function of the excitation wavelength at 0.55 V_{RHE}. The inset shows the IPCE ratio between the pristine TiO_2 and laser-irradiated TiO_{2-x} photoanodes.

This would stimulate charge separation at the reaction interface, which is linked to the enhancement of water-splitting. Moreover, the characteristic cross-sectional morphology in the laser-irradiated samples would contribute to the enhancement of the PEC property. The structural and physical properties of the pristine and laser-irradiated photoanodes are summarized in Figure 7.23. The crystallite size and stoichiometric oxygen content are uniform in the pristine TiO_2 layer; therefore, the number of grain boundaries, conductivity, and reduced absorbance for visible light are also homogeneous. In contrast, the laser-irradiated samples would have a gradient distribution of crystallite sizes and oxygen deficiency; the crystallite size and oxygen deficiency increased toward the top surface, although the oxygen deficiency at the top surface was eliminated by reoxidation. The crystallite growth reduced the number of grain boundaries and the oxygen deficiency enhanced the bulk conductivity. Both would increase the majority carrier diffusion length, which is crucial for improving the PEC activity.[56,57] Furthermore, because visible light penetrates deeper into the photoanode than UV light, the improved

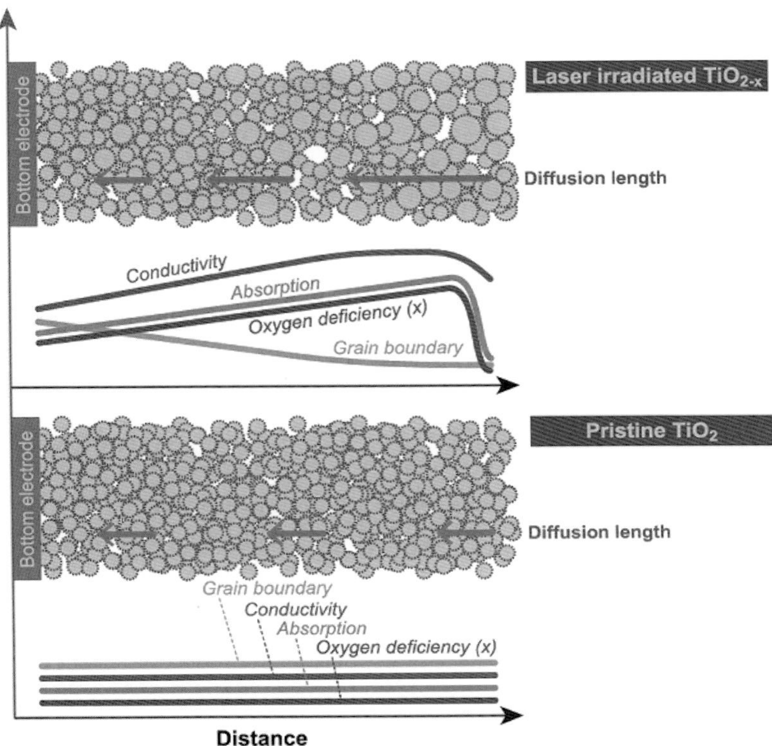

Figure 7.23. Schematic of the structural and physical properties of the pristine TiO_2 and laser-irradiated TiO_{2-x} photoanodes.

absorption in the visible range means that the photoanode interior is used more efficiently. The extended diffusion length of the excited carriers and efficient use of the excitatory light improve the solar water-splitting properties of the laser-irradiated photoanodes.

Thus, we have demonstrated that pulsed UV laser irradiation of TiO_2 films is a simple, effective process for preparing oxygen-deficient black titania (TiO_{2-x}) films. The oxygen-deficient TiO_{2-x} photoanodes were fabricated by KrF laser irradiation under low vacuum conditions for several minutes. The resulting TiO_{2-x} film had a larger absorbance for visible light than pristine TiO_2, and showed a higher PEC property for solar water-splitting. The STH efficiency of the TiO_{2-x} photoanode was 0.52%, which was 2.6-fold higher than that of the pristine TiO$_2$ photoanode. Our rapid fabrication process for black titania photoanodes has great potential for industrial solar H$_2$ production.

7.4 Concluding Remarks

Solar energy is a key component for creating a sustainable society. The breakthrough of black titania has opened up new research pathways for the application of TiO$_2$ in solar H$_2$ production. In recent years, developed blackening processes of TiO$_2$ coatings, such as high/atmospheric pressure H$_2$ annealing, ion implantation, H$_2$ plasma, annealing with oxygen getters, electron beam/pulsed laser irradiation, and electrochemical methods, have provided not only straightforward production of black titania coating but also many insights into TiO$_2$ modification for improving photocatalytic/PEC activity for solar H$_2$ production. Efficient oxygen-deficient black titania film fabrication is possible using pulsed UV laser irradiation. The blackening process for the preparation of TiO$_{2-x}$ photoanodes was completed using KrF laser irradiation under low vacuum conditions for only a few minutes. The TiO$_{2-x}$ film showed greater absorbance than pristine TiO$_2$ in the visible light region and exhibited a substantial increase in PEC water-splitting under simulated solar light. The STH efficiency of the TiO$_{2-x}$ photoanode reached 0.52%, which was 2.6-fold higher than that of the pristine TiO$_2$ photoanode. This rapid fabrication process for black titania photoanodes is promising for industrial solar H$_2$ production. The significant enhancement of these properties of black titania is strongly related to the defects and coordination environment surrounding Ti^{3+} ions in many cases. Precise control of these points in a nanoscale environment is possible by tiny variations in the fabrication conditions for thin films/small nanomaterials, e.g. defect/void formations contribute not only to the enhancement of water splitting but also to additional effects such as reduction of noble metal cocatalysts. Both the phenomenological understanding and the development of a straightforward fabrication process of black titania coatings will accelerate the realization of a solar-powered H$_2$ supply for the future.

References

(1) Alexander, B. D.; Kulesza, P. J.; Rutkowska, I.; Solarska, R.; Augustynski, J. *J. Mater. Chem.* **2008**, *18*, 2298.

(2) Nowotny, J.; Bak, T.; Nowotny, M. K.; Sheppard, L. R. *Int. J. Hydrog. Energy* **2007**, *32*, 2609.

(3) Kronawitter, C. X.; Vayssieres, L.; Shen, S.; Guo, L.; Wheeler, D. A.; Zhang, J. Z.; Antoun, B. R.; Mao, S. S. *Energy Environ. Sci.* **2011**, *4*, 3889.

(4) Osterloh, F. E. *Chem. Soc. Rev.* **2013**, *42*, 2294.

(5) Park, Y.; McDonald, K. J.; Choi, K-S. *Chem. Soc. Rev.* **2013**, *42*, 2321.

(6) Hisatomi, T.; Formal, F. L.; Cornuz, M.; Brillet, J.; Tétreault, N.; Sivula, K.; Grätzel, M. *Energy Environ. Sci.* **2011**, *4*, 2512.

(7) Liu, X.; Wang, F.; Wang, Q. *Phys. Chem. Chem. Phys.* **2012**, *14*, 7894.

(8) Saito, R.; Miseki, Y.; Sayama, K. *Chem. Commun.* **2012**, *48*, 3833.

(9) Pinaud, B. A.; Vesborg, P. C. K.; Jaramillo, T. F. *J. Phys. Chem. C* **2012**, *116*, 15918.

(10) Grätzel, M. *Nature* **2001**, *414*, 338.

(11) Fujishima, A.; Zhang, X.; Tryk, D. A. *Surf. Sci. Rep.* **2008**, *63*, 515.

(12) Hoffmann, M. R.; Martin, S. T.; Choi, W.; Bahnemann, D. W. *Chem. Rev.* **1995**, *95*, 69.

(13) Chen, X.; Liu, L.; Yu, P. Y.; Mao, S. S. *Science* **2011**, *331*, 746.

(14) Leshuk, T.; Parviz, R.; Everett, P.; Krishnakumar, H.; Varin, R. A.; Gu, F. *ACS Appl. Mater. Interfaces* **2013**, *5*, 1892.

(15) Yang, C.; Wang, Z.; Lin, T.; Yin, H.; Lü, X.; Wan, D.; Xu, T.; Zheng, C.; Lin, J.; Huang, F.; Xie, X.; Jiang, M. *J. Am. Chem. Soc.* **2013**, *135*, 17831.

(16) Wang, Z.; Yang, C.; Lin, T.; Yin, H.; Chen, P.; Wan, D.; Xu, F.; Huang, F.; Lin, J.; Xie, X.; Jiang, M. *Energy Environ. Sci.* **2013**, *6*, 3007.

(17) Khan, M. M.; Ansari, S. A.; Pradhan, D.; Ansari, M. O.; Han, D. H.; Lee, J.; Cho, M. H. *J. Mater. Chem. A* **2014**, *2*, 637.

(18) Cronemeyer, D. C. *Phys. Rev.* **1959**, *113*, 1222.

(19) Janotti, A.; Varley, J. B.; Rinke, P.; Umezawa, N.; Kresse, G.; Van de Walle, C. G. *Phys. Rev. B* **2010**, *81*, 085212.

(20) Y. H. Hu, *Angew. Chem. Int. Ed.* **2012**, *51*, 12410.

(21) Lin, T.; Yang, C.; Wang, Z.; Yin, H.; Lü, X.; Huang, F.; Lin, J.; Xie, X.; Jiang, M. *Energy Environ. Sci.* **2014**, *7*, 967.

(22) Binetti, E.; El Koura, Z.; Patel, N.; Dashora, A.; Niotello, A. *Appl. Catal. A* **2015**, *500*, 69.

(23) Cui, H.; Zhao, W.; Yang, C.; Yin, H.; Lin, T.; Shan, Y.; Xie, Y.; Gu, H.; Huang, F. *J. Mater. Chem. A* **2014**, *2*, 8612.

(24) Wang, G.; Wang, H.; Ling, Y.; Tang, Y.; Yang, X.; Fitzmorris, R. C.; Wang, C.; Zhang, J. Z.; Li, Y. *Nano Lett.* **2011**, *11*, 3026.

(25) Liu, N.; Schneider, C.; Freitag, D.; Hartmann, M.; Venkatesan, U.; Müller, J.; Spiecker, E.; Schmuki, P. *Nano Lett.* **2014**, *14*, 3309.

(26) Xie, S.; Li, M.; Wei, W.; Zhai, T.; Fang, P.; Qiu, R.; Lu, X.; Tong, Y. *Nano Energy* **2014**, *10*, 313.

(27) Nakajima, T.; Nakamura, T.; Shinoda, K.; Tsuchiya, T. *J. Mater. Chem. A* **2014**, *2*, 6762.

(28) Liu, N.; Häublein, V.; Zhou, X.; Venkatesan, U.; Hartmann, M.; Mačković, M.; Nakajima, T.; Spiecker, E.; Osvet, A.; Frey, L.; Schmuki, P. *Nano Lett.* **2015**, *15*, 6815.

(29) Khan, M. M.; Ansari, S. A.; Pradhan, D.; Ansari, M. O.; Han, D. H.; Lee, J.; Cho, M. H. *J. Mater. Chem. A* **2014**, *2*, 637.

(30) Wang, Z.; Yang, C.; Lin, T.; Yin, H.; Chen, P.; Wan, D.; Xu, F.; Huang, F.; Lin, T.; Xie, X.; Jiang, M. *Adv. Funct. Mater.* **2013**, *23*, 5444.

(31) Page, Y. L.; Strobel, P. *J. Solid State Chem.* **1983**, *47*, 6.

(32) Han, W.-Q.; Zhang, Y. *Appl. Phys. Lett.* **2008**, *92*, 203117.

(33) Wang, Z.; Yang, C.; Lin, T.; Yin, H.; Chen, P.; Wan, D.; Xu, F.; Huang, F.; Lin, T.; Xie, X.; Jiang, M. *Energy Environ. Sci.* **2013**, *6*, 3007.

(34) Yang, C.; Wang, Z.; Lin, T.; Yin, H.; Lü, X.; Wan, D.; Xu, T.; Zheng, C.; Lin, J.; Huang, F.; Lin, J.; Xie, X.; Jiang, M. *J. Am. Chem. Soc.* **2013**, *135*, 17831.

(35) Gurylev, V.; Su, C.-Y.; Perng, T.-P. *J. Catal.* **2015**, *330*, 177.

(36) Zhu, G.; Yin, H.; Yang, C.; Cui, H.; Wang, Z.; Xu, J.; Lin, T.; Huang, F. *ChemCatChem* **2015**, *7*, 2614.

(37) Li, H.; Chen, J.; Xia, Z.; Xing, J. *J. Mater. Chem. A* **2015**, *3*, 699.

(38) Xu, C.; Song, Y.; Lu, L.; Cheng, C.; Liu, D.; Fang, X.; Chen, X.; Zhu, X.; Li, D. *Nanoscale Res. Lett.* **2013**, *8*, 391.

(39) Wang. C.; Hu, Q.; Huang, J.; Wu, L.; Deng, Z.; Liu, Z.; Liu, Y.; Cao, Y. *Appl. Surf. Sci.* **2013**, *283*, 188.

(40) Wang. G.; Xiao, X.; Li, W.; Lin, Z.; Zhao, Z.; Chen, C.; Wang C.; Li, Y.; Huang, X.; Miao, L.; Jiang, C.; Huang, Y.; Duan, X. *Nano Lett.* **2015**, *15*, 4692.

(41) Bruel, M. *Nucl. Instrum. Methods Phys. Res., Sect. B* **1996**, *108*, 313.

(42) Stefanov, B. I.; Niklasson, G. A.; Granqvist, C. G.; Österlund, L. *J. Mater. Chem. A* **2015**, *3*, 17369.

(43) Nakajima, T.; Shinoda, K.; Tsuchiya, T. *Chem. Soc. Rev.* **2014**, *43*, 2027.

(44) Nakajima, T.; Tsuchiya, T.; Ichihara, M.; Nagai, H.; Kumagai, T. *Chem. Mater.* **2008**, *20*, 7344.

(45) Nakajima, T.; Tsuchiya, T.; Ichihara, M.; Nagai, H.; Kumagai, T. *Appl. Phys. Express* **2009**, *2*, 023001.

(46) Tsuchiya, T.; Watanabe, A.; Imai, Y.; Niino, H.; Yamaguchi, I.; Manabe, T.; Kumagai, T.; Mizuta, S. *Jpn. J. Appl. Phys.* **1999**, *38*, L823.

(47) Nakajima, T.; Shinoda, K.; Tsuchiya, T. *Phys. Chem. Chem. Phys.* **2013**, *15*, 14384.

(48) Joya, Y. F.; Liu, Z. *Scr. Mater.* **2009**, *60*, 467.

(49) Nakajima, T.; Tsuchiya, T.; Kumagai, T. *J. Solid State Chem.* **2009**, *182*, 2560.

(50) Nakajima, T.; Kitamura, T.; Tsuchiya, T. *Appl. Catal. B-Environ.* **2011**, *108–109*, 47.

(51) Kačiulis, S.; Mattogno, G.; Napoli, A.; Bemporad, E.; Ferrari, F.; Montenero, A.; Gnappi, G. *J. Electron Spectrosc. Relat. Phenom.* **1998**, *95*, 61.

(52) Liu, G.; Yang, H. G.; Wang, X.; Cheng, L.; Lu, H.; Wang, L.; Lu, G. Q.; Cheng, H-M. *J. Phys. Chem. C* **2009**, *113*, 21784.

(53) Pan, X.; Yang, M-Q.; Fu, X.; Zhang, N.; Xu, Y.-J. *Nanoscale* **2013**, *5*, 3601.

(54) Jacob, K. T.; Gupta, S. *JOM* **2009**, *61*, 56.

(55) Prasai, B.; Cai, B.; Underwood, M. K.; Lewis, J. P.; Drabold, D. A. *J. Mater. Sci.* **2012**, *47*, 7515.

(56) Södergren, S.; Hagfeldt, A.; Olsson, J.; Lindquist, S.-E. *J. Phys. Chem.* **1994**, *98*, 5552.

(57) Leng, W. H.; Barnes, P. R. F.; Juozapavicius, M.; O'Regan, B. C.; Durrant, J. R. *J. Phys. Chem. Lett.* **2010**, *1*, 967.

CHAPTER EIGHT

Hydrogen-Treated TiO$_2$ Nanowires for Charge Storage and Photoelectrochemical Water Splitting

Gongming Wang[*], *Xihong Lu*[†] *and Yat Li*[‡]

[*]*Department of Chemistry,*
University of Science and Technology of China,
Hefei, Anhui 230026, People's Republic of China

[†]*KLGHEI of Environment and Energy Chemistry,*
MOE of the Key Laboratory of Bioinorganic and Synthetic Chemistry,
School of Chemistry and Chemical Engineering,
Sun Yat-Sen University,
Guangzhou 510275,
People's Republic of China

[‡]*Department of Chemistry and Biochemistry,*
University of California, Santa Cruz,
California 95064, United States

8.1 Introduction of Hydrogen-Treated TiO$_2$ (Black TiO$_2$)

As one of the most widely studied metal oxides, TiO$_2$ has received considerable attention in the past several decades due to its excellent photocatalytic activity under ultraviolet (UV) light, low cost, and excellent chemical stability.[1–10] In light of these excellent properties, TiO$_2$ has been widely used as photoanode for photoelectrochemical water splitting to generate solar hydrogen.[5,6,11,13–16] However, a major drawback for TiO$_2$ photoelectrode is its large bandgap of 3.0–3.2 eV, which limits its light absorption capability in visible light and thus solar conversion efficiency. Various strategies have been designed to extend its light absorption to visible light region.[6,17–21] For example, doping with non-metal element such as nitrogen can introduce impurity states in the band structure of TiO$_2$ and lead to visible light absorption.[7,22–24] Although nitrogen doped TiO$_2$ can extend its light absorption edge to about 550 nm, the reported visible light photoactivity efficiencies are still relatively low.[7,22–24] This can be attributed to the low efficiency of utilizing solar light. Therefore, it is highly desirable to develop new strategies for preparation of TiO$_2$ with broader light absorption range and better solar energy conversion efficiency.

In 2011, Chen *et al.* reported the synthesis of "black TiO$_2$" nanoparticles by surface hydrogenation in hydrogen atmosphere, which can extend the light absorption of TiO$_2$ from UV region to the infrared light region.[12] Compared to conventional doping strategies which introduce impurity states into TiO$_2$, Chen *et al.* proposed a conceptually different mechanism and attributed the significantly enhanced solar light absorption to the creation of surface disorder of TiO$_2$ nanoparticle induced by hydrogen treatment.[12] Figure 8.1(a) (upper) shows the proposed structure formed with a crystalline TiO$_2$ nanoparticle as a core and a highly disordered surface layer. Large amount of lattice disorder in semiconductor can yield mid-gap states with energy distributions, which differ from that of single defect in a crystal. Instead of forming discrete donor states, these mid-gap states can form continuous bands and narrow the bandgap, as shown in Figure 8.1(a).[12] Meanwhile white TiO$_2$ nanoparticle powder converts to black TiO$_2$ powder, the black TiO$_2$ exhibits excellent photocatalytic activities for dye degradation and solar hydrogen generation in the presence of sacrificial chemicals under white light irradiation.[12] Wang *et al.* further demonstrated that the photoelectrochemical performance of rutile and anatase TiO$_2$ nanos- tructures for water oxidation can be substantially enhanced through hydrogen treatment.[11] Hydrogen is a reducing gas that reduces Ti^{4+} in pristine TiO$_2$ to Ti^{3+}, with the charge balanced by creating oxygen vacancies (V$_{Os}$). Figure 8.1(b) (upper) shows the simplified band structure of hydrogen-treated TiO$_2$, with discrete oxygen vacancy

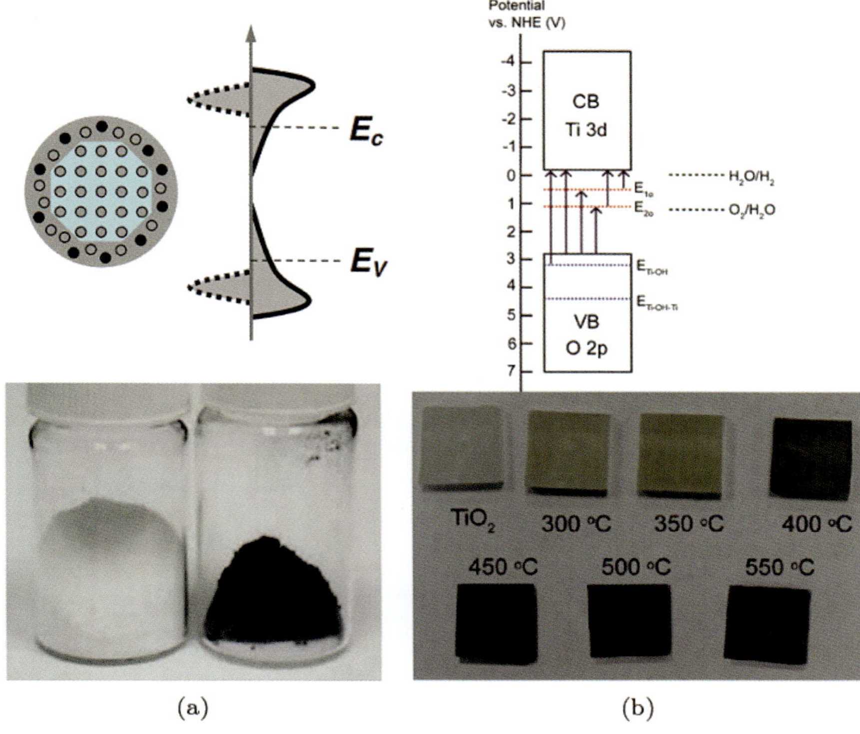

(a)   (b)

Figure 8.1. (a) Upper: the schematic illustration of structure and electronic DOS of black TiO_2 with created surface disordered layer. Lower: digital picture of pristine TiO_2 and black TiO_2 powder. Reproduced from Ref. [12]. © AAAS. (b) Upper: the schematic illustration of the band structure of hydrogen-treated TiO_2. Dashed lines highlight the states of oxygen vacancies. Lower: digital image of pristine TiO_2 and TiO_2 hydrogenated at the temperatures between 300–550°C. Reproduced from Ref. [11]. © American Chemical Society.

(V_O) states. By annealing in hydrogen gas at different temperatures from 300–550°C, the color of rutile TiO_2 nanowire arrays can change from white to yellow, and eventually black.[11] Although the color change indicates that the hydrogen-treated TiO_2 absorbs visible light, incident photon-to-current efficiency (IPCE) measurements confirmed that the visible light photoactivity of hydrogen-treated TiO_2 is negligible.[11] The enhanced photoactivity was attributed to the improved charge separation and electrical conductivity induced by the increased donor density as a result of creation of V_{Os}.[11] Since the demonstration of hydrogen-treated TiO$_2$ with excellent photocatalytic performance, efforts have been made to extend the hydrogen treatment strategy to different metal oxides and explore the hydrogen-treated metal oxides for various applications.[25–37]

Hydrogen-treated TiO$_2$ with improved electrical conductivity has also been studied as electrochemical electrode for charge storage such as lithium-ion batteries.[38–40] and supercapacitors.[27,28,41] TiO$_2$ is considered as an anode alternative for lithium-ion battery, because its high lithiation potential can avoid lithium plating and diminishing electrolyte decomposition rate.[42,43] Importantly, TiO$_2$ electrode only has minor structural change during lithiation, which is promising for its implementation as lithium-ion battery anode.[42,43] However, TiO$_2$ is not considered to be a good electrode material for electrochemical applications. The major limitation of TiO$_2$ for electrochemical charge storage is its poor electrical conductivity. Therefore, the discovery of hydrogen-treated TiO$_2$ provides new opportunities to explore TiO$_2$ for electrochemical charge storages. The conductivity of TiO$_2$ can be tuned from semiconducting to semimetallic properties through controlled variation of hydrogen treatment conditions (time and temperatures).[44] In the past few years, significant efforts have been made to develop black TiO$_2$ electrodes for battery and supercapacitor applications.[27,28,38–41] In this book chapter, we will highlight the recent progress of using hydrogen-treated TiO$_2$ nanomaterials for photoelectrochemical water splitting and supercapacitors.

8.2 Hydrogen-Treated TiO$_2$ Nanowire Arrays for PEC Water Splitting

8.2.1 *Hydrogen-Treated Rutile TiO$_2$ Nanowire Arrays*

Single crystal rutile TiO$_2$ nanowire arrays with large electrolyte/ semiconductor interfacial area, short diffusion length for minority carriers, excellent photostability represent an excellent electrode material for photoelectrochemical reactions.[22,45] Importantly, Wang *et al.* demonstrated that hydrogen treatment can considerably increase the performance of TiO$_2$ nanowire arrays for photoelectrochemical water oxidation.[11] Vertically aligned rutile TiO$_2$ nanowire arrays were fabricated on conducting fluorine-doped tin oxide (FTO) glass substrate via a hydrothermal method.[11,46] Hydrogen treatment was conducted in a home built tube furnace filled with ultrahigh purity hydrogen gas at a temperature range of between 200°C and 550°C. Upon hydrogenation, the color of TiO$_2$ changed from white to light yellow, yellowish green, and then black, with the increase of annealing temperature. The gradual color change suggests that hydrogenation modifies the optical absorption properties of TiO$_2$. Linear sweep voltammograms were collected for TiO$_2$ and hydrogen-treated TiO$_2$ nanowires prepared at the hydrogenation temperatures of 350°C, 400°C, and 450°C, in the dark and under 100 mW/cm^2 AM 1.5G simulated solar light illumination (Figure 8.2(a)). Hydrogen-treated TiO$_2$ samples exhibit enhanced photocurrent densities compared to pristine TiO$_2$ sample at the same potential.[11] Figure 8.2(b) shows the temperature dependent photocurrent density at the potential of 0.23 V vs. Ag/AgCl.[11] The photocurrent densities gradually increase with the increase of annealing temperature from 200°C to 350°C, and then decrease with further increase of annealing temperature. The hydrogen-treated TiO$_2$ at 350°C exhibits the optimal photocurrent density of $\sim$2.5 mA/cm^2, which is almost four times higher than the pristine TiO$_2$.

IPCE measurement is a powerful way to evaluate the enhanced photoactivity as a function of incident light wavelength. IPCE can be expressed by the equation:

$$\text{IPCE} = (1240 \times I)/(\lambda \times J_{\text{light}}),$$

where I is the photocurrent density at specific wavelength, λ is the wavelength, and J_{light} is the power density of the light.[11] The IPCE spectra of pristine TiO$_2$ and hydrogen-treated TiO$_2$ nanowires collected at -0.6 V vs. Ag/AgCl show that hydrogen-treated TiO$_2$ nanowire electrodes exhibit significantly enhanced photoactivity over the entire UV region, but negligible photoactivity in the visible light region (Figure 8.2(c)). The IPCE values decrease gradually from $\sim$95% at 370 nm to $\sim$1% at 420 nm, which is consistent with the bandgap energy (3.0 eV) of rutile TiO$_2$. The IPCE values in the

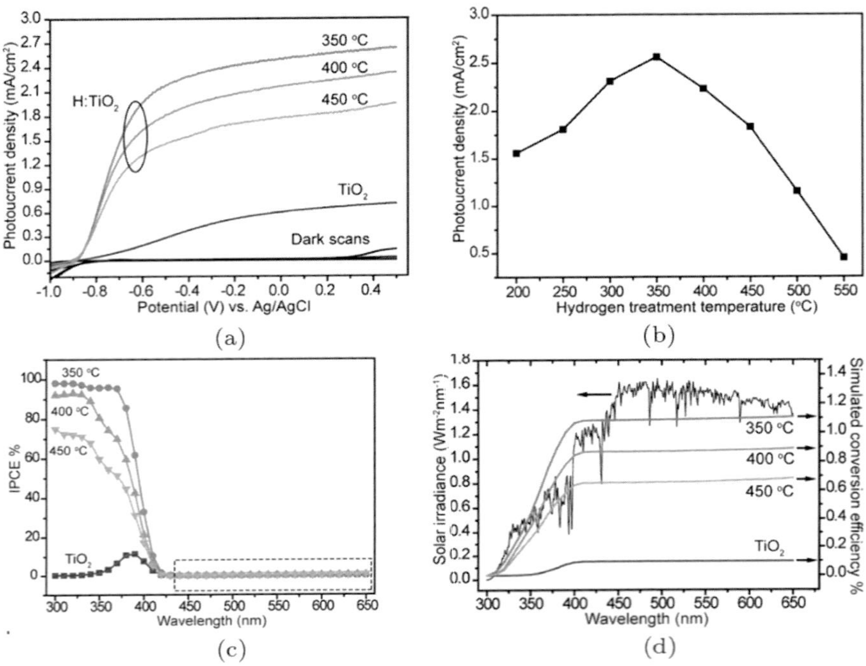

Figure 8.2. (a) Linear sweep voltammogram curves of TiO₂, H:TiO₂ at 350°C, 400°C, and 450°C under AM 1.5G 100 mW/cm² light irradiation. (b) Temperature dependent photocurrent density collected at 0.23 V vs. Ag/AgCl. (c) IPCE spectra of TiO₂, H:TiO₂ at 350°C, 400°C, and 450°C at −0.6 V vs. Ag/AgCl. (d) Simulated STH conversion efficiency for the pristine TiO₂ and H:TiO₂ nanowires as a function of wavelength by integrating their IPCE spectra with a standard AM 1.5G solar spectrum. Reproduced from Ref. [11]. © American Chemical Society.

range of 450–650 nm is minimal compared to the values in UV region. This result indicates the improved photoactivity for hydrogen-treated TiO₂ is due to improved efficient of charge collection, instead of increased visible light photoactivity. Figure 8.2(d) shows the simulated solar-to-hydrogen (STH) conversion efficiency as a function of wavelength in the range from 300 nm to 650 nm by integrating IPCE spectra with standard AM 1.5G solar spectrum. The pristine white TiO₂, and hydrogen-treated TiO₂ nanowires prepared at 350°C, 400°C, and 450°C achieve solar to hydrogen conversion efficiencies of ~0.1%, 1.1%, 0.89%, and 0.68% at −0.6V vs. Ag/AgCl.[11] From the simulated curve, the STH conversion efficiency level off at wavelength beyond 420 nm, again confirming the visible light contribution to the photoactivity is tiny.

Wang *et al.* also demonstrated that the same hydrogen treatment approach can be applied to improve the photoelectrochemical performance of anatase TiO$_2$ nanotube arrays.[11] The hydrogen-treated anatase TiO$_2$ nanotube electrode showed at least two-fold enhancement in photocurrent.[11] Hydrogen treatment at higher temperature may damage the FTO glass substrate and increase its resistance. To eliminate the substrate effect, rutile TiO$_2$ nanowire, anatase TiO$_2$ nanowire, and mixed-rutile/anatase phase TiO$_2$ nanowire were prepared on Ti metal foil using different solution methods. Ti foil treated under various hydrogenation temperatures can maintain equally good conductivity, which offers a platform to systematically study the effects of crystal phase and hydrogen treatment temperature on the photoactivity of TiO$_2$.[47] The optimal performance of rutile and anatase hydrogenated TiO$_2$ photoanodes was achieved at the hydrogenation temperature of 350°C and 550°C, respectively.[47] The difference in optimal hydrogenation temperature was believed to be related to the degree of enhancement in carrier density of different TiO$_2$ phases upon hydrogenation. This work provides further insights into the effect of hydrogenation temperature on the performance of TiO$_2$ materials for PEC water oxidation. Other types of hydrogen-treated TiO$_2$ nanostructures have been demonstrated and shown enhanced performance for dye degradation and solar hydrogen generation.[31,33,37,48]

8.2.2 *Hydrogen-Treated TiO$_2$ Shows Improved Efficiency of Charge Separation*

There is an increasing attention for using hydrogen-treated TiO$_2$ photocatalyst for various applications. It is important to understand the underlying mechanism for the enhanced performance of hydrogen-treated TiO$_2$. Time-resolved spectroscopy techniques have been employed to probe the charge carrier dynamics of hydrogen-treated TiO$_2$ and pristine TiO$_2$. Wheeler *et al.* employed ultrafast laser absorption spectroscopy and time-resolved fluorescence spectroscopy

(TRFS) to study hydrogen-treated TiO$_2$,[49] aiming to probe the dynamics of the charge carrier in the V$_O$ related states and understand the constitution of the mid-band states and their relation with the improved PEC performance. In the ultrafast transient absorption (TA) measurements, a short, usually femtosecond duration, optical pulse (the "pump") excites the sample and, after a temporal delay, a second short pulse (the "probe" pulse) is used to interrogate the population of excited charge carriers. The detected signal of the probe pulse as a function of time profiles provide information of the dynamics of the excited carriers. Ultrafast TA were collected for TiO$_2$ and hydrogen-treated TiO$_2$ using UV pump (380 nm, 475 nm probe) and visible pump (450 nm, 505 nm probe) wavelength.[49] Under UV excitation, pristine TiO$_2$ exhibits a fast initial decay followed by a medium decay and a long-lived slow decay (Figure 8.3(a)). The decay profile was fitted to a tripe exponential with time constant of 30 ps, 100 ps, and >1 ns.[49] On the contrary, the UV-pumped hydrogen-treated TiO$_2$ displays a similar fast initial decay but with slower-medium decay than the pristine TiO$_2$, indicating the existence of long-lived electron–hole pairs. UV-pumped hydrogen-treated TiO$_2$ was also fitted well to a trip exponential with time constants of 60 ps, 200 ps, and >1 ns.[49] Under visible wavelength

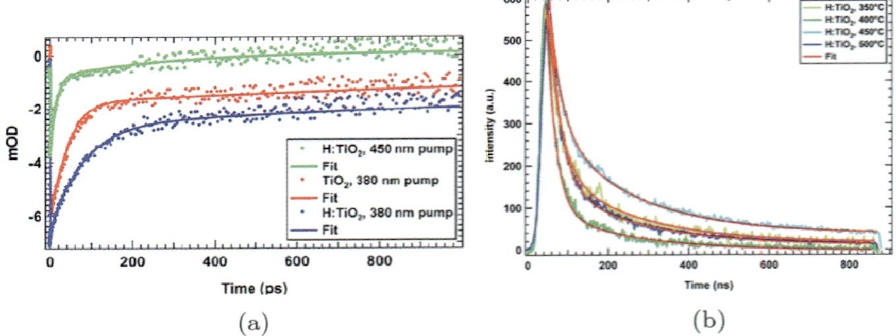

Figure 8.3. (a) Time-resolved fluorescence (TRF) Intensity vs. time of H:TiO$_2$ NWs prepared at different hydrogen annealing temperatures, along with fitting curves. (b) Transient bleach relaxation traces of TiO$_2$ and H:TiO$_2$ nanowires following 450 nm and 380 nm excitation. Reproduced from Ref. [49]. © American Chemical Society.

excitation, there was no noticeable signal on pristine TiO$_2$, because the excitation wavelength is beyond the pristine TiO$_2$ bandgap absorption. However, the visible light pumped (450 nm) H–TiO$_2$ can be detected and it exhibits a fast initial decay followed by a nearly complete relaxation to baseline.[49] The fast decay indicated very efficient electron–hole recombination under visible light irradiation, which explains the reason why visible light photoactivity of hydrogen-treated TiO$_2$ is minimal.

TRFS is a TRT which is complementary of TA spectroscopy. TRFS is based on the measurements of lifetime of photoluminescence. Figure 8.3(b) shows the TRF of hydrogen-treated TiO$_2$ prepared at 350°C, 400°C, 450°C, and 500°C.[49] The decay profile of TiO$_2$ sample was fitted to time constants of 35 ps, 120 ps, and >1 ns regardless of pump fluence.[49] The H–TiO$_2$ samples displayed a faster decay for the fast component with increasing annealing temperature. The middle component fit well to a 200 ps lifetime, and the long component had a lifetime of greater than 1 ns for all samples under 0.4 μJ/cm^2 and 1.3 μJ/cm^2 fluence runs. The global fitting reveals that the relative intensity of the fast component of the decays increased from 68% to 77% as the annealing temperature increases from 350°C to 550°C.[49] The faster decay of the fast component for hydrogen-treated and untreated samples suggests that hydrogen treatment fundamentally alters the quantity and energetic position or depth of trap states within the bandgap since a higher density of states within the bandgap is expected to quench PL and reduce the lifetime observed. The fact that the medium component becomes slower from TiO$_2$ (120 ps) to H–TiO$_2$ (200 ps) indicates that the PL decay process is slowed on this timescale due to some longer-lived trap states after hydrogen treatment, likely deeper trap states.

Based on the ultrafast TA spectroscopy and time resolved fluorescence spectroscopy, Wheeler *et al.* proposed a model to understand the energy levels in hydrogen-treated and untreated TiO$_2$ nanowires, as shown in Figure 8.4.[49] Based on the PL peak position, there is one fluorescent trap state estimated to be ~0.3 eV below the CB (the bandgap of TiO$_2$ is 3.0 eV). The V$_O$ state was estimated to lie ~0.75 eV below the CB. The slower decay of the TA in the H–TiO$_2$

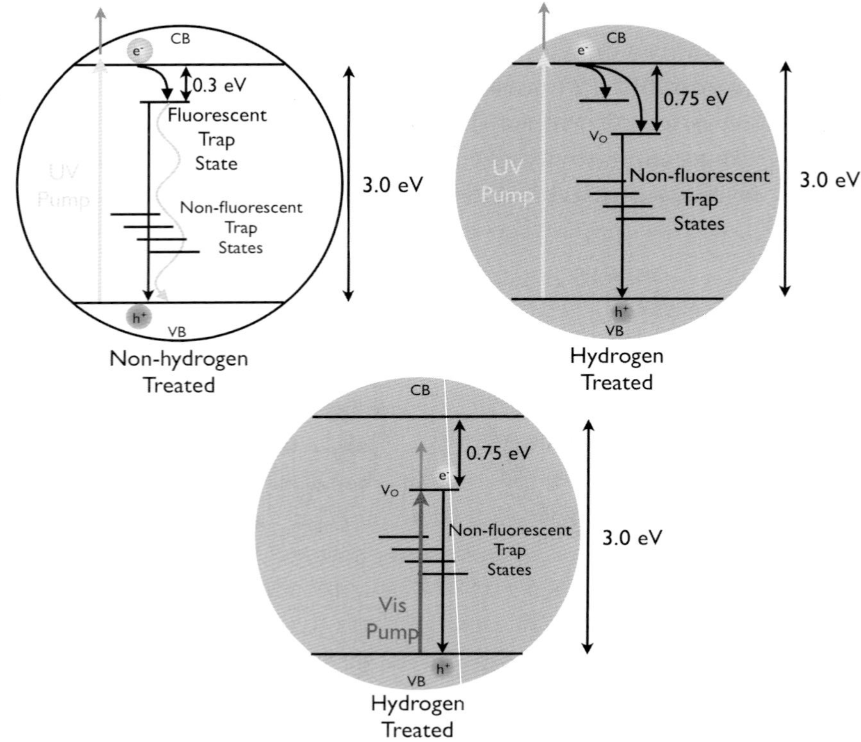

Figure 8.4. Proposed model for energy levels related to the optical properties and dynamic studies. Reproduced from Ref. [49]. © American Chemical Society.

sample with UV excitation was attributed that V_O state induced by hydrogen treatment lengthened the lifetime of the charge carriers. For visible-pumped H–TiO$_2$, the visible excitation takes the electron from the VB to the V_O.[49] The overall fast decay of the charge carriers with visible excitation is likely due to a dense manifold of states under the V_O. The lack of a long-lived component is consistent with negligible visible light photoactivity.

TA spectroscopy can also be used for *in situ* measurement of the variation of photoexcited carrier concentration as a function of time in a PEC cell under different applied bias. The *in situ* measurement can provide important insights into the bias-dependent rates of charge carrier recombination, transport and transfer. Pesci *et al.*

reported *in situ* TA measurement in a complete PEC cell.[36] In their bias dependent TA studies, they found that the photoelectron yield increased with the increase of applied bias, for both pristine TiO$_2$ and hydrogen-treated TiO$_2$. However, the yield of photogenerated charge carrier in hydrogen-treated TiO$_2$ is more sensitive to the applied bias than pristine TiO$_2$, which is a significant factor behind the enhanced activity in the hydrogen-treated TiO$_2$. Importantly, the decoupling of the electron and hole kinetics show non-identical decay traces on the microseconds–seconds timescales, indicating the processes apart from direct electron–hole recombination are able to occur. Figure 8.4 shows the TA decay of photoelectron and photohole measured at -0.6 V vs. Ag/AgCl. The photoelectrons in hydrogen-treated TiO$_2$ at -0.6 V decays by more than 50% between $10\,\mu$s and 1 ms, whereas the photohole concentration remains almost unchanged (Figure 8.5).[36] The hole signal of hydrogen-treated TiO$_2$ was fitted to a single stretched exponential function with a lifetime of 0.15 ± 0.03 s, which is a much slower decay than pristine TiO$_2$. This indicates electron–hole recombination is effectively blocked on hydrogen-treated TiO$_2$ on the microseconds–seconds time scales at -0.6 V vs. Ag/AgCl.[36] However, the average rate of hole transfer into solution on hydrogen-treated TiO$_2$ is not sufficiently different from that of pristine TiO$_2$. Therefore, the suppression of recombination is believed to be the key factor for the improved photocatalytic activity of the hydrogen-treated TiO$_2$.[36]

8.2.3 *Synergistic Effect Between Hydrogen Treatment and Element Doping*

Hydrogen treatment has been demonstrated to be a general strategy for improving the separation efficiency of photoexcited charge carriers by increasing the carrier density of TiO$_2$. Importantly, this method can couple with other chemical modification strategies such as element doping to improve the PEC performance of TiO$_2$ in a synergistic way. Incorporation of non-metal ions such as C[15, 20, 50] and N,[17, 22, 24] into TiO$_2$ to achieve visible light photoactivity have been extensively studied. Typically, the p state of the non-metal foreign

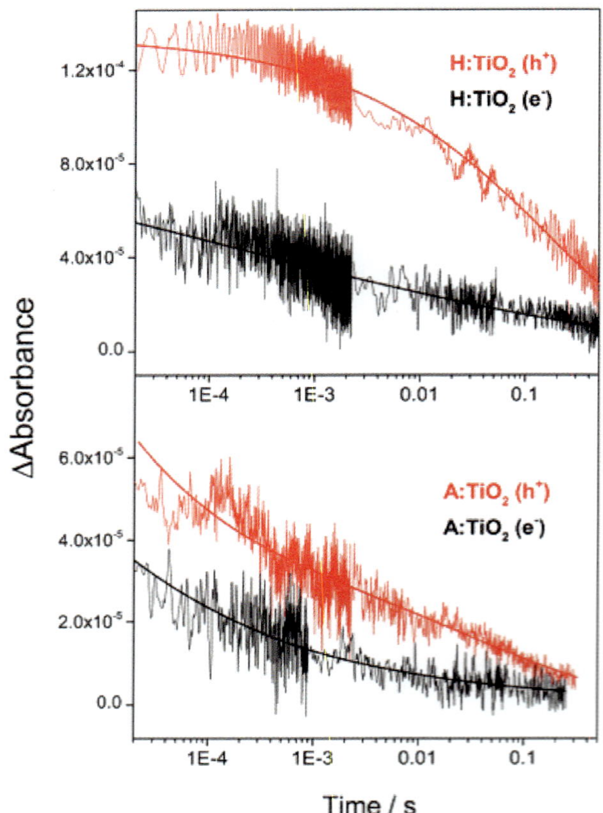

Figure 8.5. TA decay traces of photoholes ($\lambda = 500\,\text{nm}$, red curve) and electrons ($\lambda = 800\,\text{nm}$, black curve) collected after excitation of hydrogen-treated (H:TiO$_2$, upper) and pristine (A:TiO$_2$, lower) in 1 M NaOH at $-0.6\,\text{V}$ vs. Ag/AgCl. The excitation wavelength is 355 nm with power density of $70\,\mu\text{J/cm}^2$ of 0.33 Hz laser repetition rate. Reproduced from Ref. [36]. © American Chemical Society.

elements normally form impurity states above the valence band or hybridize with O 2p states.[17] Among these dopants, N-doping is the most widely studied and has achieved some success toward visible light photoactivity. Hoang *et al.* combined hydrogen treatment and nitrogen doping strategies to prepare H, N–TiO$_2$ nanowire arrays and demonstrated a synergistic effect between V$_{\text{Os}}$ and nitrogen dopant to enhance the visible light PEC water-splitting performance.[17] Figure 8.6(a) shows the linear sweep voltammetry

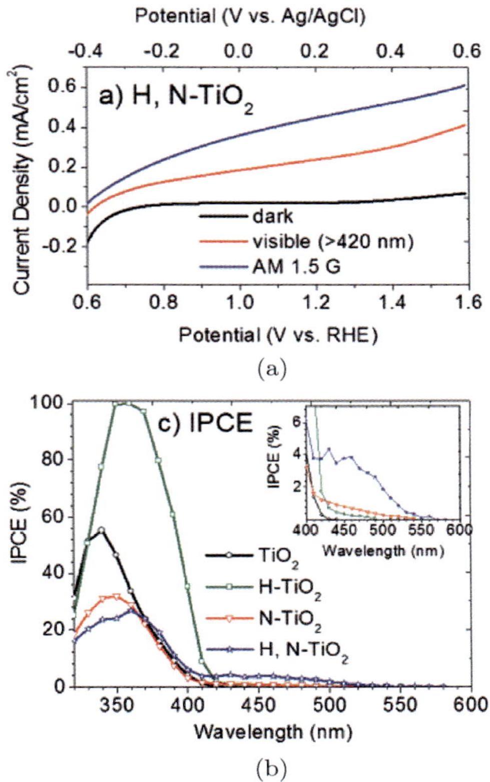

Figure 8.6. (a) Linear sweep voltammograms of H, N–TiO$_2$ sample at a scan rate of 5 mV/s. (b) IPCE spectra measured at 1.23 V vs. RHE. The inset is the magnified curve at 400–600 nm. Reproduced from Ref. [17]. © American Chemical Society.

results of H, N–TiO$_2$ nanowire at a scan rate of 5 mV/s under white and visible light irradiation.[17] The H, N–TiO$_2$ exhibits a remarkable visible light water oxidation performance. The photocurrent density measured at 1.23 V vs. RHE was ∼0.16 mA/cm^2 under visible light illumination.[17] Figure 8.6(b) shows the IPCE curves of pristine TiO$_2$, hydrogen-treated TiO$_2$ (H–TiO$_2$), nitrogen-doped TiO$_2$ (N–TiO$_2$), and nitrogen doped and hydrogen-treated TiO$_2$ (H, N–TiO$_2$) nanowire arrays. H–TiO$_2$ exhibits significant enhanced IPCE values in UV region, but negligible photoactivity in visible light region. N–TiO$_2$ shows minimal IPCE values in the visible region.

In contrast, the H, N–TiO$_2$ sample shows considerably enhanced photoactivity visible light region.[17] The interaction between V$_{Os}$ and Ti^{3+} is believed to be the key to the extension of the active region and superior visible light driven PEC water oxidation performance in H, N–TiO$_2$ nanowires.

Yang *et al.* also observed that sulfur-doped black TiO$_2$ has both remarkable UV and visible light PEC performance for water oxidation.[50] The black TiO$_2$ was prepared by molten Al reduction, not hydrogen treatment. Sulfur doping was achieved by annealing in H$_2$S gas. The IPCE of sulfur-doped black TiO$_2$ at the wavelength of 450 nm can reach $\sim$5%, which is similar to the IPCE values obtained from H, N–TiO$_2$ photoelectrode.[50] These successful examples demonstrated that the coupling between hydrogen treatment and other conventional chemical modification strategies could open up new opportunities for further improving the performance of TiO$_2$-based photoelectrodes for PEC water splitting.

8.3   Hydrogen-Treated TiO$_2$ Nanowire Arrays for Supercapacitor Applications

8.3.1   *Hydrogen-Treated TiO$_2$ as Supercapacitor Electrode*

Introducing V$_{Os}$ (or Ti^{3+} sites) into TiO$_2$ lattice has been demonstrated to be a promising strategy to improve its photoelectrochemical performance. In fact, the same approach can also be used to change the electronic property and surface functionality of TiO$_2$, and thus, their electrochemical properties. One-dimensional (1D) TiO$_2$ nanowire/tube arrays are believed to be promising electrodes materials for supercapacitors (SCs), owing to their ordered structure, large ion accessible surface area, and excellent mechanical and electrochemical stability.[51,52] Nevertheless, the reported specific capacitances for 1D TiO$_2$ nanoarrays are considerably lower than other metal oxides like V$_2$O$_5$,[53–55] MnO$_2$,[28,56,57] and RuO$_2$.[58] The poor capacitive behavior of TiO$_2$ was believed to be due to the low electrochemical activity and poor electrical conductivity of TiO$_2$. Controlled introduction of V$_{Os}$ (or Ti^{3+} sites) into the TiO$_2$

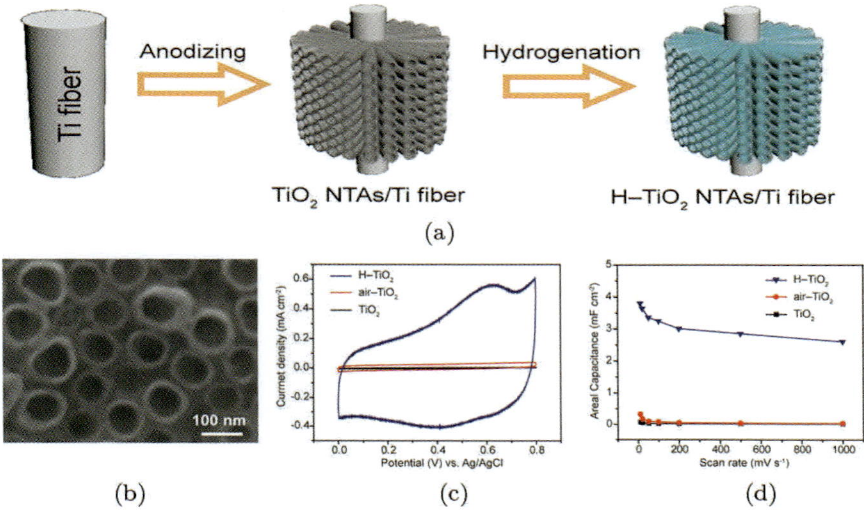

Figure 8.7. (a) A schematic diagram showing the fabrication of H–TiO$_2$ NTAs. (b) Scanning electron microscope (SEM) image of H–TiO$_2$ NTAs. (c) CV curves of the untreated TiO$_2$, air-TiO$_2$, and H–TiO$_2$ NTAs obtained at a scan rate of 100 mV s^{-1}. (d) Areal capacitance of TiO$_2$ samples measured as a function of scan rate. Reproduced from Ref. [27]. © American Chemical Society.

was found to be an effective approach to improve the electrical conductivity and capacitive properties of TiO$_2$. In 2012, Lu *et al.* demonstrated for the first time the feasibility of using hydrogenated TiO$_2$ as SC electrode.[27] Figure 8.7(a) shows the fabrication process of hydrogenated TiO$_2$ nanotube arrays (H–TiO$_2$ NTAs).[27] Highly ordered TiO$_2$ nanotube arrays (TiO$_2$ NTAs) were fabricated on Ti fibers with diameters of 1 mm via an anodizing method in glycerol aqueous solution containing 0.75% NH$_4$F. Then the obtained TiO$_2$ NTAs were directly annealed under hydrogen atmosphere with a temperature of 400°C. The H–TiO$_2$ NTAs are vertically aligned on the Ti fiber with a uniform diameter of ∼100 nm (Figure 8.7(b)).[27]

The cyclic voltammetric (CV) curves of TiO$_2$, air–TiO$_2$, H–TiO$_2$ electrode collected at 100 mV s^{-1} in 0.5 M Na$_2$SO$_4$ aqueous solution.[27] As shown in Figure 8.7(c), H–TiO$_2$ electrode exhibits a considerably larger area when compared to the other two electrodes, implying that H–TiO$_2$ has substantially enhanced capacitance. The areal capacitance of H–TiO$_2$ was determined to be $3.24\,\mathrm{mF\,cm^{-2}}$, which is 124- and 40-fold higher than the values obtained from TiO$_2$ and air–TiO$_2$ electrodes (Figure 8.7(d)). Meanwhile, the redox peaks observed in CV curve suggest the H–TiO$_2$ electrode has pseudo-capacitive behavior.[27] Pseudocapacitance was believed to originate from surface hydroxyl groups introduced during the hydrogenation process. Moreover, H–TiO$_2$ electrode also exhibited excellent cycling stability with almost no decay after 10000 charge/discharge cycles at $100\,\mathrm{mV\,s^{-1}}$.

Hydrogen plasma treatment method has been demonstrated to be another effective method to produce V$_{Os}$ in TiO$_2$ lattice.[41,59] Plasma technique is well known as a powerful method for surface modification (e.g. cleaning, etching, doping). This method can not only reduce the reaction temperature and duration, but also realize the microscale treatment of samples. For instance, Li and cowork-ers systematically investigated the plasma-assisted hydrogenated anodic titanium oxide nanotubes (ATO-H) for SCs application.[59] Figure 8.8(a) displayed the SEM images of TiO$_2$ NTAs prepared on Ti foil after hydrogen plasma treatment. Unlike traditional hydrogen gas treatment, the side-facets of nanotubes treated with hydrogen plasma are very rough due to the plasma etching effect.[59] TEM image of hydrogenated TiO$_2$ nanotubes (Figure 8.8(b)) reveals a core–shell structure, in which an amorphous shell of 1–2 nm thickness is uniformly coated on TiO$_2$ crystalline core.[59] The electrochemical performance of these TiO$_2$ NTAs was investigated through a symmetric two-electrode configuration. In comparison to ATO electrode, ATO-H electrode has substantially longer discharge time, indicating it has larger specific capacitance (Figure 8.8(c)). Furthermore, CV curves of ATO-H electrode retain *quasi*-rectangular shape even at a high scan rate of $10\,\mathrm{V\,s^{-1}}$, demonstrating its excellent rate capability (Figure 8.8(d)). Nyquist plots collected for ATO-H electrode reveals that the charge transfer resistance of the electrode is small.[59] The superior electrochemical performance of ATO-H electrode was ascribed to the highly rough surface and the large amounts of oxygen-Ti^{3+} vacancies serving as active sites in the amorphous layer.

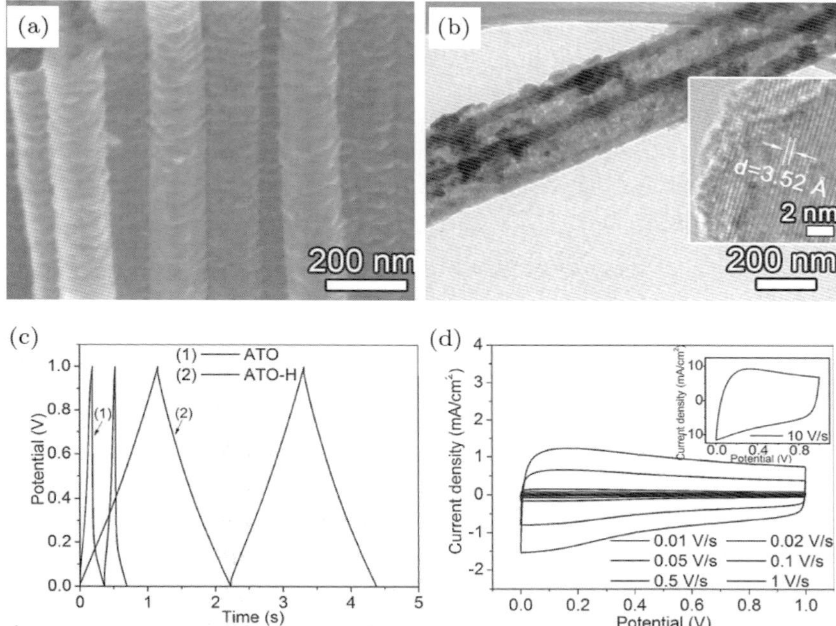

Figure 8.8. FESEM images of highly oriented (a) ATO-H films. (b) TEM images of ATO-H nanotubes. (c) Galvanostatic charge/discharge curves at the current density of $1\,\text{mA}\,\text{cm}^{-2}$ for ATO and ATO-H SCs with the voltage window of $1\,\text{V}$, respectively. (d) CV curves of the ATO-H SC recorded at the scan rate from 0.01 to $10\,\text{V}\,\text{s}^{-1}$, respectively. Reproduced from Ref. [59]. © Wiley-VCH Verlag GmbH & Co. KGaA, Weinheim.

Other strategies have also been employed to improve the capacitive performance of 1D TiO_2 electrodes, such as annealing in oxygen-deficient environment,[51,52] and electrochemical reduction method.[60–63] For example, Salari *et al.* reported the controlled phase transformation of TiO_2 by thermal treatment under an argon atmosphere.[52] The optimized TiO_2 electrode delivered an excellent areal capacitance of $2.6\,\text{mF}\,\text{cm}^{-2}$ at $1\,\text{mV s}^{-1}$. Zhou *et al.* used a simple cathodic polarization treatment on TiO_2 NTAs to improve their conductivity and capacitance.[60,61] The electrochemically treated TiO_2 NTA exhibited five orders of magnitude improved carrier density and 39 times enhancement in capacitance compared to the values obtained from pristine TiO_2 NTAs.[60,61] While different methods have been used to improve the electrochemical performance of TiO$_2$, a common strategy used in all these methods were to improve the electrical conductivity of TiO$_2$ through controlled creation of V$_{Os}$ (or Ti^{3+} sites).

8.3.2   *Hydrogen-Treated TiO$_2$ Nanowire Arrays as Supports for Hybrid SC Electrodes*

Although the electrical conductivity and specific capacitance of TiO$_2$ can be greatly improved through hydrogenation, its specific capacitance is still relatively low compared to pseudocapacitive materials like RuO$_2$, MnO$_2$, polyaniline, polypyrrole, etc. In this context, using hydrogenated TiO$_2$ nanoarchitectures as support to construct core–shell or composited electrodes for high-performance SCs has attracted some attention.[28,64,65] For instance, hydrogenated TiO$_2$ 1D support can offer a large surface area and efficient electron transport pathway for other pseudocapacitive materials. Lu *et al.* prepared hydrogen-treated TiO$_2$ nanowires (H–TiO$_2$ NWs) on a carbon cloth through hydrothermal method, followed by hydrogenation.[28] H–TiO$_2$ NWs were used as support to fabricate H–TiO$_2$–MnO$_2$ and H–TiO$_2$–C core–shell electrodes. As shown in Figures 8.9(a) and (b), both MnO$_2$ and C were uniformly coated on the surface of H–TiO$_2$ NWs with average thickness of 7.0 nm and 18 nm, respectively.[28] CV curves collected at 100 mV s^{-1} revealed that the H–TiO$_2$–MnO$_2$ electrode exhibited substantially higher areal capacitance than those of H–TiO$_2$, MnO$_2$, and TiO$_2$–MnO$_2$ electrodes, which benefits from the highly electrochemical activity of MnO$_2$, increased surface area, and improved charge transport. The H–TiO$_2$–MnO$_2$ achieved an excellent specific capacitance of 449.6 F g^{-1}, which is considerably higher than the values obtained from MnO$_2$ (325.8 F g^{-1}) and TiO$_2$–MnO$_2$ (359.7 F g^{-1}) electrodes (Figure 8.9(c)). Likewise, H–TiO$_2$–C exhibited better capacitive performance compared to H–TiO$_2$, C, and TiO$_2$–C electrodes (Figure 8.9(d)).[28] H–TiO$_2$–MnO$_2$ and TiO$_2$–C were used respectively as positive and negative electrode to assemble a *quasi*-solid-state flexible asymmetric SCs (ASCs). Significantly, the stable operation window of the H–TiO$_2$–MnO$_2$//H–TiO$_2$–C-ASC device was extended to 1.8 V (Figure 8.9(e)). The

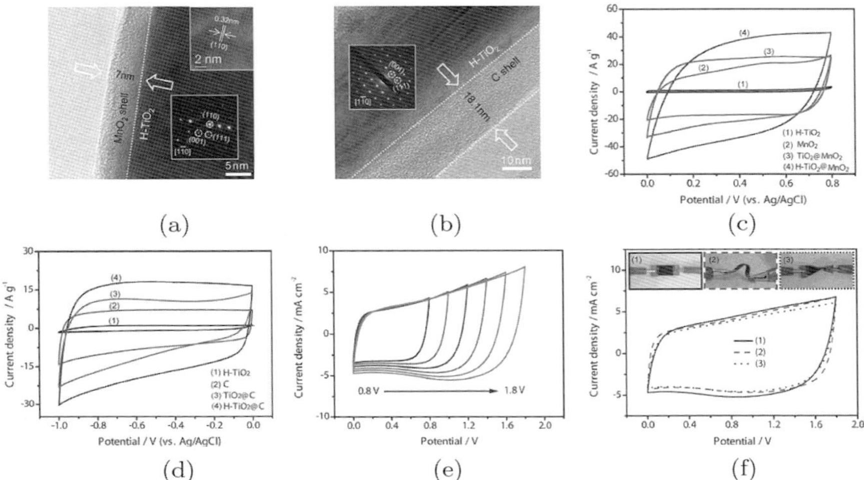

Figure 8.9. TEM image of (a) H–TiO$_2$–MnO$_2$ and (b) H–TiO$_2$–C NWs. (c) CV curves collected for H–TiO$_2$, MnO$_2$, TiO$_2$–MnO$_2$, and H–TiO$_2$–MnO$_2$ electrodes at the scan rate of 100 mV s^{-1}. (d) CV curves collected for H–TiO$_2$, C, TiO$_2$–C, and H–TiO$_2$–C electrodes at the scan rate of 100 mV s^{-1}. (e) CV curves of the solid-state ASC device collected in different scan voltage windows. (f) CV curves collected at a scan rate of 100 mV s^{-1} for the ASC device under flat, bent, and twisted conditions. Insets are the device pictures under test conditions. Reproduced from Ref. [28]. © Wiley-VCH Verlag GmbH & Co. KGaA, Weinheim.

ASC device showed excellent mechanical stability with almost no change of capacitance under flat, bent and twisted conditions (Figure 8.9(f)).[28]

Besides metal oxides, conducting polymers are another promising class of pseudocapacitive electrode materials for SCs due to their large specific capacitance, flexibility and low cost. The relatively low electrical conductivity is one of the major limitations for polymer-based electrodes. 1D H–TiO$_2$ nanoarrays has also been used to support conducting polymers for developing high-performance SC electrodes. Xia and coworkers deposited a layer of PANI on H–TiO$_2$ nanotubes (NTs) via a facile electrochemical method.[33] The disordered PANI nanowires tightly grafted on the surface of H–TiO$_2$ NTs. PANI/H–TiO$_2$ NTs exhibited better capacitive behavior in comparison with air–TiO$_2$, H–TiO$_2$, and PANI/air–TiO$_2$.

The negligible IR drop (1.9 mV) observed in the charge-discharge curve at 1.2 A g^{-1} again confirms the important role of H–TiO$_2$ for improving the conductivity of electrode.[33] These results have clearly demonstrated that hydrogenation is an efficient method to boost the capacitive performance of 1D nanostructured TiO$_2$. The H–TiO$_2$ nanomaterials with significantly improved electrical conductivity can also be employed as support to other pseudocapacitive materials for constructing high performance hybrid electrodes for electrochemical capacitors.

8.4 Conclusion and Perspective

In this chapter, we have summarized the recent advancements of hydrogen-treated TiO$_2$ nanostructures for photoelectrochemical water splitting and electrochemical supercapacitor applications. The outstanding photoelectrochemical and electrochemical performance of hydrogen-treated TiO$_2$ has been attributed to the existences of V$_{Os}$ and/or surface layer disorder of TiO$_2$. Based on the understanding of the role of V$_{Os}$ and surface layer disorder, a number of different synthetic/treatment methods have been developed to obtain high-performance TiO$_2$.[5,66] For example, Wang *et al.* found that V$_{Os}$ can also be created by annealing TiO$_2$ in oxygen-deficient condition without hydrogen gas, the prepared oxygen-deficient TiO$_2$ can achieve similar performance as hydrogen-treated TiO$_2$.[32] Reduction methods also have been developed to prepare reduced TiO$_2$, such as electrochemical reduction,[60,61] flame reduction,[67] and chemical reduction using reducing agent such as NaBH$_4$.[68,69] These synthetic strategies developed for TiO$_2$ have also been expanded to other metal oxides such as WO$_3$,[70] BiVO$_4$,[29] ZnO,[30,71] and Fe$_2$O$_3$.[72,73] Although significant processes have been achieved in the past few years, there still are outstanding problems needed to be addressed. For instance, it is challenging to increase the capability of hydrogen-treated TiO$_2$ for utilization of solar light in visible and infrared region. The specific capacitance of hydrogen-treated TiO$_2$ is still relatively low, as there is a lack of pseudocapacitive properties.

To increase the pseudocapacitance of hydrogen-treated TiO$_2$ via doping should be an interesting research direction in future. It is also important to improve the quantitative understanding on the effect of hydrogenation on the material properties of TiO$_2$. For example, the effect of creating V$_{Os}$ on the charge carrier mobility, the rate of electron–hole pair recombination, the density of surface states and how it affects the charge transfer at TiO$_2$/electrolyte interface. Additionally, it is critical to understand the interplay between these effects and the concentration as well as the distribution of V$_{Os}$ in TiO$_2$ materials. These fundamental understanding of the relation between V$_{Os}$ and TiO$_2$ material properties should open up a lot of exciting opportunities for applications.

References

(1) Schneider, J.; Matsuoka, M.; Takeuchi, M.; Zhang, J. L.; Horiuchi, Y.; Anpo, M.; Bahnemann, D. W. *Chem. Rev.* **2014**, *114*(19), 9919–9986.

(2) Abe, R. *J. Photochem. Photobiol. C-Photochem. Rev.* **2010**, *11*(4), 179–209.

(3) Chen, X. B.; Liu, L.; Huang, F. Q. *Chem. Soc. Rev.* **2015**, *44*(7), 1861–1885.

(4) Chen, X.; Mao, S. S. *Chem. Rev.* **2007**, *107*(7), 2891–2959.

(5) Wang, G. M.; Ling, Y. C.; Li, Y. *Nanoscale* **2012**, *4*(21), 6682–6691.

(6) Wang, G. M.; Ling, Y. C.; Wang, H. Y.; Lu, X. H.; Li, Y. *J. Photochem. Photobiol. C-Photochem. Rev.* **2014**, *19*, 35–51.

(7) Asahi, R.; Morikawa, T.; Ohwaki, T.; Aoki, K.; Taga, Y. *Science* **2001**, *293*(5528), 269–271.

(8) Fujishima, A.; Honda, K. *Nature* **1972**, *238*(5358), 37–38.

(9) Gratzel, M. *Nature* **2001**, *414*(6861), 338–344.

(10) Linsebigler, A. L.; Lu, G. Q.; Yates, J. T. *Chem. Rev.* **1995**, *95*(3), 735–758.

(11) Wang, G. M.; Wang, H. Y.; Ling, Y. C.; Tang, Y. C.; Yang, X. Y.; Fitzmorris, R. C.; Wang, C. C.; Zhang, J. Z.; Li, Y. *Nano Lett.* **2011**, *11*(7), 3026–3033.

(12) Chen, X. B.; Liu, L.; Yu, P. Y.; Mao, S. S. *Science* **2011**, *331*(6018), 746–750.

(13) Li, Y.; Zhang, J. Z. *Laser Photon. Rev.* **2010**, *4*(4), 517–528.

(14) Bard, A. J.; Fox, M. A. *Accounts Chem. Res.* **1995**, *28*(3), 141–145.

(15) Khan, S. U. M.; Al-Shahry, M.; Ingler, W. B. *Science* **2002**, *297*(5590), 2243–2245.

(16) Hwang, Y. J.; Hahn, C.; Liu, B.; Yang, P. D. *ACS Nano* **2012**, *6*(6), 5060–5069.

(17) Hoang, S.; Berglund, S. P.; Hahn, N. T.; Bard, A. J.; Mullins, C. B. *J. Am. Chem. Soc.* **2012**, *134*(8), 3659–3662.

(18) Zhang, Z. H.; Zhang, L. B.; Hedhili, M. N.; Zhang, H. N.; Wang, P. *Nano Lett.* **2013**, *13*(1), 14–20.

(19) Pu, Y. C.; Wang, G. M.; Chang, K. D.; Ling, Y. C.; Lin, Y. K.; Fitzmorris, B. C.; Liu, C. M.; Lu, X. H.; Tong, Y. X.; Zhang, J. Z.; Hsu, Y. J.; Li, Y. *Nano Lett.* **2013**, *13*(8), 3817–3823.

(20) Park, J. H.; Kim, S.; Bard, A. J. *Nano Lett.* **2006**, *6*(1), 24–28.

(21) Asahi, R.; Morikawa, T.; Irie, H.; Ohwaki, T. *Chem. Rev.* **2014**, *114*(19), 9824–9852.

(22) Wang, G. M.; Xiao, X. H.; Li, W. Q.; Lin, Z. Y.; Zhao, Z. P.; Chen, C.; Wang, C.; Li, Y. J.; Huang, X. Q.; Miao, L.; Jiang, C. Z.; Huang, Y.; Duan, X. F. *Nano Lett.* **2015**, *15*(7), 4692–4698.

(23) Cong, Y.; Zhang, J. L.; Chen, F.; Anpo, M. *J. Phys. Chem. C* **2007**, *111*(19), 6976–6982.

(24) Hoang, S.; Guo, S. W.; Hahn, N. T.; Bard, A. J.; Mullins, C. B. *Nano Lett.* **2012**, *12*(1), 26–32.

(25) Naldoni, A.; Allieta, M.; Santangelo, S.; Marelli, M.; Fabbri, F.; Cappelli, S.; Bianchi, C. L.; Psaro, R.; Dal Santo, V. *J. Am. Chem. Soc.* **2012**, *134*(18), 7600–7603.

(26) Liang, Z.; Zheng, G. Y.; Li, W. Y.; Seh, Z. W.; Yao, H. B.; Yan, K.; Kong, D. S.; Cui, Y. *ACS Nano* **2014**, *8*(5), 5249–5256.

(27) Lu, X. H.; Wang, G. M.; Zhai, T.; Yu, M. H.; Gan, J. Y.; Tong, Y. X.; Li, Y. *Nano Lett.* **2012**, *12*(3), 1690–1696.

(28) Lu, X. H.; Yu, M. H.; Wang, G. M.; Zhai, T.; Xie, S. L.; Ling, Y. C.; Tong, Y. X.; Li, Y. *Adv. Mater.* **2013**, *25*(2), 267–272.

(29) Wang, G. M.; Ling, Y. X.; Lu, X. H.; Qian, F.; Tong, Y. X.; Zhang, J. Z.; Lordi, V.; Leao, C. R.; Li, Y. *J. Phys. Chem. C* **2013**, *117*(21), 10957–10964.

(30) Lu, X. H.; Wang, G. M.; Xie, S. L.; Shi, J. Y.; Li, W.; Tong, Y. X.; Li, Y. *Chem. Commun.* **2012**, *48*(62), 7717–7719.

(31) Zheng, Z. K.; Huang, B. B.; Lu, J. B.; Wang, Z. Y.; Qin, X. Y.; Zhang, X. Y.; Dai, Y.; Whangbo, M. H. *Chem. Commun.* **2012**, *48*(46), 5733–5735.

(32) Wang, H. Y.; Wang, G. M.; Ling, Y. C.; Lepert, M.; Wang, C. C.; Zhang, J. Z.; Li, Y. *Nanoscale* **2012**, *4*(5), 1463–1466.

(33) Xia, T.; Chen, X. B. *J. Mater. Chem. A* **2013**, *1*(9), 2983–2989.

(34) Zhou, W.; Li, W.; Wang, J. Q.; Qu, Y.; Yang, Y.; Xie, Y.; Zhang, K. F.; Wang, L.; Fu, H. G.; Zhao, D. Y. *J. Am. Chem. Soc.* **2014**, *136*(26), 9280–9283.

(35) Liu, N.; Schneider, C.; Freitag, D.; Hartmann, M.; Venkatesan, U.; Muller, J.; Spiecker, E.; Schmuki, P. *Nano Lett.* **2014**, *14*(6), 3309–3313.

(36) Pesci, F. M.; Wang, G. M.; Klug, D. R.; Li, Y.; Cowan, A. J. *J. Phys. Chem. C* **2013**, *117*(48), 25837–25844.

(37) Lin, T. Q.; Yang, C. Y.; Wang, Z.; Yin, H.; Lu, X. J.; Huang, F. Q.; Lin, J. H.; Xie, X. M.; Jiang, M. H. *Energy Environ. Sci.* **2014**, *7*(3), 967–972.

(38) Lu, Z. G.; Yip, C. T.; Wang, L. P.; Huang, H. T.; Zhou, L. M. *ChemPlusChem* **2012**, *77*(11), 991–1000.

(39) Liu, Y.; Liu, C. J.; Li, J. L. *J. Mater. Chem. A* **2014**, *2*(38), 15746–15751.

(40) Zhang, Z. H.; Zhou, Z. F.; Nie, S.; Wang, H. H.; Peng, H. R.; Li, G. C.; Chen, K. Z. *J. Power Sources* **2014**, *267*, 388–393.

(41) Wu, H.; Xu, C.; Xu, J.; Lu, L. F.; Fan, Z. Y.; Chen, X. Y.; Song, Y.; Li, D. D. *Nanotechnology* **2013**, *24*(45), 7.

(42) Sondergaard, M.; Shen, Y. B.; Mamakhel, A.; Marinaro, M.; Wohlfahrt-Mehrens, M.; Wonsyld, K.; Dahl, S.; Iversen, B. B. *Chem. Mat.* **2015**, *27*(1), 119–126.

(43) Ren, Y.; Liu, Z.; Pourpoint, F.; Armstrong, A. R.; Grey, C. P.; Bruce, P. G. *Angew. Chem.-Int. Edit.* **2012**, *51*(9), 2164–2167.

(44) Janotti, A.; Varley, J. B.; Rinke, P.; Umezawa, N.; Kresse, G.; Van de Walle, C. G. *Phys. Rev. B* **2010**, *81*(8), 7.

(45) Feng, X. J.; Shankar, K.; Varghese, O. K.; Paulose, M.; Latempa, T. J.; Grimes, C. A. *Nano Lett.* **2008**, *8*(11), 3781–3786.

(46) Liu, B.; Aydil, E. S. *J. Am. Chem. Soc.* **2009**, *131*(11), 3985–3990.

(47) Yang, Y.; Ling, Y. C.; Wang, G. M.; Li, Y. *Eur. J. Inorg. Chem.* **2014**, 2014(4), 760–766.

(48) Leshuk, T.; Parviz, R.; Everett, P.; Krishnakumar, H.; Varin, R. A.; Gu, F. *ACS Appl. Mater. Interfaces* **2013**, *5*(6), 1892–1895.

(49) Wheeler, D. A.; Ling, Y. C.; Dillon, R. J.; Fitzmorris, R. C.; Dudzik, C. G.; Zavodivker, L.; Rajh, T.; Dimitrijevic, N. M.; Millhauser, G.; Bardeen, C.; Li, Y.; Zhang, J. Z. *J. Phys. Chem. C* **2013**, *117*(50), 26821–26830.

(50) Yang, C. Y.; Wang, Z.; Lin, T. Q.; Yin, H.; Lu, X. J.; Wan, D. Y.; Xu, T.; Zheng, C.; Lin, J. H.; Huang, F. Q.; Xie, X. M.; Jiangl, M. H. *J. Am. Chem. Soc.* **2013**, *135*(47), 17831–17838.

(51) Salari, M.; Aboutalebi, S. H.; Konstantinov, K.; Liu, H. K. *Phys. Chem. Chem. Phys.* **2011**, *13*(11), 5038–5041.

(52) Salari, M.; Konstantinov, K.; Liu, H. K. *J. Mater. Chem.* **2011**, *21*(13), 5128–5133.

(53) Wang, G. M.; Lu, X. H.; Ling, Y. C.; Zhai, T.; Wang, H. Y.; Tong, Y. X.; Li, Y. *ACS Nano* **2012**, *6*(11), 10296–10302.

(54) Lu, X. H.; Yu, M. H.; Zhai, T.; Wang, G. M.; Xie, S. L.; Liu, T. Y.; Liang, C. L.; Tong, Y. X.; Li, Y. *Nano Lett.* **2013**, *13*(6), 2628–2633.

(55) Zhai, T.; Lu, X. H.; Wang, H. Y.; Wang, G. M.; Mathis, T.; Liu, T. Y.; Li, C.; Tong, Y. X.; Li, Y. *Nano Lett.* **2015**, *15*(5), 3189–3194.

(56) Lu, X. H.; Zhai, T.; Zhang, X. H.; Shen, Y. Q.; Yuan, L. Y.; Hu, B.; Gong, L.; Chen, J.; Gao, Y. H.; Zhou, J.; Tong, Y. X.; Wang, Z. L. *Adv. Mater.* **2012**, *24*(7), 938–944.

(57) Zhai, T.; Xie, S. L.; Yu, M. H.; Fang, P. P.; Liang, C. L.; Lu, X. H.; Tong, Y. X. *Nano Energy* **2014**, *8*, 255–263.

(58) Zhang, J. T.; Ma, J. Z.; Zhang, L. L.; Guo, P. Z.; Jiang, J. W.; Zhao, X. S. *J. Phys. Chem. C* **2010**, *114*(32), 13608–13613.

(59) Xu, J.; Wu, H.; Lu, L. F.; Leung, S. F.; Chen, D.; Chen, X. Y.; Fan, Z. Y.; Shen, G. Z.; Li, D. D. *Adv. Funct. Mater.* **2014**, *24*(13), 1840–1846.

(60) Zhou, H.; Zhang, Y. R. *J. Power Sources* **2013**, *239*, 128–131.

(61) Zhou, H.; Zhang, Y. R. *J. Phys. Chem. C* **2014**, *118*(11), 5626–5636.

(62) Kim, C.; Kim, S.; Lee, J.; Kim, J.; Yoon, J. *ACS Appl. Mater. Interfaces* **2015**, *7*(14), 7486–7491.

(63) Li, Z.; Ding, Y. T.; Kang, W. J.; Li, C.; Lin, D.; Wang, X. Y.; Chen, Z. W.; Wu, M. H.; Pan, D. Y. *Electrochim. Acta* **2015**, *161*, 40–47.

(64) Cao, X. Y.; Xing, X.; Zhang, N.; Gao, H.; Zhang, M. Y.; Shang, Y. C.; Zhang, X. T. *J. Mater. Chem. A* **2015**, *3*(7), 3785–3793.

(65) Chen, J. Q.; Xia, Z. B.; Li, H.; Li, Q.; Zhang, Y. J. *Electrochim. Acta* **2015**, *166*, 174–182.

(66) Xiaoyang, P.; Min-Quan, Y.; Xianzhi, F.; Nan, Z.; Yi-Jun, X. *Nanoscale* **2013**, *5*(9), 3601–3614.

(67) Cho, I. S.; Logar, M.; Lee, C. H.; Cai, L. L.; Prinz, F. B.; Zheng, X. L. *Nano Lett.* **2014**, *14*(1), 24–31.

(68) Fang, W. Z.; Xing, M. Y.; Zhang, J. L. *Appl. Catal. B-Environ.* **2014**, *160*, 240–246.

(69) Huang, H. B.; Leung, D. Y. C.; Ye, D. Q. *J. Mater. Chem.* **2011**, *21*(26), 9647–9652.

(70) Wang, G. M.; Ling, Y. C.; Wang, H. Y.; Yang, X. Y.; Wang, C. C.; Zhang, J. Z.; Li, Y. *Energy Environ. Sci.* **2012**, *5*(3), 6180–6187.

(71) Cooper, J. K.; Ling, Y. C.; Longo, C.; Li, Y.; Zhang, J. Z. *J. Phys. Chem. C* **2012**, *116*(33), 17360–17368.

(72) Ling, Y. C.; Wang, G. M.; Reddy, J.; Wang, C. C.; Zhang, J. Z.; Li, Y. *Angew. Chem.-Int. Edit.* **2012**, *51*(17), 4074–4079.

(73) Mettenborger, A.; Singh, T.; Singh, A. P.; Jarvi, T. T.; Moseler, M.; Valldor, M.; Mathur, S. *Int. J. Hydrog. Energy* **2014**, *39*(10), 4828–4835.

CHAPTER NINE

Rationalizing the Efficiency of Hydrogen-Treated TiO$_2$ Nanomaterials in Light Driven Water-Splitting Applications

Mark Forster and Alexander J. Cowan

Department of Chemistry and Stephenson Institute for Renewable Energy, University of Liverpool, L69 7ZF, UK

9.1 Introduction

Light driven water splitting provides a potential solution to the problem of gathering and storing solar energy on the global scale. In an approach comparable to photosynthesis, solar energy is captured and stored in the form of chemical bonds to yield fuels such as H$_2$, which can be utilized whenever required. The splitting of water to produce H$_2$ and O$_2$ is an endothermic process (237 kJ mol^{-1}, Eq. (1–3)), corresponding to 1.23 eV per electron transferred.[1] Numerous approaches to utilizing solar energy to drive the half reactions for water oxidation (Eq. (1)) and hydrogen evolution (Eq. (2)) have been explored,[2] including, but not limited to the use of coupled photovoltaic-electrolyzers, photobiological systems, photothermal/photochemical materials, molecular photocatalysts, and light absorbing semiconductor materials. Amongst these, the use of semiconducting photocatalysts and photoelectrodes are of particular interest due to the potential of developing materials that meet both cost and efficiency requirements.[3] The focus of this chapter is on the application and mechanism of operation of hydrogen-treated or "black" titanium dioxide (TiO$_2$) materials; we do not aim to provide a comprehensive introduction to the use of semiconductor photoelectrodes and photocatalysts as these have been discussed in detail elsewhere.[4,5] Nonetheless, it is necessary to give a brief introduction to provide a framework for the discussions in Sections 2 and 3.

$$\text{H}_2\text{O} \rightarrow \frac{1}{2}\text{O}_2 + 2\text{H}^+ + 2\text{e}^-, \quad \beta \qquad \text{(OER)} \tag{1}$$

$$2\text{H}^+ + 2\text{e}^- \rightarrow \text{H}_2, \quad p \qquad \text{(HER)} \tag{2}$$

$$\text{H}_2\text{O} \rightarrow \frac{1}{2}\text{O}_2 + \text{H}_2 \quad \Delta G = +237\,\text{kJ}\,\text{mol}^{-1}. \tag{3}$$

Fundamental to all of the studies discussed here is the initial absorption of a photon by the semiconductor and the subsequent generation of holes and electrons, oxidizing and reducing equivalents, which are able to drive the half reactions Eqs. (1) and (2), respectively. In order for a photon to be absorbed, the energy must exceed the semiconductor bandgap energy (E_g). Unlike molecular absorbers, semiconductors have a continuum of states in both their highest occupied levels (the valence band, VB) and their lowest unoccupied levels (the conduction band, CB), so that the absorption coefficient generally does not decline with increasing excitation energy but instead continually increases above the bandgap of the semiconductor.[5] Once generated, the electron–hole pairs can either be directly utilized in a wireless particulate photocatalytic system (Figure 9.1(a)) or in a wired photoelectrochemical cell (Figure 9.1(b)). Both approaches have been explored using hydrogen-treated TiO$_2$ and they are briefly discussed below.

9.1.1   *Wireless Particulate Photocatalysis*

The simplest type of photocatalysts are in the form of particles or powders suspended in aqueous solution where the photogenerated holes in the VB and electrons in the CB directly oxidize and reduce water at the semiconductor surface (Figure 9.1). In order for this to

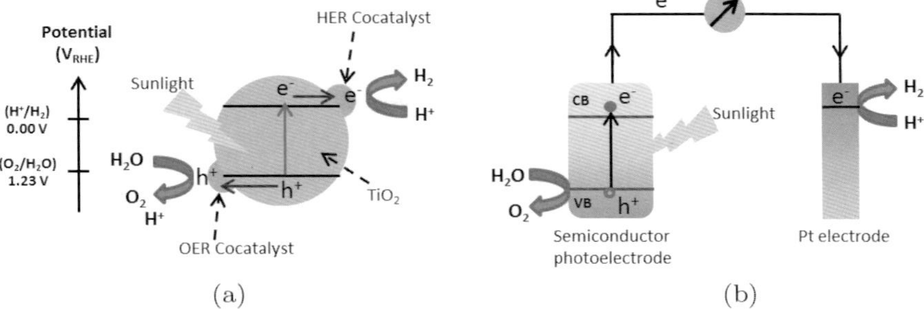

Figure 9.1. The basic photoelectrochemical setup under illumination for (a) a single particle in a particle reactor and (b) a photoelectrode cell setup with working electrode connected to a metal counter electrode. Catalysts can be attached to the surface to assist the reaction kinetics.

occur using a single light absorber, the potential of the VB needs to be positive of the H$_2$O/O$_2$ couple (four electrons, 1.23 V$_{RHE}$) and the CB needs to be sufficiently negative to enable hydrogen production (H$^+$/H$_2$, two electrons, 0 V$_{RHE}$). Although this may appear to make the minimum bandgap requirement 1.23 eV, the need to have a sufficient overpotential for the water oxidation and proton reduction reactions to enable sufficiently fast utilization of the photogenerated charges, coupled to potential losses due to surface trapping at lower energy defect sites makes the generally accepted minimum bandgap closer to 2 eV.[4,5] A further consideration is that the material must have sufficient stability against photocorrosion, a feature of many of the wider bandgap metal oxide semiconductors. Because of these requirements, the number of suitable semiconductor materials for direct water splitting is limited and critically the need to have a material with a relatively large bandgap limits the ability to achieve high solar energy to hydrogen conversion efficiencies, as only the higher energy portion (typically ultraviolet (UV)) of the solar spectrum can be harvested.[4]

In addition to challenges associated with poor-light harvesting, achieved quantum yields for single particle photocatalysts have tended to be relatively low[6] due to excessive electron–hole recombination (Figure 9.2(a)). Following initial electron–hole generation, it is necessary to separate and migrate the charges to the required surface

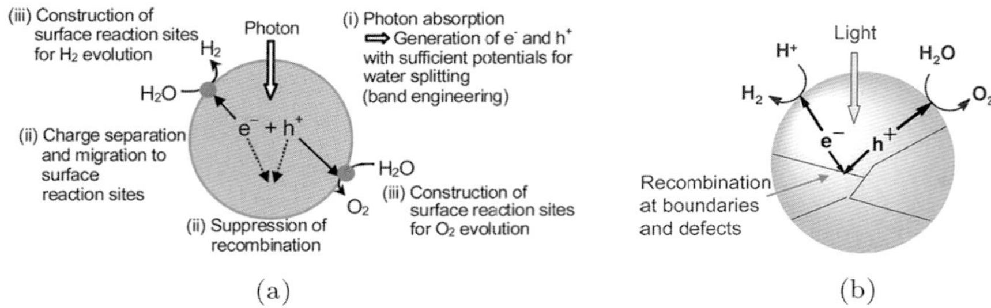

Figure 9.2. (a) Key processes in photocatalytic water splitting using a single light absorbing semiconductor. (b) Electron–hole recombination is expected to occur at both surface sites and defect/boundary trap sites. Reprint from Ref. [6]. © The Royal Society of Chemistry.

reaction sites for oxygen and hydrogen evolution. During migration, bulk trap-mediated recombination may occur (Figure 9.2(b)), and if the hole and electron surface reaction sites are not sufficiently separated, surface recombination may also become a significant loss pathway on the particulate photocatalyst, further reducing device efficiency. In light of these difficulties, numerous studies have not actually examined complete water splitting with a single light absorber and have instead concentrated on the optimization of a single half reaction, either photocatalytic hydrogen or oxygen production in the presence of a sacrificial reagent. For hydrogen production, high energy sacrificial reagents such as an alcohol or a sulfide ion are added to the solution.[6] These reagents require less oxidizing holes, relaxing the bandgap requirements of the material, in addition to being typically easier to oxidize which leads to rapid hole removal and hence a decrease in the level of electron–hole recombination. Clearly the use of a high energy sacrificial reagent is impractical for use in a solar energy system, nonetheless these studies are of great value as once optimized the individual photocatalysts for hydrogen production and water oxidation can be coupled together in a two-step Z-scheme system.[7]

9.1.2 *Photoelectrochemical Water Splitting*

Here we briefly introduce the energetics of an n-type semiconductor used as a photoanode in a photoelectrochemical (PEC) cell for water splitting (Figure 9.1(b)). In such a device, only the OER occurs at the semiconductor surface, making it necessary for the semiconductor VB to be at a potential positive of the H$_2$O/O$_2$ couple. Whilst it may also be desirable for the CB to lie negative of the H$^+$/H$_2$ couple, the use of an additional electrical energy input means that this is no longer a requirement, enabling the use of a wider range of semiconductors with narrower bandgaps and hence better light harvesting properties.

An energy level diagram for an *n*-type semiconductor brought into contact with a liquid electrolyte containing a redox couple is shown in Figure 9.3. Typically, the Fermi level of an *n*-type semiconductor lies above the potential of the redox couple in solution and charge flows between the semiconductor and the electrolyte until equilibrium of the Fermi levels is established (Figure 9.3(b)). The value of the Fermi level in the semiconductor changes much more than that of the electrolyte as the solution has far more states per unit energy than the semiconductor's bandgap region. Because of the interfacial charge flow, the electrode at equilibrium possesses

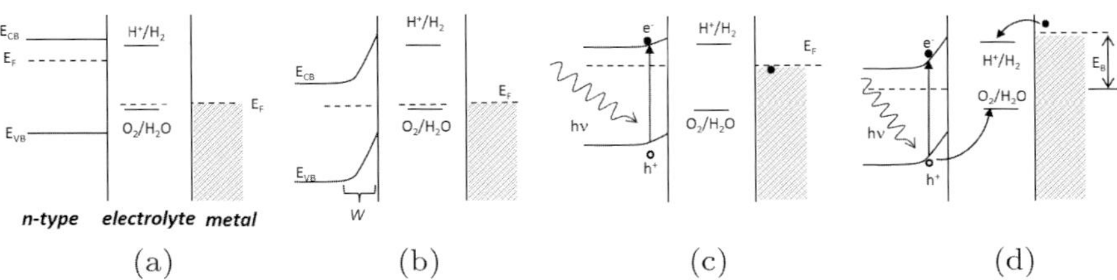

Figure 9.3. Energetics of an n-type semiconductor metal photoelectrochemical cell with an aqueous electrolyte. (a) Prior to contact being made between the semiconductor and the electrolyte in the dark, where E_{cb}, E_{vb}, and E_F are the CB, VB, and Fermi-level energies, respectively. (b) After contact in the dark, electrons have flowed from the semiconductor to the solution such that the Fermi level of the semiconductor, E_F, matches that of the electrolyte. In a semiconductor where the surface states pin the energy level, band bending occurs, with the resultant region of positive charge in the semiconductor distributed over the depletion width (W). (c) Under strong illumination when the two electrodes are shorted together, the Fermi level rises flattening the bands. (d) Under illumination, an applied bias (E_B) can be used to raise the metal Fermi level above the potential of the H$^+$/H$_2$ couple and to maintain the band bending in the semiconductor, aiding charge separation.

excess positive charge, arising from the ionized dopant atoms in the semiconductor, and the solution possesses excess negative charge. This positive charge is spread out over a width of the semiconductor known as the depletion width, W, and the negative charge is spread over a much narrower region close to the electrode. The strength of the resultant electric field across the semiconductor depends on the potential dropped in the solid, which, in turn, is a function of the initial difference in the Fermi level of the semiconductor relative to the electrolyte and the depletion width. Under the influence of this electric field, charge carriers (electrons or holes) generated within the depletion width are swiftly separated with holes being swept to the semiconductor surface and electrons to the bulk (Figure 9.3(c)). Under intense illumination, when the electrodes are short circuited, the high rate of charge generation can lead to a raising of the Fermi level and a reduction of the band bending with often a subsequent decrease in charge separation yields. Application of an external bias (E_B) across the n-type semiconductor can help to maintain the band bending in the semiconductor aiding charge separation, and in many PEC cells it is also required to raise the metal Fermi level above the potential of the H$^+$/H$_2$ couple (Figure 9.3(d)). The use of an external electrical energy input does limit the efficiency of the photon-to-current efficiency (PCE) device, which is often reported as an applied bias PEC (ABPE, Eq. (4))[8] making it desirable to develop materials that require a lower electrical energy input.

$$\text{ABPE} = \left[\frac{|j_{\text{ph}}(\text{mA cm}^{-2})| \times |(1.23 - V_B)(V)|}{P_{\text{total}}(\text{mW cm}^{-2})} \right]. \qquad (4)$$

Another important metric of efficiency that will be described here is the incident photon-to-current efficiency (IPCE, Eq. (5)) which is a measure of current collected (j_{ph}) at a particular photon flux ($\varphi(\lambda)$) as a function of wavelength. IPCEs are reported at stated applied biases but do not provide any information on the bias dependence of the PEC itself. The efficiency of the electrode is often described in terms of three fundamental processes: (i) the efficiency of light harvesting at a particular wavelength ($\eta_{\text{LH}}(\lambda)$), (ii) the efficiency of initial charge separation and transport to the semiconductor liquid junction (SCLJ, hole) and the semiconductor current collector junction (electron) ($\eta_{\text{sep}}(\lambda)$) and (iii) the efficiency of the hole transfer reaction at the interface ($\eta_{\text{int}}(\lambda)$), Eq. (5).[9]

$$\text{IPCE}(\lambda) = \eta_{\text{LH}}(\lambda) \times \eta_{\text{sep}}(\lambda) \times \eta_{\text{int}}(\lambda). \tag{5}$$

Electron–hole recombination occurs in competition with the charge separation, transport, and transfer reactions and it is the kinetic balance of these processes which is a significant factor in controlling the efficiency of a device. Recombination can occur via a number of different pathways (Figure 9.4).[5] If absorption of the photon occurs in the bulk of the semiconductor, outside of the depletion region, charge carriers can rapidly recombine prior to hole transport

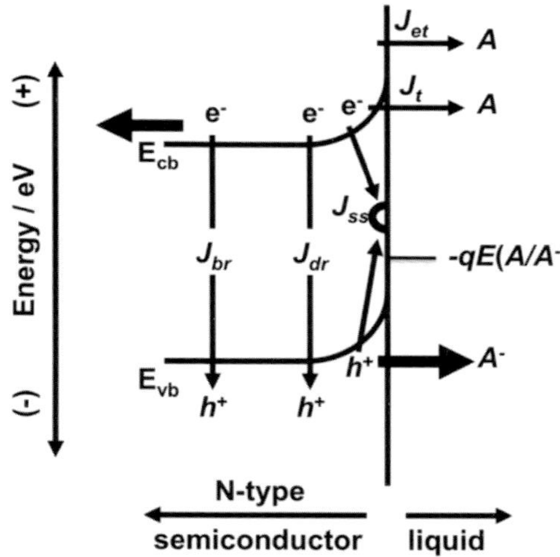

Figure 9.4. Recombination pathways for photoexcited carriers in a semiconductor PEC cell held at open circuit can be broken down into different categories, represented by the thin arrows in the diagram. The electron–hole pairs can recombine via radiative or non-radiative recombination in the bulk of the semiconductor (J_{br}), depletion-region recombination (J_{dr}), surface recombination due to defects (J_{ss}), indirect surface recombination following tunneling (J_{t}), and indirect recombination following majority carriers traversing the interfacial barrier (J_{et}). Electron collection by the back contact and hole collection by the redox couple (e.g. oxidation of water to O_2) are processes that contribute positively to device efficiency, and these are depicted by thick black arrows. Reprint from Refs. [5, 10]. © American Chemical Society.

to the SCLJ. In a similar manner, even in the depletion region, recombination can occur if charge separation and transport is not sufficiently efficient. Indirect recombination can also take place if electrons either traverse or tunnel through the potential barrier at the SCLJ, leading to reduction of O$_2$ or water-splitting intermediates. The presence of trap or defect states, often at the semiconductor surface, can also lead to enhanced recombination losses if the rate of trap-mediated recombination is sufficiently large enough to compete with the hole transfer reaction. Each of these processes can in principle respond differently to variations such as changes in dopant density, temperature, illumination intensity, electric field strength; in Section 9.4, the effect of hydrogen treatments on the kinetics of charge carriers of TiO$_2$ photoelectrodes will be explored in an attempt to understand which are the key factors controlling efficiency so that future design directions can be identified.

9.2 Titanium Dioxide

TiO$_2$ is the most widely studied semiconductor in light driven water-splitting applications due to its ability to act as a photocatalyst for both water oxidation and reduction, excellent thermal and photostability and low cost.[6] However, whilst the individual OER and HER reactions are achievable on the most commonly studied crystalline phases of TiO$_2$, anatase and rutile, the majority of studies utilize sacrificial electron or hole donors and complete water splitting is not commonly achieved, with the few reported efficiencies being very low.[4,5,11–13]

In addition to being widely studied as a suspension photocatalyst TiO$_2$ has also been explored as a photoanode material for over 40 years. The 1972 study of Fujishima and Honda, where a TiO$_2$ single crystal electrode is used in a PEC cell for water splitting, is widely recognized as being a breakthrough study in solar fuels research, inspiring countless researchers since.[14] Unfortunately, although TiO$_2$ photoelectrodes and photocatalysts have been reported with impressive efficiencies for the OER under UV illumination, the wide bandgap of the material, *ca.* 3.2 eV (anatase) and *ca.* 3.0 eV (rutile)[15] means that they have little activity in the visible region as the large bandgap prevents absorption of photons with wavelengths longer than *ca.* 400 nm. The fraction of photons with sufficient energy to generate electron–hole pairs by direct bandgap excitation of TiO$_2$ therefore is relatively low, $\sim$2%, which represents less than 5% of the total incident solar energy.[16] Numerous attempts to extend the absorption profile and hence photocatalytic activity of TiO$_2$ into the visible region have been made, primarily through doping with main group[4,5] and metal elements,[4,5] however to date the activities of these materials remains modest.

In 2011, Chen *et al.*[17] introduced a conceptually different approach where the absorption spectrum of anatase TiO$_2$ nanocrystals was drastically changed through a thermal treatment (200°C) under a high pressure of hydrogen (20 bar), leading to the formation of a material that is becoming commonly referred to as black TiO$_2$ (Figure 9.5). Although thermal treatments under lower pressures of hydrogen have been known to impart coloration to TiO$_2$ since the 1950s, improvements in visible light photocatalytic activity were not commonly reported due to the incorporation of oxygen vacancies (V$_{Os}$).[18] The striking feature of the 2011 study[17] was

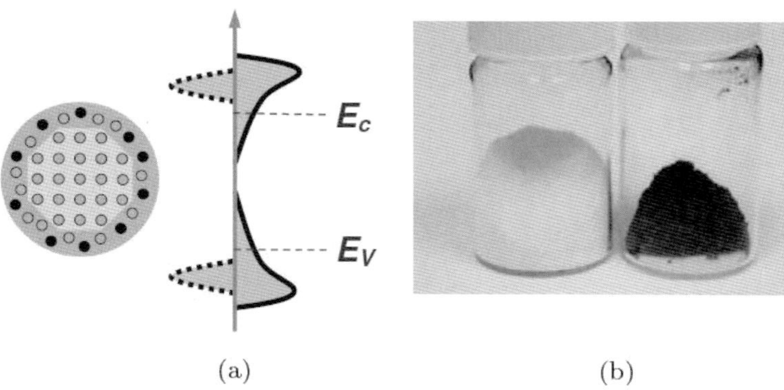

(a)   (b)

Figure 9.5. Thermal treatment of TiO$_2$ under a high pressure of TiO$_2$ can lead to the formation of H:TiO$_2$ with a black coloration (B). The TiO$_2$ nanoparticles have been shown to have a crystalline core and highly disordered surface leading to a bandgap narrowing. Reprint from Ref. [17]. © The American Association for the Advancement of Science.

that when examined for light driven HER in the presence of a sacrificial electron donor under full simulated solar irradiation a very large improvement, approximately two orders of magnitude, in activity was reported due to both increases in activity under UV and visible light. The hydrogen-treated TiO$_2$ nanocrystals were also described as possessing two phases: a crystalline TiO$_2$ quantum dot or nanocrystal as a core, and a highly disordered surface layer where dopants are introduced. This was in contrast to previous studies where solely V$_{Os}$ were present. The surface disorder led to a significant bandgap narrowing and an onset in absorption at ~1200 nm (~1 eV). This study has since been cited over 1500 times in less than 5 years[19] and numerous groups have since gone on to explore the application of hydrogen-treated TiO$_2$ materials for use as both particulate photocatalysts and photoelectrodes,[20] with many materials showing both enhanced UV and visible light activity for water splitting.[21–26] In this chapter, we aim to summarize many of the key mechanistic studies which explore the fundamental factors controlling the change in activity upon hydrogen treatments. We do not limit ourselves solely to surface disordered materials, as increasingly, wrongly or rightly, the term black TiO$_2$ is becoming applied to any TiO$_2$ that has achieved coloration following heating in a hydrogen (reducing) atmosphere. We do however endeavor to highlight whenever possible the conditions of preparation of the materials of study as it will be shown that these are critical factors in controlling the mechanism of enhancement in water-splitting activity.

9.3   Core–Shell Black TiO$_2$

In order to rationalize the change in photocatalytic activity, it is necessary to first address the nature of the structural changes upon hydrogen treatment in more detail and identify the cause of the strong visible light absorption. As noted above, Chen *et al.*[17,27] demonstrated that heating anatase TiO$_2$ nanoparticles (*ca.* 8 nm diameter) to 200°C under H$_2$ at 20 bar for 5 days led to the formation of a disordered outer shell as measured by high resolution transmission electron microscopy (HRTEM). Furthermore, Raman spectroscopy showed the broadening of the anatase modes in addition to the presence of new bands assigned to previously forbidden modes which became allowed in the disordered structure. Detailed X-ray photoelectron spectroscopy (XPS) measurements indicated that the induced disorder had led to little change in the nature of the CB states, however VB XPS showed a strong tailing of the VB leading to a bandgap narrowing to *ca.* 1.5 eV.[17] In many past studies, it has been shown that the treatment of TiO$_2$ with hydrogen at elevated temperatures can cause the formation of V$_O$ to yield Ti^{3+} centers.[20,28–31] The presence of the Ti^{3+} gives rise to mid-gap states at *ca.* 0.75 eV and 1.18 eV below the CB of TiO$_2$ and hence can cause coloration of the material, even making the TiO$_2$ appear black when present in high concentrations.[18,32] To assess if such species were also present in the core–shell H:TiO$_2$ prepared by Chen *et al.*,[17,27] Ti 2p XPS measurements were carried out which provided unambiguous evidence that Ti^{3+} was not present at high levels. Perhaps most surprising was that the controlled introduction of Ti^{3+} sites actually lead to a removal of the VB states that were proposed to be the cause of the strong visible light absorption in the hydrogen-treated TiO$_2$.[27]

In order to assess the mechanism of formation of the disordered surface layer and to identify how this causes the visible light transitions, density functional theory (DFT) calculations have been carried out by several groups, all of which appear to confirm that the presence of V$_O$ is unlikely to cause the observed optical properties in the core–shell materials.[17,33–36] Lei *et al.*[33] demonstrated that H atoms interacting on the surface of anatase were able to break up Ti$_3$O units leading to distortion of the lattice and the formation of stable ~Ti–H and ~O–H sites, surface species which were also later experimentally verified.[37] In contrast, in the bulk of the crystal the defect sites were only metastable with Ti–O bonds predicted to reform. This difference in stability of the defect sites would account for the core–shell structure observed by HRTEM. An alternative theory for the confinement of the disorder induced defect sites to the surface region is that a barrier to H and H$_2$ migration through the TiO$_2$ lattice exists.[34] The $\sim$Ti–H and $\sim$O–H sites were predicted to generate mid-bandgap states[17,33,36] with the exact energy being strongly dependent on the local environment. The high level of disorder in the shell region would therefore lead to the formation of a continuum of states that could potentially overlap with the VB edge, giving rise to the observed XPS spectrum. Furthermore, in agreement with the XPS studies, DFT calculations[17,33,36] also predicted that the presence of the mid-bandgap states would lead to no major change in the apparent CB level. In contrast to the majority of experimental and theoretical works which predicted that the core–shell materials were based on the presence of crystalline and defective TiO$_2$ regions, a recent study[38] has identified that Ti$_2$O$_3$ can form on the shell of TiO$_2$ in a reductive atmosphere. Ti$_2$O$_3$ is a narrow bandgap semiconductor and it was postulated that this may actually be the cause of the visible light absorption of core–shell TiO$_2$ treated under H$_2$. It should however be noted that the Ti$_2$O$_3$ on TiO$_2$ was observed following a thermal treatment under an argon atmosphere, in contrast to the other materials discussed here which employed hydrogen. Overall, in light of the experimental[17,22,27] and DFT studies[17,33–36] we can now propose a mechanism for the observed visible light activity of core–shell black TiO$_2$, where visible light absorption leads to excitation of mid-bandgap states to the largely unmodified CB of TiO$_2$. As the TiO$_2$ CB edge remains unchanged, the generated photoelectrons have sufficient driving force for H$_2$ production to occur. In the original study by Chen *et al.*,[17] methanol was used as a hole scavenger and hence the loss of oxidative power of the mid-bandgap hole is likely to be an insignificant factor, Figure 9.6(a).

In addition to being visible light active photocatalysts, the core–shell black TiO$_2$ particles also show greatly enhanced activity under UV light. In the original 2011 study, a H$_2$ evolution rate of 10 mmol hr^{-1} g^{-1} of photocatalyst was achieved under 1 SUN illumination (UV + Vis) which is approximately two orders of magnitude higher than achievable with reference "white" TiO$_2$ particles. Interestingly when examined using a 400 nm long pass filter to exclude the UV light, the activity drops dramatically to

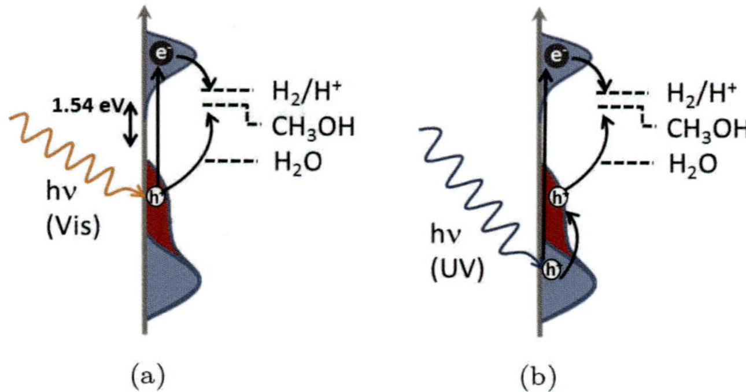

Figure 9.6. Schematic of core–shell TiO$_2$ in the presence of methanol. (a) Visible excitation of electrons from mid-bandgap states located in the shell region (red) to the largely unchanged CB of TiO$_2$ can occur in addition to (b) UV excitation of electrons from the core of the TiO$_2$ nanoparticle to the CB. Initially formed VB holes migrate to less energetic surface trap states decreasing the rate of recombination. Oxidation potentials are based on Ref. [39]. The CB and VBs of TiO$_2$ are shown in blue and these are expected to cover the whole TiO$_2$ nanoparticle, the red region shows the hydrogen treatment induced mid-bandgap states that occur in the disordered surface shell.

0.1 mmol hr^{-1} g^{-1}.[17] It is therefore apparent that the primary reason for the very high activity of the core–shell black TiO$_2$ is not its improved light harvesting properties but actually that the core–shell structure is assisting in more efficient utilization of photogenerated charges formed following UV excitation. This may appear surprising as commonly in photocatalytic materials the introduction of defects and trap states inhibits photocatalysis due to the generation of recombination centers which lower charge separated lifetimes.[6] However, as outlined above, these mid-gap states are localized at or close to the TiO$_2$ surface. This formation of two distinct regions within the TiO$_2$ nanoparticle provides a junction over which charge separation can occur. Following UV excitation, initially formed holes can rapidly migrate to the less energetic trap states at the surface, spatially separating them from CB photoelectrons, whose wave function can spread throughout both the core and shell regions. This spatial separation of the charges across the core/shell junction greatly decreases the rate of electron–hole recombination enhancing the photocatalytic efficiency as shown in Figure 9.6(b).[33]

The role of charge localization in determining activity was further developed by Lu *et al.*[36] who proposed that photogenerated electron–hole separation may also be enhanced by electron–hole flow between different facets with a predicted flow of holes from the (101) to the (001) facet due to the relative stability of the local defects. In contrast, the electrons were predicted to localize at (101) facets. In addition to the predicted differences in stability of facets, the relative distribution has also been studied experimentally using X-ray diffraction (XRD) with it being noted that this may also be a significant factor behind the very high levels of photocatalytic activity.[40]

So far the discussion has concentrated on the use of black core–shell TiO$_2$ for photocatalytic HER in the presence of a sacrificial hole scavenger such as methanol. An interesting study explored alternative oxidative pathways, looking at the thermal treatment of protonated titanate nanotubes under a H$_2$ atmosphere at 500°C. This treatment led to the synthesis of anatase nanowires ($\sim$8 nm diameter), which were proposed to have a high concentration of surface Ti–H groups leading to a disordered core–shell structure.[37] In a similar manner to previous works, the presence of the disordered TiO$_2$ was identified by Raman, XPS, and HRTEM and the materials were found to be visible light active photocatalysts for hydrogen production in the presence of a hole scavenger. In the absence of a hole scavenger in water under a Xe lamp (UV and visible wavelengths), it was observed that $^\bullet$OH radicals were formed. It was in fact shown that TiO$_2$:H had a *ca.* 13 times greater yield of $^\bullet$OH radicals when compared to P25, Figure 9.7. This is an extremely promising result as the $^\bullet$OH radical is the product of the one-electron oxidation of H$_2$O (H$_2$O $\rightarrow$ $^\bullet$OH + H$^+$ + e$^-$, E$_0$ $\sim$2.8 V$_{RHE}$),[41] which is believed to be the key step in water oxidation on TiO$_2$. Whilst this demanding oxidation reaction has been commonly observed on crystalline anatase TiO$_2$, where the VB lies several 100 mV positive of the required potential, it is to the best of our knowledge the first report on core–shell black TiO$_2$. What makes this significant is that if $^\bullet$OH generation can still occur following hole transfer to the less oxidizing surface states, then it is feasible that

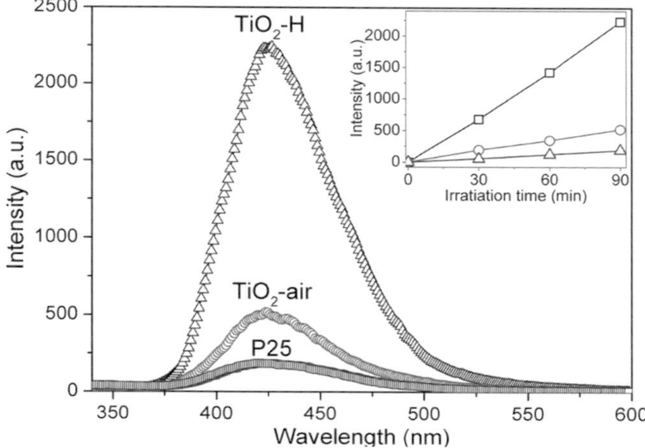

Figure 9.7. Fluorescence spectra of the visible-light irradiated TiO$_2$ samples in terephthalic acid (TA) solution. The Fluorescence is from TAOH formed by reaction of the TA with $^{\bullet}$OH radicals generated by 90 min irradiation of the TiO$_2$ samples under Xe lamp irradiation. Reprint from Ref. [37]. © The Royal Society of Chemistry.

a complete water-splitting core–shell black TiO$_2$ materials may be developed.

9.4   The Role of Oxygen Vacancies in Hydrogen-Treated TiO$_2$

Although Chen *et al.* clearly demonstrated minimal formation of V_O in their materials prepared at 20 bar H$_2$ and 200°C,[17] V_O have since been observed in several other studies of core–shell black TiO$_2$ materials prepared using modified conditions.[20, 22, 42, 43] Of particular mechanistic interest is the work of Naldoni *et al.* who presented[22] a structural and electronic analysis of TiO$_2$ treated at 500°C under H$_2$ at atmospheric pressure. Using Raman spectroscopy, XPS, and HRTEM, they confirmed the presence of the core–shell structure with VB tailing. However, synchrotron X-ray powder diffraction (SXRPD) and cathodoluminescence (CL) analyses also revealed that Ti^{3+} species were present in the nanoparticles, with an occupation factor for O of ~0.95 indicating a 5% concentration of V_O sites. It was therefore proposed that electronic transitions from (i) the tailed VB to the CB, (ii) the tailed VB to localized V_O states, and from (iii) V_O states to the CB all contributed to the strong Vis-IR absorption of the hydrogen-treated TiO$_2$, as shown in Figure 9.8. The presence of V_O in the core–shell materials prepared is likely to be due to the changes in hydrogen treatment conditions, which is carried out at a lower

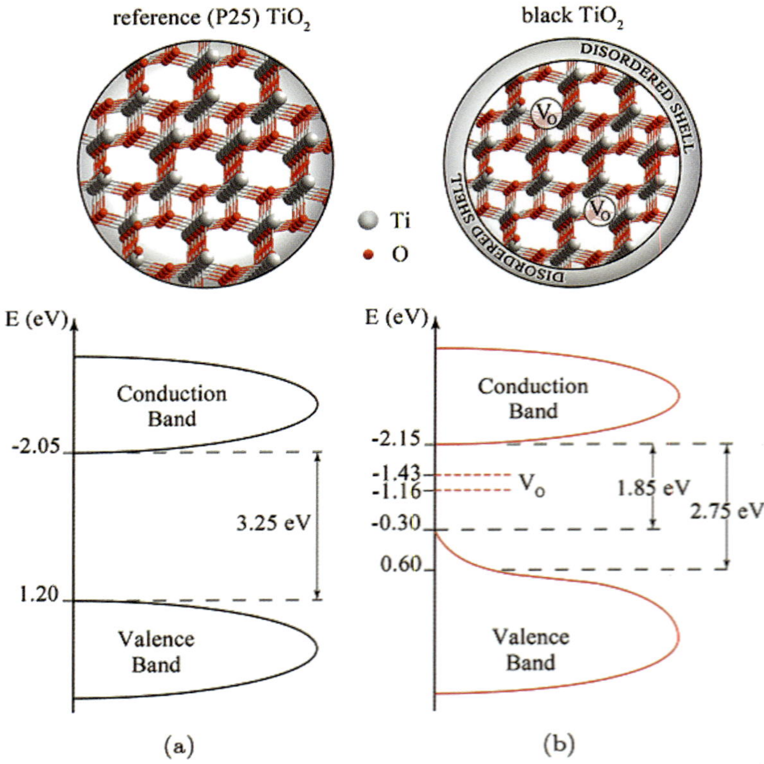

Figure 9.8. Schematic of nanoparticle's structure and DOS for (a) TiO$_2$ P25 Degussa and (b) black TiO$_2$ treated at lower pressure (*ca.* 1 atm) H$_2$ at 500°C. Both representations were built using experimental data from SXRPD, CL, UV–Vis spectroscopy, and XPS analysis. In contrast to previous models for core–shell materials prepared under high pressures of H$_2$, the presence of V_O is reported. The energy positions of V_O localized states were calculated by subtracting components at 2.36 eV and 2.63 eV of black TiO$_2$ CL band from the VB maximum in reference TiO$_2$. Reprint from Ref. [22]. © The American Chemical Society.

pressure than Chen *et al.* initially used. This is inline with theoretical works which have predicted that the nature and relative distribution of surface and sub-surface states, including Ti–H and ~O–H sites and V$_{Os}$ are temperature and H$_2$ pressure dependent.[34] It is therefore apparent that the mechanism of visible light absorption by TiO$_2$ may not be common across all materials which are labeled as "black TiO$_2$" and in this section we assess the role of V$_O$ in controlling the activity of some hydrogen-treated TiO$_2$ photocatalysts. An excellent review of black TiO$_2$ materials has now tabulated many studies documenting the conditions of preparation and reporting the presence or absence of the potential core–disordered shell structure and of V$_O$ sites.[20]

V$_O$ sites in TiO$_2$ have been studied for many years and they can be readily formed by treatment of TiO$_2$ in a reducing atmosphere, such as H$_2$.[18,44] In addition to acting as color centers, the shallow V$_O$ sites which sit *ca.* 0.75 eV and 1.18 eV below the CB edge of rutile TiO$_2$[32] are known to both modify the surface properties and act as self-dopants. The presence of V$_{Os}$ leads to the creation of Ti^{3+} to maintain the electrostatic balance, increasing the conductivity of the deficient TiO$_2$.[32,45] There are a number of techniques which can provide information on the role of V$_O$/defects in TiO$_2$. Photoluminescence (PL)[25,46] and XPS[47] spectroscopies provide information on the concentrations of Ti^{3+} sites in the TiO$_2$ lattice, whilst STM offers a route to image the TiO$_2$ surface allowing for mapping of defects/vacancies.[48] Positron annihilation lifetime spectroscopy (PALS) has also proven a particularly useful tool when assessing V$_O$ in TiO$_2$ materials prepared under a H$_2$ atmosphere.[49] PALS involves the injection of positrons into the bulk TiO$_2$. The positrons then thermalize and upon interaction with electrons are annihilated, emitting γ-rays which can be detected, on the picosecond timescale with the lifetime depending on the local electron density. Alternatively the positrons can form bound positron–electron pairs which preferentially distribute in the regions where electron density is low, e.g. vacancy type defects, vacancy clusters, and microvoids, with the slower electron–positron annihilation rate (nanosecond) providing information regarding various defect sizes and concentrations.

Pan *et al.* used PALS to investigate P25 that has been thermally treated at 400°C under H$_2$ at atmospheric pressure for 10 h.[25] The resulting TiO$_2$ was found to be strongly colored with both core–shell structure (as detected by Raman and HRTEM) and V$_O$ sites (as detected by Ti 2p XPS) being present. The presence of ∼3% V$_{Os}$ was proposed to be the cause of an increase in the positron lifetime following H$_2$ treatments, with experiments suggesting that V$_O$ existed either in isolated sites or in small clusters on the TiO$_2$. PL studies further confirmed the role of V$_O$ in suppressing the rate of electron–hole recombination, which was proposed to be a significant factor in improving the photocatalytic activity of this material. The decreased recombination losses when V$_O$ were present could be assigned to the increased concentration of surface O$_2$, which acted as a transient electron trap (Figure 9.9).[25,50]

PALS has also been used to explore the importance of the location of V$_O$ sites[48] and it has been noted that materials with a higher ratio of bulk to surface defects being more active for photocatalysis. This contrasts with several other works including Electron paramagnetic resonance (EPR) studies[23] which attributed a 10-fold increase in TiO$_2$ nanocrystal photoactivity following

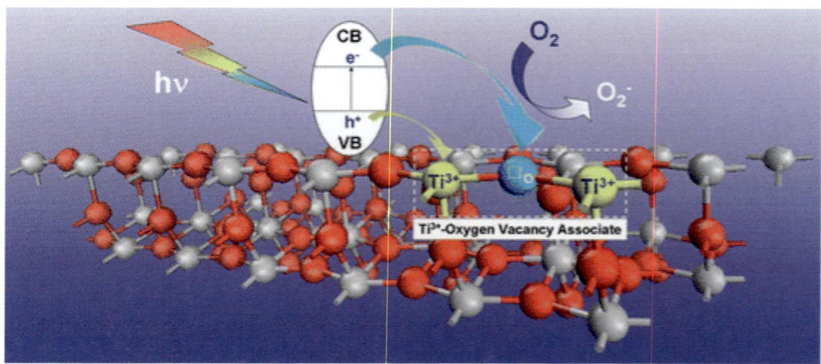

Figure 9.9. The introduction of V$_O$ into a hydrogen-treated core–shell TiO$_2$ has been proposed to lower electron–hole recombination possibly due to enhanced O$_2$ adsorption at the defect sites leading to increased electron scavenging or due to hole transfer to the Ti^{3+} sites. Reprint from Ref. [25]. © The American Chemical Society.

hydrogen treatment to a high surface-to-bulk defect ratio. In this work, EPR detected both bulk Ti^{3+} defects and O$^-$ surface sites with O$^-$ being formed by the surface interaction of O$_2$ with Ti^{3+} defects. Interestingly, it was noted that the defect distribution evolved with time during annealing and by comparing the photocatalytic activity of materials that had been treated for different time periods the relative roles of the different sites could be understood (Figure 9.10). V$_O$ formation has also been proposed to be the cause of decreased photocatalytic activity.[51] When TiO$_2$ particles were treated under high pressures of H$_2$ (20 bar) at a range of temperatures (250–450°C), a sample with a high bulk V$_O$ content was formed. Despite the improved light harvesting properties, these photocatalysts showed significantly lower levels of activity which led to the suggestion that the bulk Ti^{3+}/V$_O$ sites were acting as recombination centers, in contrast to their proposed role when generated at, or close to, the TiO$_2$ surface.[25]

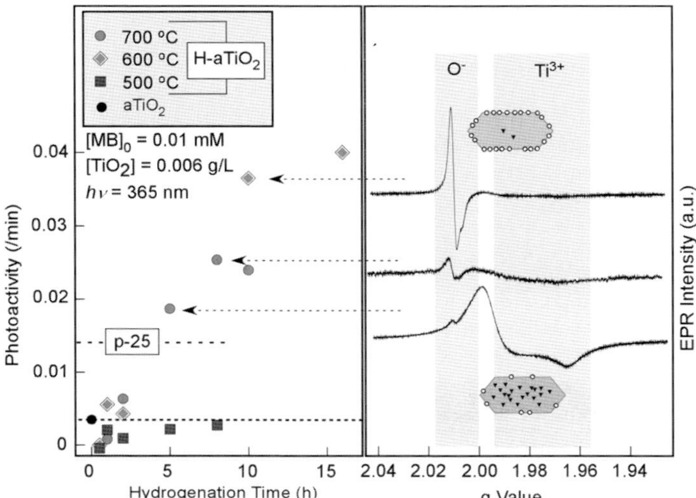

Figure 9.10. Combined EPR and photocatalytic studies explored the evolution of both surface and bulk defects during hydrogen treatment with a high level of surface O$^-$ sites correlating with increased photocatalytic activity. Reprint from Ref. [23]. © American Chemical Society.

Shortly after the report of the first core–shell materials,[17] Wang *et al.*[47] described the first hydrogen-treated TiO$_2$ 1D nanostructures for use as a water oxidation photoanode in a PEC cell. Rutile nanowires were treated at a range of different temperatures (200–550°C) under a flow of H$_2$. In agreement with previous studies, the hydrogen treatment was found to lead to visible light absorption, with materials prepared at higher temperatures turning black (Figure 9.11). Interestingly Wang *et al.* did not observe any changes in the VB XPS upon hydrogen treatment, indicating that the formation of a surface disordered layer with a shifted VB was not the cause of the visible light absorption, in contrast to previous studies on core–shell materials.[17] Instead, the Ti 2p XPS spectrum of the hydrogen-treated material showed the presence of high concentration of Ti^{3+} sites demonstrating that the NIR and visible light absorption

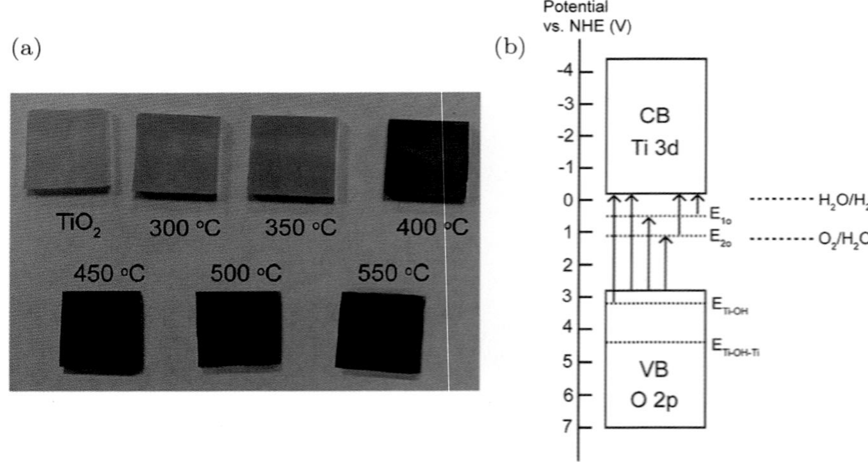

Figure 9.11. (a) Digital pictures showing the extent of coloration of TiO$_2$ photoelectrodes at varying hydrogen annealing temperatures. (b) A simplified energy diagram of H:TiO$_2$ nanowires. E_{1o} and E_{2o} refer to the V$_{Os}$ (red dashed lines) located at 0.73 eV and 1.18 eV below the TiO$_2$ CB; $E_{Ti–OH}$ and $E_{Ti–OH–Ti}$ (blue dashed lines) located at 0.7 eV and 2.6 eV below the TiO$_2$ VB represent the energy levels of surface hydroxyl group. The black dashed lines indicate the H$_2$O/H$_2$ and O$_2$/H$_2$O potentials. Arrows highlight the possible electronic transitions between the different energy levels in H:TiO$_2$. Reprint from Ref. [21]. © American Chemical Society.

was due to the presence of V$_O$ sites which were believed to sit within the bandgap (Figure 9.11).[47] The photoelectrodes treated under hydrogen were found to show dramatically improved efficiencies vs. the untreated materials with higher photocurrent densities and photocurrent saturation occurring at low applied biases under white light illumination, with a STH efficiency of 1.1% being the highest reported at that date for a TiO$_2$ electrode. Perhaps surprisingly, given that the visible light absorption was proposed to be due to the presence of V$_O$, some visible light activity for the oxidation of water to dioxygen with hydrogen evolution occurring at the Pt counter electrode was also reported for the black 450°C sample at wavelengths between 400–650 nm even at low applied biases. Nonetheless, IPCE measurements showed that the primary reason for improved activity was due to improved utilization of UV absorbed photons at lower applied potentials. A near 100% IPCE value was recorded for photons below 380 nm indicating near complete suppression of electron–hole recombination losses following generation of electrons and holes in the CB and VB of TiO$_2$ respectively (see Eq. (5)).

Visible light photocatalytic or PEC water-splitting activity due to transitions to and from the V$_O$ traps as reported in this study is rare; the V$_O$ traps lie below the H$^+$/H$_2$ reduction potential, therefore, unless a significant electrical energy input is added into the system, the trapped electrons are unable to produce H$_2$, making visible light transitions to the trap state inactive.[21] Nonetheless, hydrogen production following visible light excitation of V$_O$ rich TiO$_2$ is not without precedent, as suspension photocatalysts have been observed to be active for hydrogen production in the presence of a sacrificial hole scavenger, suggesting that under certain conditions it is possible to further promote the trapped electrons to a suitable state for proton reduction.[52] Nonetheless, the most striking feature of the V$_O$ rich TiO$_2$ nanowires is not the limited visible light activity but instead the remarkably efficient utilization of UV light generated charge carriers. Several studies have explored the mechanistic factors controlling the efficiency of these materials using transient absorption (TA) and time-resolved fluorescence (TRF) spectroscopies.[53,54] TA and TRF enable the concentrations of electrons and holes to be directly probed. When accompanied by complimentary techniques, scavenger controls and kinetic analysis, useful insight into the locations, lifetimes, and fates of these excited charges can be obtained. TiO_2 has been studied with TA spectroscopy for over 25 years.[55] Although slight differences in charge carrier spectra are observed, depending on the electrolyte and material phase, it is generally accepted that on anatase TiO_2, trapped photoholes absorb light at λ ~450–550 nm, trapped photoelectrons absorb at λ ~ 800–900 and that CB photoelectrons have an absorption profile that increases in intensity with wavelength (>900 nm).[56–58] Following absorption of a UV photon, it has been shown consistently that trapping of holes and electrons occurs within 500 ps of the laser flash.[59] In contrast, electron–hole recombination occurs across a wide range of timescales (picosecond-to-millisecond) with the rate being highly dependent on the local environment and electron density.[56,60] TA studies on the hydrogen-treated nanowires of Wang *et al.*[21] have explored both the initial trapping and fast recombination events[53] and the slower electron density dependent electron–hole recombination events.[54] Through a combination of fs–TA and TRF Wheeler *et al.*[53] investigated the chemical nature of bandgap states in hydrogen-treated TiO_2, identifying the introduction of mid-bandgap states attributed to V_O, as well as non-fluorescent trap states situated below the localized V_O states.

They fitted PL signals of both untreated and hydrogen-treated TiO_2. In each sample, they observed a fast component and a slow component to their fits of the PL decay. On comparing the faster component of the fits between the two samples, they noted that the characteristically weak PL of rutile[61,62] was further reduced following hydrogen treatment, suggesting an alteration of the quantity and energetic position or depth of trap states within the bandgap leading to PL quenching. This observation was in line with HRTEM difference spectroscopy which indicated an increase in surface defect density in the crystalline lattice of the hydrogen-treated sample. In contrast, the slow component of the PL decay was further retarded following hydrogen treatment, which may indicate the formation of deeper lying traps. TA spectroscopy of both pristine and hydrogen-treated samples also confirmed the role of the induced trap states in changing recombination kinetics. UV excitation of the pristine TiO$_2$ gave an electron–hole TA signal containing three time constants (30 ps, 100 ps, $>$1.0 ns). Following hydrogen treatment, the lifetimes increased significantly (60 ps, 200 ps, and $>$1.0 ns), with a greater yield at $>$1 ns. This increase in overall charge carrier lifetime supported the proposed trapping of some of the charges possibly in the V$_O$ sites at $\sim$0.75 eV below the CB. The fastest decay component (30 ps and 60 ps) was tentatively assigned to electron relaxation from the bottom of the CB to both the fluorescent trap state in untreated TiO$_2$ and to a mixture of the fluorescent trap state and the V$_O$ sites in the hydrogen-treated sample. Once electrons become trapped, the localized nature of the V$_O$ sites prevents recombination, increasing the efficiency of photocatalysis under UV light.[53] On the other hand under visible light the hydrogen-treated TiO$_2$ only showed very small TA signals, the majority of which decayed within 30 ps, indicating that electrons excited from the VB to a deeper V$_O$ trap state (0.75 eV below the CB) are able to rapidly relax, potentially due to the presence of a manifold of close lying states (Figure 9.12). This lack of long-lived charge carrier signals is in line with the observed low IPCE values under visible illumination in the original PEC study.[20]

In the same year, Pesci *et al.*[54] used slow (μs–s) TA spectroscopy to study the same hydrogen-treated nanowires. In contrast to the past work which focused on the nature of the electron traps and on the fast trapping of charges, this work focused on the slower dynamics of electrons and holes in the PEC cell under operating conditions. This allows for the study of the factors controlling the efficiency of charge separation (η_{sep}) and transfer at the interface (η_{int}), Eq. (5). Notably under visible light, no long-lived photohole or photoelectron signals were observed, in line with the previous TA study that indicated that recombination on the picosecond timescale dominated.[53] Under UV light, the charge carrier yield was found to be very strongly dependent on the applied bias in the

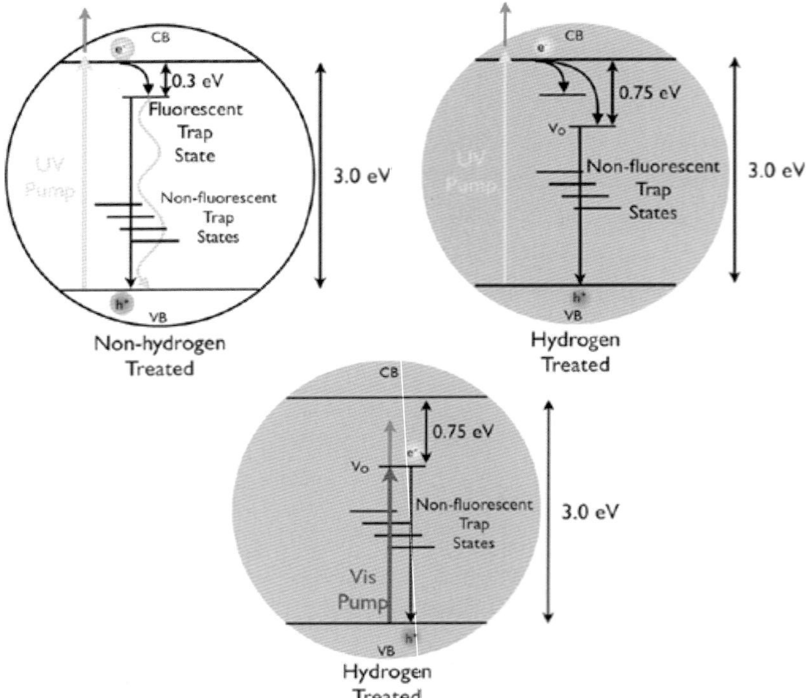

Figure 9.12. Proposed model for energy levels related to the optical properties and dynamics studies. For H:TiO$_2$ nanowires, visible pumping excites an electron to the V$_O$ state which is followed by non-radiative exciton decay mediated by a manifold of trap states. UV pumping for the same sample excites an electron across the bandgap which is followed by relaxation to the V$_O$ state and subsequent relaxation through the manifold of trap states. For TiO$_2$ nanowires, UV pumping (visible pumping not possible) likewise excites an electron across the bandgap, which is followed by relaxation into a fluorescent trap state situated ∼0.3 eV below the CB. Subsequent recombination was fast due to trap states in the bandgap. Reprint from Ref. [53]. © American Chemical Society.

hydrogen-treated material. At potentials close to the photocurrent onset it was noted that the initial yield of charge separation was much higher in hydrogen-treated materials and that the electron–hole lifetimes had decoupled, indicating a turning off of recombination (Figure 9.13). In the hydrogen-treated materials, the rate of electron transfer through the material to the external circuit could be

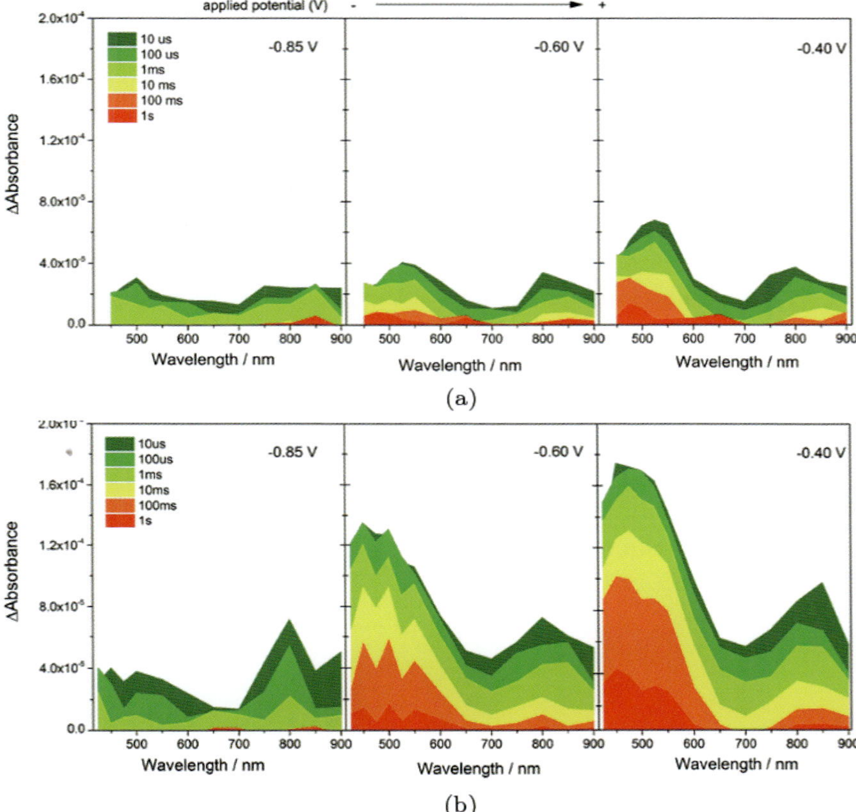

Figure 9.13. TA spectra recorded after excitation of (a) A:TiO₂ and (b) H:TiO₂ in 1 M NaOH(aq) at the bias indicated (*vs.* Ag/AgCl), 355 nm excitation, $70\,\mu\mathrm{Jcm}^{-2}$, 0.33 Hz laser repetition rate. The photohole signal can be seen at *ca.* 500 nm and the photoelectron at *ca.* 850 nm. Reprint from Ref. [54]. © American Chemical Society.

measured and the lack of recombination allowed for a direct study of the hole transfer across the interface which occurred with a lifetime on the order of $0.03\,\mathrm{s}$.[63]

The apparent turning off of electron–hole recombination in the hydrogen-treated material can be interpreted within the context of the model developed by Gartner[64] and Gerischer,[65,66] in which electron–hole pair separation is driven by the presence of a depletion layer that drives holes toward the SCLJ and electrons away from the interface, as mentioned above in Section 1.2. The maximum possible depletion layer depth depends upon the donor density, the dielectric constant, the radius of the nanowire and the distance from the substrate contact.[67,68] In highly nanostructured materials, it is common for the radial dimensions to be significantly smaller than the width of the space charge layer at a given applied potential, leading to complete depletion and limiting the degree of band bending achievable.[69] Indeed, in the study of the hydrogen-treated nanowires, it was shown that untreated TiO$_2$ wires had a depletion width which would exceed the material's dimensions under most applied biases. In the hydrogen-treated materials, the presence of the high concentration of Ti^{3+} sites leads to a large increase in the r donor density (N_d) ($\sim$4 orders of magnitude increase as measured by Mott–Schottky analysis).

The relationship between the space charge layer and the donor density is shown in Eq. (6):

$$W_{\text{SC}} = \sqrt{\frac{2\varepsilon\varepsilon_0}{qN_D}(E - E_{\text{FB}})}, \tag{6}$$

where W_{sc} is the width of the space charge layer, N_D is the donor density, ε is the dielectric constant of the surface film, ε_0 is the dielectric constant of the vacuum, q is the elementary charge, E is the potential and E_{FB} is the flat band potential. As can be seen by Eq. (6), this dramatically decreases the width of the space charge layer, making the individual nanowires thick enough to support a sufficiently large radial electric field for effective spatial charge separation, leading to both high initial charge carrier yields and the suppression of slow ($>\mu$s) recombination, Figure 9.14(b). In the untreated TiO$_2$, the inability to maintain a significantly large radial electrical field leads to higher levels of recombination losses and lower IPCE values, Figure 9.14(a). These reports are in good agreement with those initially proposed by Wang *et al.*[21]

A further development to improve the activity of the hydrogen-treated nanowires has also been reported.[70] Following treatment at atmospheric pressure at 400°C for 30 min to form

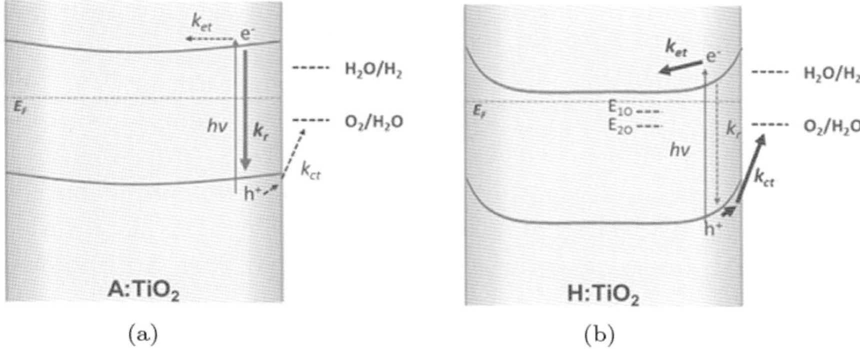

Figure 9.14. Simplified energy diagram for a cross-section of (a) A:TiO$_2$ and (b) H:TiO$_2$ under a positive applied bias at a distance away from the FTO interface showing the key kinetic processes occurring following absorption of UV light in which k_{ct}, k_r, and k_{et} correspond to the rates of charge transfer into solution, recombination, and electron transport and collection at the FTO Interface. E_{1O} and E_{2O} correspond to Vo at 0.75 eV and 1.2 eV below the CB edge. As described in the main text, A:TiO$_2$ is anticipated to be fully depleted at even moderate applied biases. Reprint from Ref. [54]. © The American Chemical Society.

gray arrays of rutile TiO$_2$, the sample can be treated with an aqueous TiCl$_4$ solution to form an anatase surface layer. After comparing efficiencies between NW photoanodes, hydrogen-treated NW photoanodes and hydrogen-treated NW photoanodes with a TiCl$_4$ treatment, it was shown that the surface treatment led to an increase in activity.[47,54] This was proposed to be due to the selective removal of surface trap sites, which may in some cases aid recombination, whilst maintaining the bulk V_O concentration which is critical for the improved electrical properties. It should however be noted that the potential role of the anatase/rutile junction acting as a charge separation promoter was also not ruled out making it important that further studies explore the potential for surface passivation treatments in enhancing activity.

9.4.1 *Surface Reaction Kinetics in Oxygen Deficient TiO$_2$*

At this stage, it is worth briefly mentioning the potential role that V_{Os} and surface disorder states could play in modifying the binding and reactivity of substrates in photocatalytic water splitting. FTIR studies of core–shell TiO_2 have been used to examine the change in surface OH groups following hydrogen treatment.[27,37,71–73] A change in the nature or quantity of a surface species such as OH groups on TiO_2 is likely to significantly affect surface reaction kinetics in water-splitting applications. The FTIR carried out by Chen *et al.*[27] showed that OH groups experienced a more varied environment following hydrogen treatment and that a disordered phase on the outside of their crystalline anatase nanoparticle cores was present, with H atoms partially bonded to O and Ti atoms. The surface interaction between TiO_2 and H_2O is also an important consideration for materials containing V_O at or near the surface. It has been reported that the chemically dissociated water molecules are energetically favored on a defective TiO_2 surface; however, water molecules are only physically adsorbed on the perfect TiO_2 surface.[74] The interaction of oxygen with TiO_2 is also of interest when considering photocatalytic applications due its potential role in electron scavenging. Thompson *et al.*[50] noted that V_O defects on the surface of TiO_2 are likely to play an essential part in governing the adsorption of O_2 molecules, which are believed to interact strongly with Ti^{3+}, offering a route to retarding charge recombination. We are aware of only a very limited number of studies on the hole transfer step into water on photocatalytic and photoelectrode materials. TA has measured the lifetime on holes as being up to ∼0.03–0.4 s on TiO_2, depending on the electrolyte pH[60,63] leading to the suggestion that this is the minimum required lifetime for charge transfer to occur. Pesci *et al.*[54] assessed the possible contribution of modified hole transfer kinetics to the activity of hydrogen-treated TiO_2, however it was found that the rate of hole transfer was not significantly modified, despite the surface of the TiO_2 material being rich in V_O sites. It would be intriguing however to see if the rate of hole transfer into water was notably modified on core–shell TiO_2 materials where the energetics of the VB had been modified, changing the thermodynamic driving force for hole transfer.

9.5 Future Directions for Hydrogen-Treated TiO$_2$

Black TiO$_2$ is clearly an exciting and ever developing field to be involved in, offering great potential in photocatalytic water-splitting applications and beyond. However, grouping what is now apparently a relatively diverse range of materials under the single name "black TiO$_2$" provides a challenge when it comes to rationalizing the efficiency of these materials. Nonetheless, common features of all materials, both oxygen deficient and core–shell, have become apparent during the writing of this chapter. In nearly all cases, the very large increase in photocatalytic activity upon hydrogen treatment can be assigned to more efficient use of the UV photons that are absorbed. Despite the materials being black in color, the visible light absorption appears to be offering only a slight increase in the photocatalytic activity. This may appear disheartening, however a two orders of magnitude increase in activity for suspension photo-catalysis[17] with core–shell materials, or a near 100% IPCE for defect rich photoelectrodes is a remarkable result.[21] Exciting new research is now trying to combine the advantages of improved charge separation and utilization upon hydrogen treatment with other more traditional treatments to impart visible light activity.[75–77] Here we conclude by considering two of these studies. Nitrogen doping of TiO$_2$ has been long explored as a route to obtaining visible light activity, however the materials developed typically have low charge separation yields in the visible. Recently[76] it was demonstrated that by treating TiO$_2$ nanowire arrays with both H$_2$ and NH$_3$ it was possible to combine the improved electrical properties of hydrogen-treated TiO$_2$ with the light harvesting properties of N-doped TiO$_2$. Photoelectrochemical water oxidation studies showed that following the dual treatments, electrodes could be prepared with significant levels of activity at wavelengths as long as 550 nm. Another co-treatment approach has been presented by Wang *et al.*[77] who coupled hydrogen treatment of TiO$_2$ nanowires with CdS quantum dot sensitization to enhance photoactivities in the wavelength region from 350 nm to 550 nm. Interestingly as a consequence of this, the UV activity of the CdS/hydrogen-treated samples was reduced compared to the sample

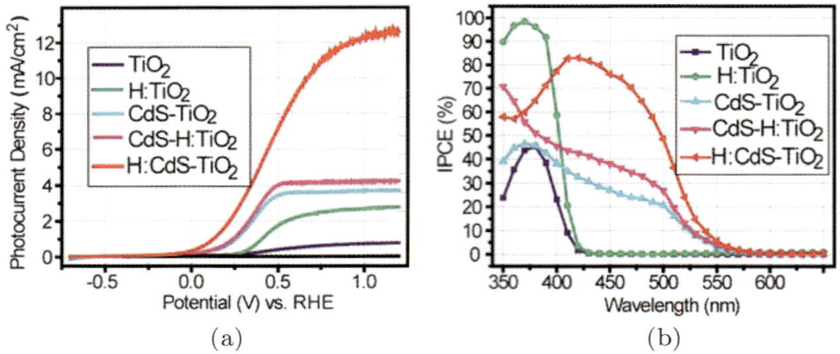

Figure 9.15. (a) Linear sweep voltammograms of the pristine TiO_2, $H{:}TiO_2$, $CdS{-}TiO_2$, $CdS{-}H{:}TiO_2$ and $H{:}CdS{-}TiO_2$ nanowires, measured at a scan rate of $20~mV\,s^{-1}$ under $100~mW\,cm^{-2}$ white light illumination. (b) Corresponding IPCE spectra collected at a potential of 0.5 V vs. RHE. Reprint from Ref. [77]. © The Royal Society of Chemistry.

only treated with hydrogen (Figure 9.15). It was found that by hydrogen treating after CdS sensitization, the crystallinity of the CdS was improved while simultaneously increasing the donor density of the TiO_2 by V_O formation.

The co-treatments reported to date, of which those discussed above are just a selection, show great promise as alternative routes towards further enhancing water-splitting activity of black TiO_2 and can lead to interesting property alterations. Due to the simplicity of such co-treatments and the wide range of possibilities, this could be a potential route to obtain activity from doped and sensitized materials that have previously been reported and found to be of limited use.

References

(1) Navarro Yerga, R. M.; Álvarez Galván, M. C.; del Valle, F.; Villoria de la Mano, J. A.; Fierro, J. L. G. *ChemSusChem* **2009**, *2*, 471–485.

(2) Rajeshwar, K.; McConnell, R.; Licht, S. *Solar Hydrogen Generation*, Springer New York, New York, NY, 2008.

(3) Pinaud, B. A.; Benck, J. D.; Seitz, L. C.; Forman, A. J.; Chen, Z.; Deutsch, T. G.; James, B. D.; Baum, K. N.; Baum, G. N.; Ardo, S.; Wang, H.; Miller, E.; Jaramillo, T. F. *Energy Environ. Sci.* **2013**, *6*, 1983.

(4) Maeda, K.; Domen, K. *J. Phys. Chem. Lett.* **2010**, *1*, 2655–2661.

(5) Lewis, N. S. *Inorg. Chem.* **2005**, *44*, 6900–6911.

(6) Kudo, A.; Miseki, Y. *Chem. Soc. Rev.* **2009**, *38*, 253–278.

(7) Bard, A. J. *J. Photochem.* **1979**, *10*, 59–75.

(8) Chen, Z.; Jaramillo, T. F.; Deutsch, T. G.; Kleiman-Shwarsctein, A.; Forman, A. J.; Gaillard, N.; Garland, R.; Takanabe, K.; Heske, C.; Sunkara, M.; McFarland, E. W.; Domen, K.; Miller, E. L.; Turner, J. A.; Dinh, H. N. *J. Mater. Res.* **2010**, *25*, 3–16.

(9) Cowan, A. J.; Leng, W.; Barnes, P. R. F.; Klug, D. R.; Durrant, J. R.; *Phys. Chem. Chem. Phys.* **2013**, *15*, 8772–8778.

(10) Walter, M. G.; Warren, E. L.; McKone, J. R.; Boettcher, S. W.; Mi, Q.; Santori, E. A.; Lewis, N. S. *Chem. Rev.* **2010**, *110*, 6446–6473.

(11) Hashimoto, K.; Irie, H.; Fujishima, A. *Jpn. J. Appl. Phys.* **2006**, *44*, 8269–8285.

(12) Ni, M.; Leung, M. K. H.; Leung, D. Y. C.; Sumathy, K. *Renew. Sustain. Energy Rev.* **2007**, *11*, 401–425.

(13) Szymanski, P.; El-Sayed, M. A.; *Theor. Chem. Acc.* **2012**, *131*, 1202.

(14) Fujishima, A.; Honda, K. *Nature* **1972**, *238*, 37–38.

(15) Scanlon, D. O.; Dunnill, C. W.; Buckeridge, J.; Shevlin, S. A.; Logsdail, A. J.; Woodley, S. M.; Catlow, C. R. A.; Powell, M. J.; Palgrave, R. G.; Parkin, I. P.; Watson, G. W.; Keal, T. W.; Sherwood, P.; Walsh, A.; Sokol, A. A. *Nat. Mater.* **2013**, *12*, 798–801.

(16) *http://rredc.nrel.gov/solar/spectra/am1.5/*.

(17) Chen, X.; Liu, L.; Yu, P. Y.; Mao, S. S. *Science* **2011**, *331*, 746–750.

(18) Cronemeyer, D. C. *Phys. Rev.* **1959**, *113*, 1222–1226.

(19) Reported by google scholar as of 05/01/2016.

(20) Chen, X.; Liu, L.; Huang, F. *Chem. Soc. Rev.* **2015**, *44*, 1861–1885.

(21) Wang, G.; Wang, H.; Ling, Y.; Tang, Y.; Yang, X.; Fitzmorris, R. C.; Wang, C.; Zhang, J. Z.; Li, Y. *Nano Lett.* **2011**, *11*, 3026–3033.

(22) Naldoni, A.; Allieta, M.; Santangelo, S.; Marelli, M.; Fabbri, F.; Cappelli, S.; Bianchi, C. L.; Psaro, R.; Dal Santo, V.; *J. Am. Chem. Soc.* **2012**, *134*, 7600–7603.

(23) Yu, X.; Kim, B.; Kim, Y. K. *ACS Catal.* **2013**, *3*, 2479–2486.

(24) Huo, J.; Hu, Y.; Jiang, H.; Li, C. *Nanoscale* **2014**, *6*, 9078.

(25) Jiang, X.; Zhang, Y.; Jiang, J.; Rong, Y.; Wang, Y.; Wu, Y.; Pan, C. *J. Phys. Chem. C* **2012**, *116*, 22619–22624.

(26) Yang, C.; Wang, Z.; Lin, T.; Yin, H.; Lu, X.; Wan, D.; Xu, T.; Zheng, C.; Lin, J.; Huang, F.; Xie, X.; Jiang, M. *J. Am. Chem. Soc.* **2013**, *135*, 17831–17838.

(27) Chen, X.; Liu, L.; Liu, Z.; Marcus, M. A.; Wang, W.-C.; Oyler, N. A.; Grass, M. E.; Mao, B.; Glans, P.-A. Yu, P. Y.; Guo, J.; Mao, S. S. *Sci. Rep.* **2013**, *3*, 1510.

(28) Di Valentin, C.; Pacchioni, G.; Selloni, A. *J. Phys. Chem. C* **2009**, *113*, 20543–20552.

(29) Liu, H.; Ma, H. T.; Li, X. Z.; Li, W. Z.; Wu, M.; Bao, X. H.; *Chemosphere* **2003**, *50*, 39–46.

(30) Heller, A.; Degani, Y.; Johnson, D. W.; Gallagher, P. K. *J. Phys. Chem.* **1987**, *91*, 5987–5991.

(31) Haerudin, H.; Bertel, S.; Kramer, R. *J. Chem. Soc. Faraday Trans.* **1998**, *94*, 1481–1487.

(32) Cronemeyer, D. C.; Gilleo, M. A. *Phys. Rev.* **1951**, *82*, 975–976.

(33) Liu, L.; Yu, P. Y.; Chen, X.; Mao, S. S.; Shen, D. Z. *Phys. Rev. Lett.* **2013**, *111*, 065505.

(34) Raghunath, P.; Huang, W. F.; Lin, M. C. *J. Chem. Phys.* **2013**, *138*, 154705.

(35) T. S. Bjørheim, Kuwabara, A.; Norby, T. *J. Phys. Chem. C* **2013**, *117*, 5919–5930.

(36) Lu, J.; Dai, Y.; Jin, H.; Huang, B. *Phys. Chem. Chem. Phys.* **2011**, *13*, 18063.

(37) Zheng, Z.; Huang, B.; Lu, J.; Wang, Z.; Qin, X.; Zhang, X.; Dai, Y.; Whangbo, M.-H. *Chem. Commun.* **2012**, *48*, 5733.

(38) Tian, M.; Mahjouri-Samani, M.; Eres, G.; Sachan, R.; Yoon, M.; Chisholm, M. F.; Wang, K.; Puretzky, A. A.; Rouleau, C. M.; Geohegan, D. B.; Duscher, G. *ACS Nano* **2015**, *9*, 10482–10488.

(39) Berr, M. J.; Wagner, P.; Fischbach, S.; Vaneski, A.; Schneider, J.; Susha, A. S.; Rogach, A. L.; Jäckel, F.; Feldmann, J. *Appl. Phys. Lett.* **2012**, *100*, 223903.

(40) Xia, T.; Chen, X. *J. Mater. Chem. A* **2013**, *1*, 2983.

(41) Bard, A. J. *J. Am. Chem. Soc.* **2010**, *132*, 7559–7567.

(42) Zeng, L.; Song, W.; Li, M.; Zeng, D.; Xie, C. *Appl. Catal. B Environ.* **2014**, *147*, 490–498.

(43) Teng, F.; Li, M.; Gao, C.; Zhang, G.; Zhang, P.; Wang, Y.; Chen, L.; Xie, E. *Appl. Catal. B Environ.* **2014**, *148–149*, 339–343.

(44) Pan, X.; Yang, M.-Q.; Fu, X.; Zhang, N.; Xu, Y.-J. *Nanoscale* **2013**, *5*, 3601.

(45) Kim, W.-T.; Kim, C.; Choi, Q. W. *Phys. Rev. B* **1984**, *30*, 3625–3628.

(46) Riley, R. *J. Phys. Chem. C* **2012**, *116*, 22619–22624.

(47) Wang, G.; Wang, H.; Ling, Y.; Tang, Y.; Yang, X.; Fitzmorris, R. C.; Wang, C.; Zhang, J. Z.; Li, Y. *Nano Lett.* **2011**, *11*, 3026–3033.

(48) Kong, M.; Li, Y.; Chen, X.; Tian, T.; Fang, P.; Zheng, F.; Zhao, X. *J. Am. Chem. Soc.* **2011**, *133*, 16414–16417.

(49) Fong, C.; Dong, A. W.; Hill, A. J.; Boyd, B. J.; Drummond, C. J. *Phys. Chem. Chem. Phys.* **2015**, *17*, 17527–17540.

(50) Thompson, T. L.; Yates, J. T. *Top. Catal.* **2005**, *35*, 197–210.

(51) Leshuk, T.; Parviz, R.; Everett, P.; Krishnakumar, H.; Varin, R. A.; Gu, F. *ACS Appl. Mater. Interfaces* **2013**, *5*, 1892–1895.

(52) Zuo, F.; Wang, L.; Wu, T.; Zhang, Z.; Borchardt, D.; Feng, P. *J. Am. Chem. Soc.* **2010**, *132*, 11856–11857.

(53) Wheeler, D. A.; Ling, Y.; Dillon, R. J.; Fitzmorris, R. C.; Dudzik, C. G.; Zavodivker, L.; Rajh, T.; Dimitrijevic, N. M.; Millhauser, G.; Bardeen, C.; Li, Y.; Zhang, J. Z. *J. Phys. Chem. C* **2013**, *117*, 26821–26830.

(54) Pesci, F. M.; Wang, G.; Klug, D. R.; Li, Y.; Cowan, A. J. *J. Phys. Chem. C* **2013**, *117*, 25837–25844.

(55) Bahnemann, D.; Henglein, A.; Lilie, J.; Spanhel, L. *J. Phys. Chem.* **1984**, *88*, 709–711.

(56) Tang, J.; Durrant, J. R.; Klug, D. R. *J. Am. Chem. Soc.* **2008**, *130*, 13885–13891.

(57) Yoshihara, T.; Katoh, R.; Furube, A.; Tamaki, Y.; Murai, M.; Hara, K.; Murata, S.; Arakawa, H. *J. Phys. Chem. B* **2004**, *108*, 3817–3823.

(58) Yoshihara, T.; Tamaki, Y.; Furube, A.; Murai, M.; Hara, K.; Katoh, R.; *Chem. Phys. Lett.* **2007**, *438*, 268–273.

(59) Tamaki, Y.; Furube, A.; Murai, M.; Hara, K.; Katoh, R.; Tachiya, M. *Phys. Chem. Chem. Phys.* **2007**, *9*, 1453.

(60) Cowan, A. J.; Tang, J.; Leng, W.; Durrant, J. R.; Klug, D. R. *J. Phys. Chem. C* **2010**, *114*, 4208–4214.

(61) Ghosh, J. Amal K.; Wakim, F. G.; Addiss, R. R. *Phys. Rev.* **1969**, *184*, 979.

(62) Hachiya, K.; Kondoh, J. *Phys. B Condens. Matter* **2003**, *334*, 130–134.

(63) Cowan, A. J.; Barnett, C. J.; Pendlebury, S. R.; Barroso, M.; Sivula, K.; Grätzel, M.; Durrant, J. R.; Klug, D. R. *J. Am. Chem. Soc.* **2011**, *133*, 10134–10140.

(64) Gärtner, W. W. *Phys. Rev.* **1959**, *116*, 84–87.

(65) Genscher, H. *J. Electrochem. Soc.* **1966**, *113*, 1174.

(66) Södergren, S.; Hagfeldt, A.; Lindquist, S. *J. Phys. Chem.* **1994**, *98*, 5552–5556.

(67) Bisquert, J.; Garcia-Belmonte, G.; Fabregat-Santiago, F. *J. Solid State Electrochem.* **1999**, *3*, 337–347.

(68) Barnes, P. R. F.; Miettunen, K.; Li, X.; Anderson, A. Y.; Bessho, T.; Gratzel, M.; O'Regan, B. C. *Adv. Mater.* **2013**, *25*, 1881–1922.

(69) Law, M.; Greene, L. E.; Johnson, J. C.; Saykally, R.; Yang, P. *Nat. Mater.* **2005**, *4*, 455–459.

(70) Wang, D.; Zhang, X.; Sun, P.; Lu, S.; Wang, L.; Wang, C.; Liu, Y.; *Electrochim. Acta* **2014**, *130*, 290–295.

(71) Wang, Z.; Yang, C.; Lin, T.; Yin, H.; Chen, P.; Wan, D.; Xu, F.; Huang, F.; Lin, J.; Xie, X.; Jiang, M. *Adv. Funct. Mater.* **2013**, *23*, 5444–5450.

(72) Xia, T.; Zhang, C.; Oyler, N. A.; Chen, X. *Adv. Mater.* **2013**, *25*, 6905–6910.

(73) Xia, T.; Cao, Y.; Oyler, N. A.; Murowchick, J.; Liu, L.; Chen, X. *ACS Appl. Mater. Interfaces* **2015**, *7*, 10407–10413.

(74) Pan, X.; Yang, M.-Q.; Fu, X.; Zhang, N.; Xu, Y.-J. *Nanoscale* **2013**, *5*, 3601–3614.

(75) Rahman, M. A.; Bazargan, S.; Srivastava, S.; Wang, X.; Abd-Ellah, M.; Thomas, J. P.; Heinig, N. F.; Pradhan, D.; Leung, K. T. *Energy Environ. Sci.* **2015**, *8*, 3363–3373.

(76) Hoang, S.; Berglund, S. P.; Hahn, N. T.; Bard, A. J.; Mullins, C. B. *J. Am. Chem. Soc.* **2012**, *134*, 3659–3662.

(77) Wang, H.; Wang, G.; Ling, Y.; Lepert, M.; Wang, C.; Zhang, J. Z.; Li, Y.; *Nanoscale* **2012**, *4*, 1463–1466.

CHAPTER TEN

Black TiO$_2$ Nanomaterials for Lithium-Ion Batteries

Laifa Shen[*,†], *Shengyang Dong*[*], *Xiaogang Zhang*[*]
and Guozhong Cao[†]

[*]*College of Material Science and Engineering,
Nanjing University of Aeronautics and Astronautics,
Nanjing 210016, China*

[†]*Department of Materials Science and Engineering,
University of Washington, Seattle, WA 98195, USA*

10.1 Introduction of Lithium Ion Batteries

Lithium Ion Batteries (LIBs) have achieved great success in commercialization and are the most widely used in mobile electronics and considered one of the most promising energy storage devices for electric vehicles (EVs), hybrid electric vehicles (HEVs) and other electric utilities.[1–3] This is because LIBs have a high voltage as well as a large charge–discharge capacity, and less likely to have unfavorable influence caused by a memory effect and the like, compared to nickel cadmium batteries and the like.[4,5] In a typical LIBs configurations: anode, cathode, and electrolyte are the three major components. The LIBs function by converting a chemical potential into electrical energy via Faradaic reactions, accompanied by insertion/extraction of Li$^+$ ions into/from the structure of an electronic and ionic conductive solid electrode. In general, capacity, cyclability, and rate capability are the three most important performance indicators for the LIBs, which are strongly dependent on the properties of the active electrode materials.[6,7] As far as the electrodes are concerned, the composition, crystal structure, and morphology can influence the reaction rate and transfer processes and can be manipulated to alter the overall electrochemical performance of the device.

Commercial LIBs are usually composed of a graphitic anode coupled with a layered type $LiCoO_2$ cathode in the presence of an intervening electrolyte of $LiPF_6$ in a mixture of organic solvents. The working mechanism for a typical conventional LIB is as follows.[8] On charging, Li ions are deintercalated from the $LiCoO_2$ cathodes, pass across the electrolyte, and intercalated into graphite anodes. The above process is reversed during discharge. Li ions would diffuse through solid state diffusion while electrons move in the opposite direction and pass around the external circuit. Graphite has been widely used as anode material in commercial LIBs. Unfortunately, using graphitic anode-based LIBs for high power applications such as EVs and HEVs are not possible due to the following two problems: first, the lithium acceptance and lithium removal process were accompanied by the severe volume expansion and shrinkage of the graphite particles, resulting in the cracking of the active materials and degrading the electrical contact between the particles, which will lead to the decrease of the capacity retention upon cycling; second, the working potential for the graphite is in the region of 0–0.5 V (vs. Li/Li$^+$) as they are employed as anodes in LIBs. In such a low operating voltage region, the electrolyte is prone to decompose, resulting in the release of gases and accumulates pressure for possible explosion.[9,10] On the other hand, it will also give rise to the formation of lithium dendrite and solid–electrolyte interface (SEI) layer on the anodes' surface, which in turn will promote the decomposition of electrolyte.[11] The commercial LIBs are inadequate to meet high power applications, advanced materials with better safety and high rate capability are critical for next-generation LIBs.

Among various electrode materials, titanium-based materials, including various TiO_2 polymorphs and lithium titanate, are considered alternatives to graphitic anodes in LIBs because of their inherently safe, high abundance, non-toxicity, and structural integrity over many charge/discharge cycles.[7, 12] However, the Li-ion storage performance of TiO$_2$-based batteries is severely limited by poor chemical diffusion of lithium ions in all polymorphs resulting from both (i) slow lithium-ion diffusion and (ii) poor electronic conductivity. Tailoring the particle size from micrometer to nanometer is a key to improved Li$^+$ diffusion, because of the reduced diffusion length.[13–15] Introducing dopants or defects into the TiO$_2$ matrix is an effective way to enhance its electric conductivity owing to the presence of Ti^{3+} species and a large amount of oxygen vacancies.[13, 16] Recently, highly conductive black TiO$_2$ with disordered surface layer has attracted enormous attention for fast lithium storage in LIBs.[13, 16–18] For example, annealing TiO$_2$ nanotubes in CO has been found to significantly improve lithium-ion intercalation capacity and rate capability than N$_2$ annealed samples.[19] Hydrogenated anatase TiO$_2$ nanoparticles prepared by a H$_2$ plasma treatment show improved fast lithium storage capability compared to the pristine TiO$_2$ nanoparticles.[20] In this chapter, the recent progress on the black TiO$_2$ nanomaterials for LIBs has been reviewed here, and special emphasis have been given on the deeply understanding of the mechanism of the enhanced lithium-ion storage of black TiO$_2$ nanomaterials.

10.2   Mechanism and Advantages of Black TiO$_2$ for LIBs

10.2.1   *Mechanism of the Lithium-Ion Storage in TiO$_2$*

TiO$_2$ polymorphs reported to date include rutile, anatase, brookite, TiO$_2$–B (bronze), TiO$_2$–R (ramsdellite), TiO$_2$–H (hollandite), TiO$_2$–II (columbite), and TiO$_2$–III (baddeleyite). Among all the TiO$_2$ polymorphs, rutile, anatase, brookite, and TiO$_2$–B have been reported suitable as Li-insertion host. It is generally considered that at low pressures only rutile has a true field of stability; anatase and brookite form a metastable structure. The structure of TiO$_2$ polymorphs.[12]

The following typical lithium intercalation–deintercalation reaction occurs during the electrochemical process on the TiO$_2$

polymorphs anode in the potential range of 1.4–1.8 V vs. Li/Li$^+$,

$$TiO_2 + xLi^+ + xe^- = Li_xTiO_2 \quad (0 \leq x \leq 1). \tag{1}$$

This electrochemical reaction implies not only the insertion of xLi^+, but also the creation of charge compensating xTi^{3+} cations in the Ti^{4+} sublattice, as observed in X-ray photoelectron spectroscopy (XPS) experiments and supported by theoretical calculations,[21–23] with sources of structural strain and relaxation associated with both types of induced defects.[24] The capacity of TiO$_2$ polymorphs to undergo this redox reaction, and relative phase stabilities as a function of Li$^+$ content, have been examined closely with both computational molecular modeling and experiment.[25–29] For modeling studies, both quantum mechanical and empirical potential atomistic modeling have been used to predict relative phase stabilities, but not without occasional contradiction or disagreement with experiment. In one of the earliest molecular modeling studies for this purpose, Mackrodt performed periodic Hartree–Fock (HF) structure optimizations for a number of TiO$_2$ and LiTiO$_2$ polymorphs with great success.[30] Predicted relative stabilities of TiO$_2$ polymorphs include rutile > anatase > brookite > ramsdellite > spinel, with calculated energy difference between rutile and anatase to be 0.02–0.06 eV, in excellent agreement with density functional theory (DFT) calculation at the LDA level[12] and the measured H of Navrotsky and Kleppa.[31] Because of low packing density, anatase (density, $\rho = 3.89 \, \text{g cm}^{-3}$) and TiO$_2$–B ($\rho = 3.73 \, \text{g cm}^{-3}$) phases show higher lithium storage capacity as compared to densely packed brookite ($\rho = 4.13 \, \text{g cm}^{-3}$) and rutile ($\rho = 4.25 \, \text{g cm}^{-3}$) phases, both anatase TiO$_2$ and TiO$_2$–B have been extensively investigated as potential alternative anode material for LIBs.[32,33] Specially, considering the safety concerning overcharging and the stable voltage plateau at 1.78 V, anatase TiO$_2$, a typical Li-ion intercalation compound, is often used as the negative electrode in LIBs.[34,35] TiO$_2$–B is the least dense polymorph of TiO$_2$, and has a theoretical capacity of 335 mA h g^{-1}. TiO$_2$–B has an open structure with freely accessible channels for Li-ion transport perpendicular to the (010) face, which allows easy Li-ion transport. It has been demonstrated that the kinetics of lithium ion storage in

TiO$_2$–B is governed by a pseudocapacitive Faradaic process, which is not limited by solid state diffusion of Li ions.[36,37] Apart from the TiO$_2$, the spinel Li$_4$Ti$_5$O$_{12}$, a white non-conductive crystal, has been viewed as one promising anode material for high power LIBs. The Li$_4$Ti$_5$O$_{12}$ can reversibly accommodate three Li ions per formula unit with a two-phase equilibrium for 175 mA h g^{-1} at a flat potential of 1.55 V vs. Li/Li$^+$. Additionally, as a zero-strain insertion material, the spinel Li$_4$Ti$_5$O$_{12}$ possesses excellent reversibility and structural stability during the charge–discharge process.[38]

Unfortunately, titanium-based materials suffer from low electrical conductivity due to the empty Ti 3d state with a wide bandgap energy (3.2 eV) and small lithium-ion diffusivity throughout the bulk TiO$_2$ (10^{-11}–10^{-13} cm^2 s^{-1}),[39] which seriously hindered the exploration of the full advantages of TiO$_2$-based LIBs. The most commonly used strategies are to reduce the electroactive particle size to nanoscale and to coat conductive layers on the TiO$_2$ particle surface.[38,40,41] Usually, carbon-based coating via chemical vapor deposition or a hydro/solvothermal process is the most widely used.[33,42] Because of safety issues and side reactions concerning carbon materials, more attention has been directed to self-structure modifications of TiO$_2$ with controlled electronic properties.

10.2.2 *The Advantages of Black TiO$_2$ for LIBs*

Recently, the highly conductive TiO$_2$ nanostructure called "black TiO$_2$" has attracted a lot of research for both fundamental understanding and for applications particularly in energy conversion and storage including photovoltaics, photocatalysis, and fuel cells, as addressed in this book, because of its narrow bandgap of approximately 2.2 eV and relatively high electrical conductivity (10^{-2}–10^{-5} S cm^{-1}).[43–46] Black TiO$_2$ is simply prepared via thermal treatment under vacuum or in a reducing atmosphere, and Ti^{3+} or/and oxygen vacancies that cause the color changed were formed and introduced in titania crystals during such thermal treatment.[16,18,20,47,48] Self-structural modification of TiO$_2$ is a very versatile approach, which can be obtained by any type of synthesis during the post-synthesis treatment. In 2011, for the first time,

Shin *et al.* reported black TiO$_{2-\delta}$ nanoparticles prepared by hydrogen reduction that can achieve significantly enhanced capacity and better rate capability.[16] This discovery has triggered worldwide research interest in black TiO$_2$ nanomaterials for LIBs. The effect of structural properties of black TiO$_2$ on the improved lithium storage properties are briefly summarized in the following sections.

(1) Enhanced electric conductivity

It is generally believed that the donor density of black TiO$_2$ nanomaterials proportional to electrical conductivity is significantly improved by the formation of oxygen vacancies/Ti^{3+} sites,[49] thus, leading to greatly enhanced lithium storage performance. The existence of oxygen vacancies and Ti^{3+} ions in the black TiO$_2$ nanomaterials was found from hydrogen treatment,[16,43,50,51] and chemical reduction.[52] For example, by annealing at 350°C for 3 h in H$_2$ atmosphere, a larger amount of Ti^{3+} sites and oxygen vacancies were introduced into the structures of TiO$_2$(B), which dramatically increase the electronic conductivity of TiO$_2$(B) by several orders of magnitude, up to $2.79 \times 10^{-3}\,\mathrm{S\,cm^{-1}}$ compared to that of pristine TiO$_2$(B) nanostructures ($8.34 \times 10^{-10}\,\mathrm{S\,cm^{-1}}$).[53] The influence of hydrogen time on the electrical conductivity is studied in detail by Shin *et al.*[16] As shown in Figure 10.1, after hydrogen thermal treatment for 1 h (H$_2$-1 h-TiO$_{2-\delta}$), the conductivity (at 723 K) was increased by around 1 order of magnitude, showing $\approx 1.5 \times 10^{-3}\,\mathrm{S\,cm^{-1}}$, whereas more than 2 orders of magnitude higher conductivity was obtained upon 7 h treatment (H$_2$-7 h-TiO$_{2-\delta}$) ($\approx 4.4 \times 10^{-2}\,\mathrm{S\,cm^{-1}}$).[16]

(2) Enhanced diffusion coefficient of Li-ions

Different from the white TiO$_2$, amorphous or disordered surface layer in a few nanometers can be observed in black TiO$_2$ nanoparticles. After introducing oxygen vacancies and hydroxyl groups in this layer, the overlap of the titanium and oxygen orbitals decreases and the interaction between the transferred charge Li$^+$ and the host matrix becomes weaker. This will lead to easier charge transport within the modified surface layer, hence enabling the fast lithium storage ability.

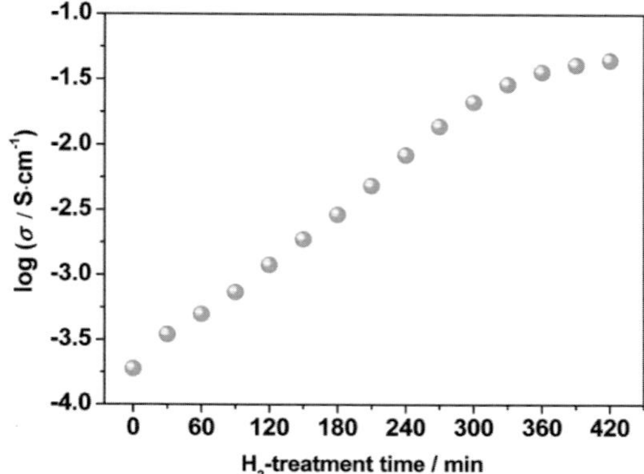

Figure 10.1. Conductivity relaxation plot of TiO_2 at 723 K under 5% H_2/95% Ar atmosphere. Reprinted from Ref. [16]. © 2012 American Chemical Society.

Zhang *et al.* calculated that the lithium-ion diffusion coefficient (D_{Li}) value of the hydrogenated TiO_2(B) electrode is 6.81×10^{-13} cm^2 s^{-1}, which is about two times higher than that of pristine TiO_2(B) electrode (3.50×10^{-13} cm^2 s^{-1}).[53] Shin *et al.* revealed that the D_{Li} of black TiO_2 via 1 h hydrogen thermal treatment is effective increased to $\approx 9.6 \times 10^{-18}$ cm^2 · s^{-1}, which is 2 orders of magnitude higher diffusivity than for the pristine (stoichiometric) TiO_2.[16] The effect of hydrogenation on the lithium-ion diffusion coefficient was analyzed by Qiu *et al.*[54] Their electrochemical impedance study revealed that the Li-ion diffusion coefficient is *ca.* 1.20×10^{-14} cm^2 s^{-1} in the blue TiO_2 while it is about *ca.* 4.73×10^{-15} cm^2 s^{-1} in the white TiO_2, demonstrating that the hydrogenation process has enhanced the mass transport within the TiO_2 lattice by a factor of three times.[54] The effect of oxygen vacancy on the mass transport of Li ions within the crystalline structure was investigated by DFT study and the model is shown in Figure 10.2(a). Their DFT calculation results show that the diffusion barrier is more than 2.00 eV in pristine TiO_2, which can be reduced to 1.23 eV (Figure 10.2(a)) if oxygen vacancies are presented between the two octahedral interstitial voids (indicated as a black circle in Figure 10.2(b)). The reason may be that the vacancies are very close to the Ti^{3+} cores that repel the positive charged Li ions immediately, which act as a driving force to

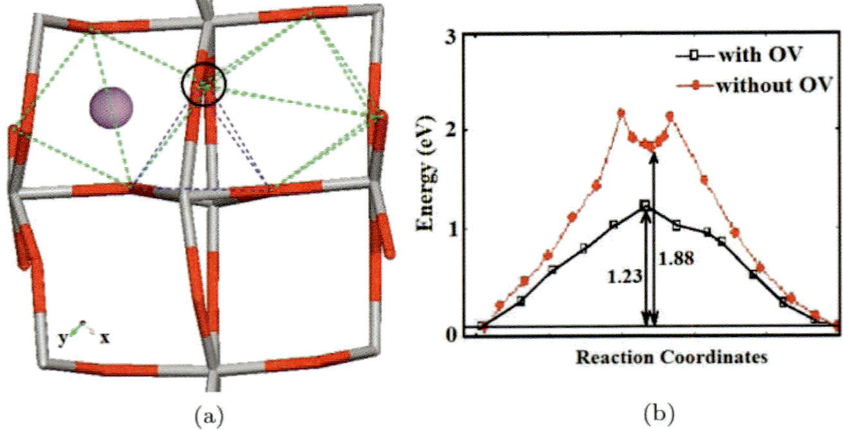

Figure 10.2. Computational study of Li-ion diffusion in the TiO_2 structure with and without oxygen vacancies. (a) Model used in the calculation of energy profiles for Li-ion diffusion; (b) energy profiles for Li-ion diffusion from one octahedral interstitial void to its neighboring one. Reprinted from Ref. [54]. © American Chemical Society, 2014.

pass the Li ions to next Ti^{3+} cores to accelerate the mass transport of Li ions within the crystalline structure.[54]

(3) Pseudocapacitive contribution

There are two different lithium storage mechanisms in TiO_2, one is surface lithium adsorbed accompanied by charge transfer, another is lithium stored by intercalation in nanocrystal. The pseudocapacitive energy storage mechanism is different from intercalation capacity, which dominated by surface redox properties rather than Faradaic diffusion-controlled insertion processes, making the high rate charge transfer become possible. The disordered surface layer in black TiO_2 may provide more active sites for Faradaic reactions than the crystalline surface in pristine-TiO_2, and induce more surface pseudocapacitive lithium storage which has much faster kinetics than those of a bulk insertion especially at high rates.[20,55,56] This phenomenon has been confirmed by Yan *et al.*,[20] in which the scan-rate dependence of the cyclic voltammetry (CV) analysis reveals that the improved rate capability of hydrogenated TiO_2 results from the

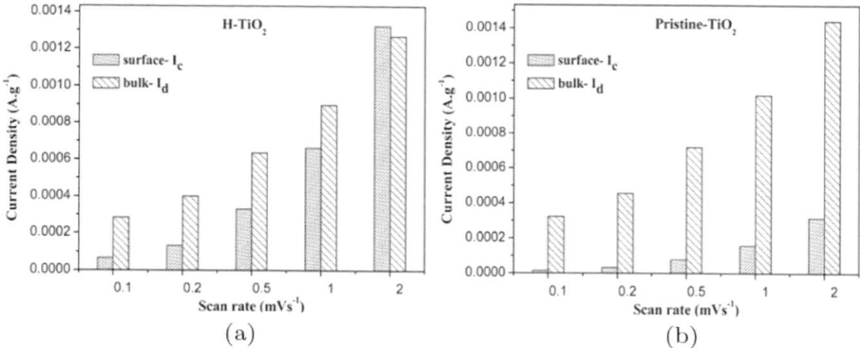

Figure 10.3. (a) Calculated surface pseudocapacitive and bulk insertion discharge currents of H–TiO$_2$ electrodes. (b) Calculated surface pseudocapacitive and bulk intercalative discharge currents of pristine-TiO$_2$ electrodes. Reprinted from Ref. [20]. © Royal Society of Chemistry, 2013.

enhanced contribution of pseudocapacitive lithium storage on the particle surface (Figure 10.3).

10.3   Black TiO$_2$ Nanomaterials for LIBs

10.3.1   *Vacuum Assisted Synthesis of Black TiO$_2$ Nanomaterials for LIBs*

Creating surface defects on the semiconductor is thought to be the most simple, effective way of enhancing the ion and/or electron conductivity and a method believed to work for most anode and cathode materials.[57] Vacuum assisted synthesis technology is an efficient method for creating surface defects. Xia *et al.*[58] found that the electrochemical performance of TiO$_2$ nanocrystals can be successfully enhanced by a facile low-temperature vacuum process. Vacuum-treated TiO$_2$ nanocrystals displayed longer stability, higher Coulombic efficiency, and better rate performance. For example, after 200 cycles, the discharge capacity of vacuum-treated TiO$_2$ nanocrystals was $140\,\mathrm{mAh\,g^{-1}}$ at 1°C rate, 34.6% higher than that of pristine TiO$_2$ nanocrystals ($104\,\mathrm{mAh\,g^{-1}}$). Moreover, at 50°C charging rate, the discharge capacity of vacuum-treated TiO$_2$ nanocrystals was $44\,\mathrm{mAh\,g^{-1}}$, 26% higher than that of pristine TiO$_2$ nanocrystals. The excellent electrochemical performance was attributed to the fact that formation of the oxygen vacancies led to increased electronic conductivity. Xia *et al.*[48] also reported that surface-amorphized titania nanocrystals were formed by a high-temperature vacuum process by placing the TiO$_2$ nanoparticles in a vacuum chamber at 500°C for 4 h. Surface-amorphized TiO$_2$ nanocrystals greatly improve lithium-storage performance: 20 times rate and 340% capacity improvement over crystalline TiO$_2$ nanocrystals. This improvement is benefited from the built-in electric field (BIEF) within the nanocrystals that induces much lower lithium-ion diffusion resistance and facilitates its transport in both insertion and extraction processes, as shown in Figure 10.4. The strategy of creating BIEF within the nanocrystal electrode materials by reconstructing the surface of crystalline electrode materials into amorphous structures at the nanometer scale is shown as a new method for improving the electrochemical performance.[48]

Apart from TiO$_2$, the electrochemical performance of Li$_4$Ti$_5$O$_{12}$ can also been improved through vacuum treatment. For example, Shi *et al.*[59] reported that oxygen-deficient Li$_4$Ti$_5$O$_{12-x}$ was synthesized through a ball-milling assisted solid state reaction under vacuum using LiOH·H$_2$O, TiO$_2$ and graphene oxide (GO) as reactants. In this reaction system, GO did not only serve as a carbon source but also as a reducing agent, leading to the formation of oxygen vacancies. The lithium-ion diffusion coefficient of Li$_4$Ti$_5$O$_{12-x}$ is calculated to be $1.02 \times 10^{-12}\,\mathrm{cm^2\,s^{-1}}$, about 16 times higher than $1.61 \times 10^{-13}\,\mathrm{cm^2\,s^{-1}}$ for pure Li$_4$Ti$_5$O$_{12}$. As a result, Li$_4$Ti$_5$O$_{12-x}$ exhibits excellent lithium-ion storage performance, presenting an initial discharge capacity of $172.4\,\mathrm{mAh\,g^{-1}}$ at 0.5°C with a capacity retention of 96.9% after 100 cycles.[59]

10.3.2 *Hydrogenation Synthesis of Black TiO$_2$ Nanomaterials for LIBs*

The hydrogenation of TiO$_2$ nanomaterials has shown to increase reactivity and surface disorder of the semiconductor lattice, which in turn allows the material to produce a heightened electrical conductivity and more flexible structure in the distorted layer for greater diffusion and kinetic rates.[60,61] Several methods, including

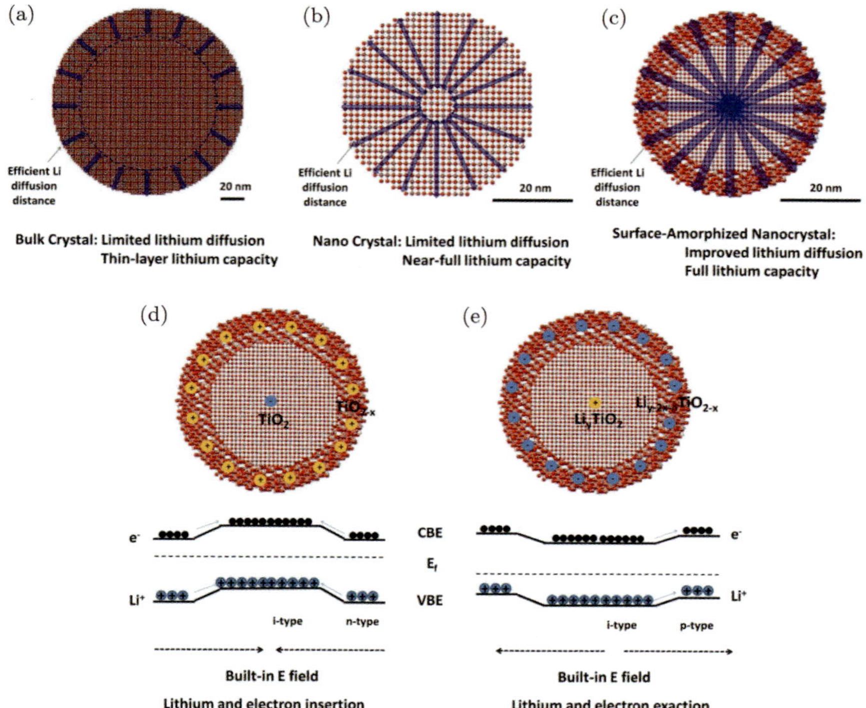

Figure 10.4. Comparison of charge diffusion in electrode materials made of (a) bulk crystals, (b) nanocrystals, (c) surface-amorphized nanocrystals, and illustration of the facilitation of charge transport under the BIEF during discharge (d) and charge (e) processes. The length and the width of the arrows illustrate the relative penetration depth and the charge transfer/transport coefficients of the lithium-ion in the material, respectively. Reprinted from Ref. [48]. © American Chemical Society, 2013.

annealing in a hydrogen/argon mixture gases, H-plasma exposure, and H-ion implantation, have been used to perform hydrogenation. Shin *et al.*[16] reported that hydrogenated black TiO$_2$ nanoparticles treated at 450°C (723 K) for 1–7 h under a 5% H$_2$/95% Ar gas flow exhibited excellent rate capability for lithium-ion storage. The reversible capacity of hydrogenated black TiO$_2$ electrode at 1°C was 161 mAh g^{-1}. Even under fast lithium insertion/extraction condition (10°C), rather nearly the same reversible capacity (159 mAh g^{-1}, 20th cycle) was maintained. The high electrochemical lithium storage properties were attributed to the well-balanced Li^+/e^- diffusion in the hydrogenated TiO$_2$ nanoparticles from the oxygen vacancies and increased electronic conductivity induced by the hydrogen reduction process. Xia *et al.*[18] reported a surface-disordered TiO$_2$ nanoparticles for high performance lithium-ion storage. This TiO$_2$ nanoparticles were obtained by the hydrogenation reaction by placing the calcinated white-colored TiO$_2$ powders in a sealed sample chamber under vacuum for 1 h followed by hydrogenation under a 20.0 bar H$_2$ atmosphere at about 200°C for 5 days. The hydrogenated surface-disordered TiO$_2$ nanocrystals can facilitate the charge transfer process and capacity retention of the pure crystalline electrode, resulting in better electrochemical energy storage performance, better rate performance, larger capacity, and longer stability.[18] The improvement of the reversible capacity and high-rate charge/discharge is obviously obtained in the 500 cycles tested and the Coulombic efficiency is still close to 100% after 500 cycles without obvious degradation. Yan *et al.*[20] reported on the fast lithium-ion storage performance of hydrogenated TiO$_2$ nanoparticles (H–TiO$_2$) synthesized by a H$_2$ plasma treatment of commercial anatase TiO$_2$ nanoparticles. When used as anode material for LIBs, the obtained H–TiO$_2$ shows a superior long term and fast lithium-ion storage capability. The capacities of H–TiO$_2$ are about 154, 141, 124, and 108 mA h g^{-1} at a rate of 5°C, 10°C, 20°C, and 40°C, respectively, which is superior to pristine TiO$_2$ nanoparticles (pristine-TiO$_2$, about 126, 109, 91, and 70 mA h g^{-1} at the corresponding rates). Furthermore, H–TiO$_2$ has a capacity of 101 mA h g^{-1} at 40°C after 5000 cycles, which is almost 1.7 times that of pristine-TiO$_2$ (61 mA h g^{-1} after 5000 cycles at 40°C). The electrochemical measurements demonstrated that the improved rate capability of H–TiO$_2$ can be attributed to the enhanced contribution of pseudocapacitive lithium-ion storage on its surface; and it is suggested that the presence of Ti^{3+} species and the disordered surface layer might be the key factors for the enhanced pseudocapacitance effect.[20]

Qiu *et al.*[54] reported that blue hydrogenated rutile TiO$_2$ nanorods is prepared by treating white rutile via an enhanced hydrogenation process (high H$_2$ pressure and high reaction temperature, the hydrogen pressure was raised to 40 bar and the temperature was maintained at 450°C for 1 h). The hydrogenation reaction increases the unit cell volume of the pristine rutile crystalline, reduces the size of the rutile nanocrystalline, causes crystalline dislocation, and produces oxygen vacancy throughout the bulk and surface of the crystalline.[54] These changes facilitate fast lithium-ion transport and electron transfer during the insertion/extraction process. For example, at the rate of 0.1°C and 5°C, the discharge capacities of the blue rutile are maintained at about 179.8 mAh g^{-1} and 129.2 mAh g^{-1}, while the capacities of the white TiO$_2$ are just 119.6 mAh g^{-1} and 55.5 mAh g^{-1}, respectively. Spherical material has a high volumetric specific energy, which has important prospects in many practical and commercial applications. Li *et al.*[51] prepared hydrogenated mesoporous TiO$_2$ microspheres, which showed twice the rate capability compared to that of pristine TiO$_2$ microspheres at 20°C due to the combination of the short lithium-ion diffusion path and the high electronic conductivity. Their electrochemical impedance (EIS) study also revealed that both the electronic conductivity and the lithium-ion diffusion kinetics were improved in the H–TiO$_2$ microspheres. TiO$_2$(B), like other Ti-based materials, also suffer from poor electronic conductivity when employed as an anode for LIBs. Zhang *et al.*[53] reported that flower-like hydrogenated TiO$_2$(B) nanostructures were synthesized by a facile solvothermal process combined with hydrogenation treatment. The thin primary TiO$_2$(B) nanosheets are in a thickness of 10±1.2 nm, which is helpful for the migration of lithium. The introduction of Ti^{3+} species and/or oxygen vacancies greatly improves the electronic conductivity (up to 2.79×10^3 S cm^{-1}) and enhances surface electrochemical reaction. Hence, the hydrogenated TiO$_2$(B) exhibits enhanced electrochemical performances compared to the pristine TiO$_2$(B), including high capacity (292.3 mAh g^{-1} at 0.5°C), excellent rate capability (179.6 mAh g^{-1} at 10°C), and good cyclic stability (98.4% capacity retention after 200 cycles at 10°C).[53]

Recently, self-supported electrodes have been reported to overcome the drawbacks of mixing with conductive carbon and have fast charging/discharging.[62] Another research of electrochemical anodized TiO$_2$ nanotube arrays targeted for use as anode in LIBs by annealing anodized TiO$_2$ nanotubes or rutile nanowires at 450°C for 1 h in a reducing atmosphere (5% H$_2$ and 95% Ar) has been proposed by Lu *et al.*[63] They revealed that the improved electrochemical performance is mainly due to the electronic conductivity increase of the bulk TiO$_2$ nanotubes rather than conductive characteristics of the surface coating because hydrogenation treatment produces a high number of oxygen vacancies inside the crystal lattices that make the TiO$_2$ nanotube arrays favor a bulk *n*-type conductor.[63] Eom *et al.*[43] reported that black TiO$_{2-x}$ nanotube arrays is prepared by an electrochemical method and subsequent thermal treatment in a hydrogen atmosphere. After the thermal conversion under hydrogen atmosphere, the black TiO$_{2-x}$ NT electrodes generated reduced Ti^{3+} ions with oxygen vacancies in the structure, which enhance the electronic conductivity of the TiO$_2$ by increasing its carrier density.[43] Moreover, the partial Ti substrate can transform into a metallic TiH$_{1.5}$ phase instead of forming an anatase passive layer on the substrate, which is easily formed after thermal treatment in air. The black TiO$_{2-x}$, as a binder-free, free-standing electrode, delivered excellent cyclability for 300 cycles at a high current rate of 1 mA cm^{-2} (10°C) and superior rate capability even at 10 mA cm^{-2} (100°C) compared to an anatase TiO$_2$ NT electrode owing to its improved electronic features and kinetic properties (Figure 10.5). Liu *et al.*[50] fabricated flexible free-standing TiO$_2$ nanowire arrays and created Ti^{3+} by hydrogenation treatment to improve the electrical conductivity. Rutile TiO$_2$ nanowire arrays were first fabricated on carbon fiber substrates, using a seed mediated hydrothermal method. The obtained sample was annealed in air and H$_2$ at 550°C for 1 h. Free-standing hydrogenated-TiO$_2$ nanowire arrays grown on a flexible carbon fiber not only provide more reaction sites for lithium-ion insertion/extraction, but also provide good mechanical support and serve as a binder-free electrode. The hydrogenated-TiO$_2$ nanowire arrays deliver a high reversible capacity of 517 mAh g^{-1} at a current density of 1°C. Moreover, the hydrogenated-TiO$_2$ nanowire arrays have an outstanding rate performance as they still retain 70 mAh g^{-1} when the current density increases to 50°C.

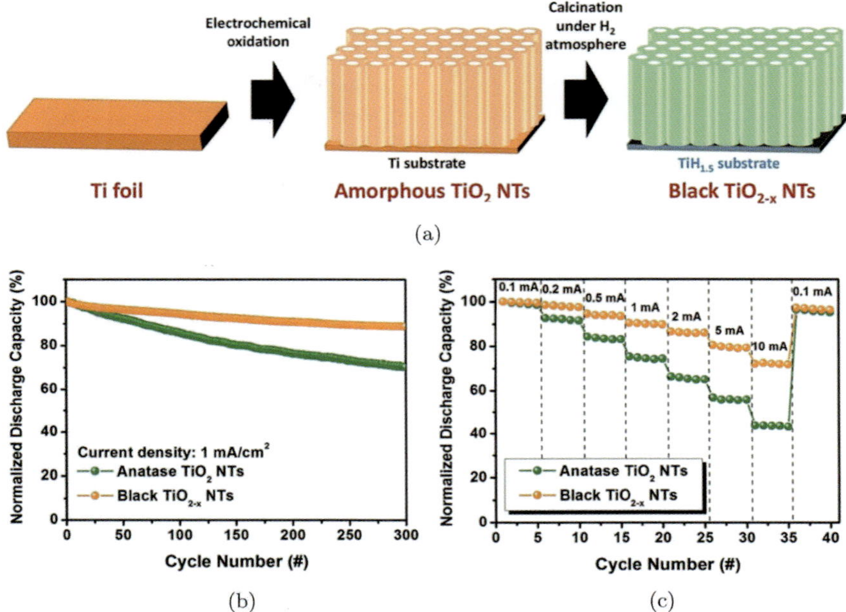

Figure 10.5. (a) Schematic illustration of the synthetic strategy for a black TiO$_{2-x}$ NT electrode. (b) Cycling performance (expressed by normalized capacities) of anatase TiO$_2$ NTs and H–TiO$_2$ NTs at a current density of 1 mA cm^{-2} for 300 cycles. (c) Rate capabilities during cycling of anatase TiO$_2$ NT and black TiO$_{2-x}$ NT electrodes at various current densities. The samples were cycled at a current density of 0.1 mA cm^{-2} for 10 cycles and then were cycled five times at each current density. Reprinted from Ref. [43]. © The Royal Society of Chemistry, 2015.

Shen *et al.*[62] developed a facile template-free strategy to fabricate Li$_4$Ti$_5$O$_{12}$ nanowire arrays (LTO NWAs) growing directly on Ti foil and improve their electronic conductivity by creating Ti^{3+} sites through hydrogenation treatment, as confirmed by XPS spectra (Figure 10.6). The hydrogenated LTO NWAs can be directly used as anodes for lithium-ion storage without any ancillary materials, which exhibited an excellent rate capability (121 mAh g^{-1} at 30°C) and a significantly enhanced cycling performance (Figures 10.6(c) and (d)). The suggested reason may be attributed to the improved electronic conductivity of hydrogenated LTO NWAs by introduction of Ti^{3+} states and the direct connection to the growth substrate.[62]

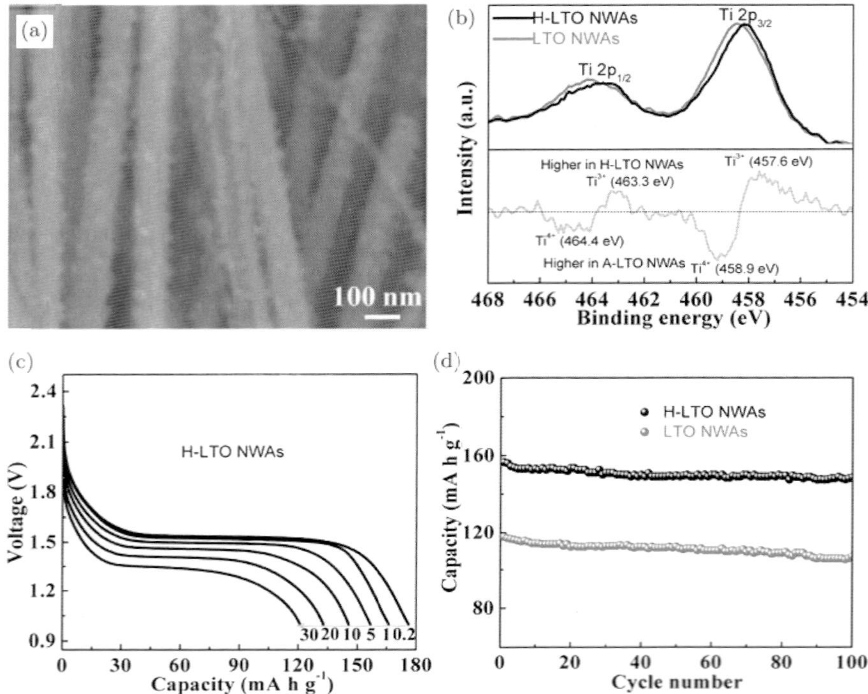

Figure 10.6. (a) Cross-sectional SEM image of H-LTO NWAs. (b) Overlay of normalized Ti 2p core level XPS spectra of H-LTO NWAs and LTO NWAs, together with the difference in their spectra ("HLTO NWAs" minus "LTO NWAs"). (c) the discharge curves of H-LTO NWAs. (d) Cycling performances at the rate of 5°C. Reprinted from Ref. [62]. © Wiley-VCH, 2012.

10.3.3 *Nitridation Synthesis of Black TiO$_2$ Nanomaterials for LIBs*

Ammonia annealing, which has been extensively used to prepare N-doped TiO$_2$ in photocatalysis,[64,65] has been shown to improve the Li-ion storage performance of titanium dioxide in terms of the stability upon cycling and the rate capability. Samiee *et al.*[66] demonstrated that the electrochemical performance of anatase TiO$_2$ nanoparticles can be improved substantially by annealing in NH$_3$ atmosphere at 450–500°C. At a high current rate of 25°C, the average discharge capacity of nanoparticles annealed in NH$_3$ at 450°C for 7 h is more than double that of the controlled nanoparticles annealed in air under the same treatment condition. This study also suggested that lower nitridation temperatures of 450–500°C for annealing in NH$_3$ atmosphere (as compared to the typical 600–650°C) can lead to better results, which are attributed to more moderate surface nitridation with thinner and less-disordered nitridated regions that enhance electronic conductivity without blocking lithium transport. Zhang *et al.*[67] reported that N-doped TiO$_2$ nanoparticles were prepared by a solvothermal method followed by annealed at 550°C for 2 h in air, and another heat treatment at 550°C under a NH$_3$ flow for 2 h. Finally, the N-doped TiO$_2$ was annealed at 550°C for 1 h in air. The N anions reside on the interstitial and substitutional sites in the material bulk, which slightly expands the crystal lattice of anatase TiO$_2$. The N-doped TiO$_2$ shows better rate capability than pristine TiO$_2$. A discharge capacity of 45 mAh g^{-1} is obtained at the current rate of 15°C, which is 80% higher than that of pristine TiO$_2$. N doping can improve the lithium-ion diffusion coefficient of TiO$_2$. Particularly, the lithium-ion diffusion coefficient is increased to about 2.14×10^{-11} cm^2 s^{-1}, which is 13 times higher than that of pristine TiO$_2$ (only about 1.65×10^{-12} cm^2 s^{-1}). Yoon *et al.*[68] prepared nitridated TiO$_2$ mesoporous spheres through hydrothermal treatment followed by post-nitridation with NH$_3$ gas at 550°C for 5 h. The nitridated TiO$_2$ mesoporous spheres exhibit a high capacity with good cyclability (above 200 mAh g^{-1} after 80 cycles at 1/10°C rate) and high rate capability (about 138 and 139 mAh g^{-1} at 5°C and 10°C rates), as the nitridated conducting layer and favorable morphology of nanostructure provides good electrical contact, accommodates cycling induced strain smoothly, and facilitates lithium-ion diffusion.[68]

Yang *et al.*[69] fabricated N-doped TiO$_2$ nanorods decorated with carbon dots with increased electrical conductivity and faster charge-transfer by a simple hydrothermal process involving TiO$_2$ powders and NaOH in the presence of carbon dots followed by ion exchange and calcinations process. Due to the advantages of the carbon dots, doping and nanostructures, the as-prepared N–TiO$_2$/C-dots nanocomposite employed as anode materials for lithium-ion storage can provide a capacity of 185 mAh g^{-1} with 91.6% retention

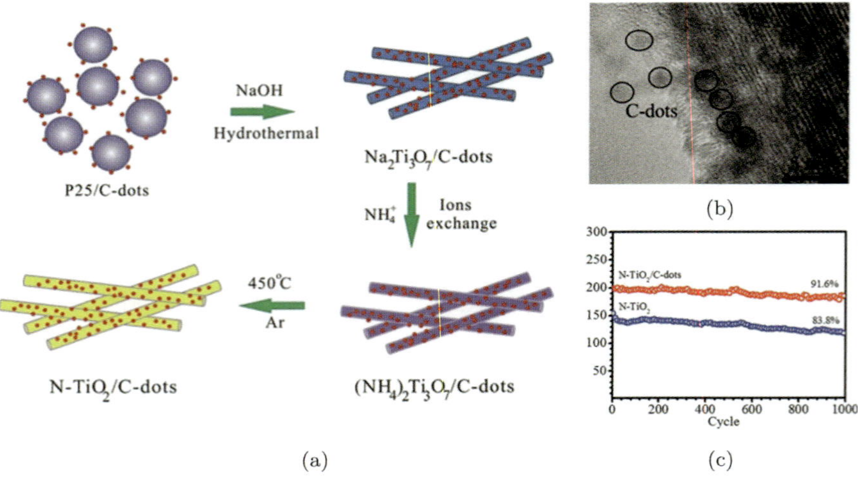

Figure 10.7. (a) Illustration for the fabrication process to produce N–TiO$_2$/C-dots composite. (b) TEM images of the N–TiO$_2$/C-dots composite. (c) Cycling performance at 10°C of lithium-ion batteries employing the N–TiO$_2$ and N–TiO$_2$/C-dots anodes. Reprinted from Ref. [69]. © The Royal Society of Chemistry, 2015.

even at a high current density of 10°C over 1000 cycles (Figure 10.7). It is very interesting to note that the ratios of capacitive charge capacity during such high rates for the N–TiO$_2$/C-dots electrodes are higher than those at low rates, which likely explains the observed excellent rate capabilities.[69] Wang *et al.*[70] reported that N-doped Li$_4$Ti$_5$O$_{12}$ nanosheets (NLTO) prepared by a simple hydrothermal process with further heat treatment at 500°C for 6 h under NH$_3$ atmosphere. When used as anode for LIBs, NLTO shows a better rate capability and cycle stability. Specially, at 10°C rate, NLTO exhibits 151.8 mAh g^{-1}, which is higher than that of LTO (139.1 mAh g^{-1}). After 200 cycles, its discharge capacity is 144 mAh g^{-1} at 10°C. The excellent rate capability can be attributed to the increased conductivity of the N-doping combined with the short ion and electron transfer lengths of nanosheet structures.

The morphology and electronic structure of metal oxides, including TiO$_2$ on the nanoscale, definitely determine their electronic or electrochemical properties, especially those relevant to application in energy devices. For this purpose, a concept for controlling the morphology and electrical conductivity in TiO_2, based on tuning by electrospinning, is proposed. Kim *et al.*[71] found that the 1D TiO_2 nanofibers interestingly gave higher cyclic retention than 0D nanopowder, and nitrogen doping in the form of TiO_2N_x also caused further improvement. TiO_2N_x nanofibers were prepared by a simple electrospinning technology with diethylenetriamine (DETA) as nitrogen source. With an increasing amount of DETA, they found a small fraction of rutile phase. The nitrogen-doped TiO_2 nanofibers show much higher cyclic retention ($185\,\mathrm{mA\,h\,g^{-1}}$) than the pure TiO_2 nanofibers ($165\,\mathrm{mA\,h\,g^{-1}}$) and nanopowder ($60\,\mathrm{mA\,h\,g^{-1}}$) after 40 cycles. This is due to the higher conductivity and faster Li^+ diffusion of TiO_2N_x nanofibers, as confirmed by electrochemical impedance spectra (EIS) and charge/discharge curve. They also demonstrated that substitutional nitrogen would be spread widely in the bulk, while interstitial nitrogen ions were mostly located near the surface in the form of TiO_2N_x, improving the electrical conductivity and changing the work function, revealed by their first principles calculations.[71] Han *et al.*[72] also reported that nitridated TiO_2 hollow nanofibers were synthesized using a simple electrospinning technology and subsequent nitridation process (Figure 10.8). They demonstrated that an outstanding rate performance of TiO_2 based

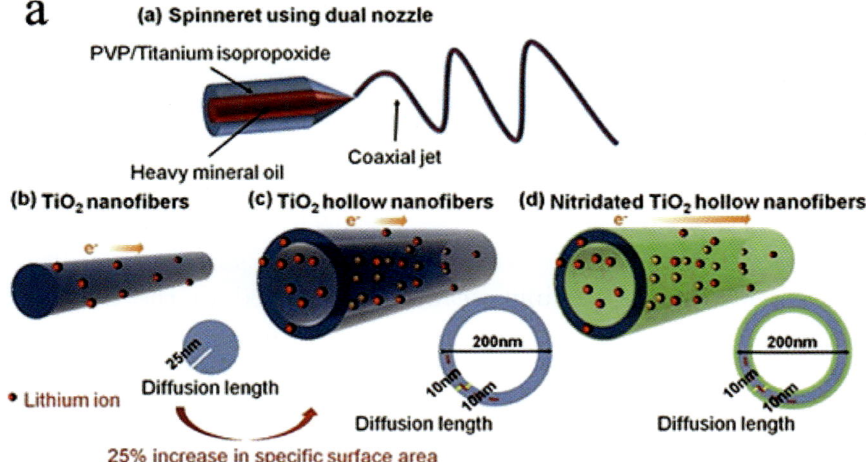

Figure 10.8. Schematic illustrations of coaxial electrospinning.

anodes for LIBs could be achieved by balanced electrode design in both morphology and composition.[72] The discharge capacity of nitridated TiO$_2$ hollow nanofibers ($85\,\mathrm{mAh\,g^{-1}}$) was almost double higher than that of pure TiO$_2$ nanofibers ($45\,\mathrm{mA\,h\,g^{-1}}$) at the rate of 2°C. The hollow structure and conducting shell layer can provide a shorter lithium-ion diffusion path and a high electronic conductivity along the surface of TiO$_2$ nanofibers, both of which lead to enormous improvement in the rate capability.[72]

10.3.4   *Sulfuration Synthesis of Black TiO$_2$ Nanomaterials for LIBs*

As we all know, anion dopants such as the most studied nitrogen tends to locate dominantly in the surface of TiO$_2$, that is largely due to the higher energy of surface than that of the bulk and the low solubility of dopants.[72] A recent research reveals that the gradient doping of substitutional N for lattice O can be achieved in anatase TiO$_2$ with the aid of interstitial boron (B).[73] However, the existence of B element in the interstitial sites will block the insertion/extraction of lithium-ions. Therefore, developing alternative preparation strategy is highly desirable for obtaining suitable anion-doped TiO$_2$ for lithium-ion storage applications. Jiao *et al.*[74] prepared anatase TiO$_2$ nanoparticles doped with nitrogen (N) and sulfur (S), where substitutional N and S atoms for lattice O, respectively, locate in the bulk and the surface of the nanocrystals. The N/S codoped anatase TiO$_2$ can be obtained by a three-step preparation route as shown in Figure 10.9(a). The resultant N/S doped nanoparticles exhibit a much higher lithium-ion storage capability at a 10°C rate than either pure TiO$_2$ or nitrogen-doped TiO$_2$ ($63.5\,\mathrm{mAh\,g^{-1}}$ vs. $3.6/17.8\,\mathrm{mAh\,g^{-1}}$) Figure 10.9(b). The as-prepared N/S codoped TiO$_2$ also has an excellent cyclic retention Figure 10.9(c). The capacity is about $60\,\mathrm{mAh\,g^{-1}}$ after 40 cycles at 10°C rate. The doping of N/S gives rise to high electric conductivity and the nanostructure reduces the transport lengths of lithium-ions and electrons, permitting fast kinetics for both transported lithium-ions and electrons, thus enabling high-power performance.[74]

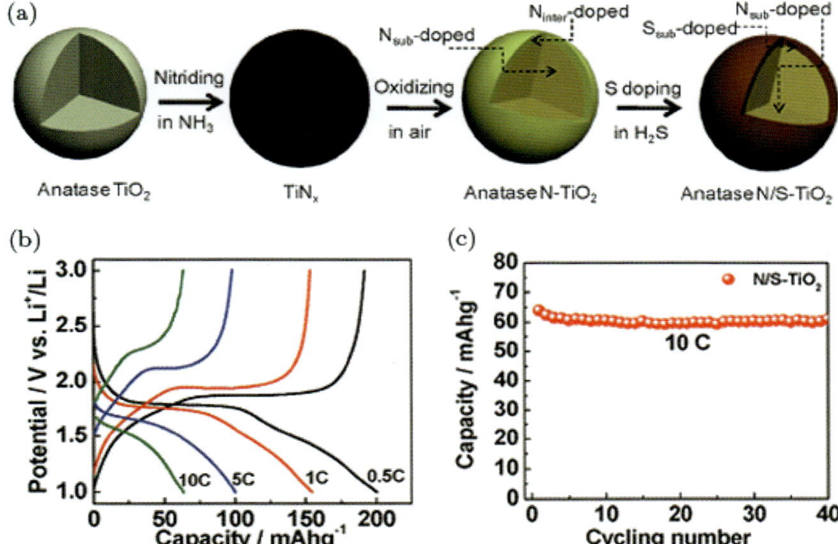

Figure 10.9. (a) Schematic of the preparation processes of N/S doped anatase TiO$_2$ crystals in three steps, namely, nitriding in gaseous NH$_3$ gas, oxidizing in air and S doping in gaseous H$_2$S. N$_{sub}$: substitutional N for lattice O; N$_{inter}$: interstitial N; S$_{sub}$: substitutional S for lattice O. (b) and (c) Electrochemical performance of N/S doped anatase TiO$_2$ crystals. Reprinted from Ref. [74]. © The Royal Society of Chemistry, 2013.

10.4 Summary and Outlook

Without doubt, titanium-based oxide has become a competitive anode material for next-generation LIBs for large-scale application, such as EVs and HEVs. But high polarization caused by slow ionic and electronic diffusion is still a big challenge to fully realize advantage of TiO$_2$ using batteries. Introducing dopants or defects into the TiO$_2$ matrix to produce highly conductive TiO$_2$ nanostructure called "black TiO$_2$" shows improved fast lithium-ion storage capability compared to the pristine TiO$_2$. After the heating treatment under vacuum or a reducing atmosphere, Ti^{3+} sites associated with oxygen vacancies will be generated in black TiO$_2$, which effectively improves the electronic conductivity of the TiO$_2$ structure by increasing its carrier density. Meanwhile, disordered layer will be formed at the surface of crystalline TiO$_2$, which accelerates the mass transport of Li ions within the crystalline structure. Also, the disordered surface layer in black TiO$_2$ induces more surface pseudocapacitive lithium-ion storage which has much faster kinetics than pristine TiO$_2$. In addition, self-structure modification of TiO$_2$ is a very versatile approach because all synthesis method need a post-annealing.

However, the fundamental understanding of the effect of structures on the improved fast lithium storage properties of TiO$_2$ is essential for future black TiO$_2$ electrochemical research. It is anticipated that this self-structure modifications approach can also be extended to improve the electrochemical performance of other titanium-based materials, such as Li–Ti–O system, Na–Ti–O system, and H–Ti–O system. Also the black titanium-based nanomaterials have potential applications in new battery systems based on intercalation reactions with abundant natural resources, such as Na-ion, Mg-ion, and Al-ion batteries. Certainly, some fundamental issues, such as a disordered layer at surface crystalline structure may induce unacceptable gas generation in Li$_4$Ti$_5$O$_{12}$ when used as anode in LIBs, should be investigated carefully.

References

(1) Tarascon, J.-M.; Armand, M. *Nature* **2001**, *414*, 359.

(2) Wang, Y.; Wang, Y.; Hosono, E.; Wang, K.; Zhou, H. *Angew. Chem. Int. Ed.* **2008**, *47*, 7461.

(3) Chan, C. K.; Peng, H.; Liu, G.; McIlwrath, K.; Zhang, X. F.; Huggins, R. A.; Cui, Y. *Nature Nanotech.* **2008**, *3*, 31.

(4) Nishi, Y. *J. Power Sources* **2001**, *100*, 101.

(5) Sasaki, T.; Ukyo, Y.; Novák, P. *Nature Mater.* **2013**, *12*, 569.

(6) Armstrong, G.; Armstrong, A. R.; Bruce, P. G.; Reale, P.; Scrosati, B. *Adv. Mater.* **2006**, *18*, 2597.

(7) Shen, L.; Li, H.; Uchaker, E.; Zhang, X.; Cao, G. *Nano Lett.* **2012**, *12*, 5673.

(8) Scrosati, B.; Garche, J. *J. Power Sources* **2010**, *195*, 2419.

(9) Guo, B.; Wang, X.; Fulvio, P. F.; Chi, M.; Mahurin, S. M.; Sun, X. G.; Dai, S. *Adv. Mater.* **2011**, *23*, 4661.

(10) Park, K.-S.; Benayad, A.; Kang, D.-J.; Doo, S.-G. *J. Am. Chem. Soc.* **2008**, *130*, 14930.

(11) Yang, S.; Feng, X.; Müllen, K. *Adv. Mater.* **2011**, *23*, 3575.

(12) Yang, Z.; Choi, D.; Kerisit, S.; Rosso, K. M.; Wang, D.; Zhang, J.; Graff, G.; Liu, J. *J. Power Sources* **2009**, *192*, 588.

(13) Myung, S.-T.; Kikuchi, M.; Yoon, C. S.; Yashiro, H.; Kim, S.-J.; Sun, Y.-K.; Scrosati, B. *Energ Environ. Sci.* **2013**, *6*, 2609.

(14) Shen, L.; Ding, B.; Nie, P.; Cao, G.; Zhang, X. *Adv. Energy Mater.* **2013**, *3*, 1484.

(15) Shen, L.; Zhang, X.; Uchaker, E.; Yuan, C.; Cao, G. *Adv. Energy Mater.* **2012**, *2*, 691.

(16) Shin, J.-Y.; Joo, J. H.; Samuelis, D.; Maier, J. *Chem. Mater.* **2012**, *24*, 543.

(17) Wang, Z. Y.; Wang, Z. C.; Liu, W. T.; Xiao, W.; Lou, X. W. *Energ Environ. Sci.* **2013**, *6*, 87.

(18) Xia, T.; Zhang, W.; Li, W.; Oyler, N. A.; Liu, G.; Chen, X. *Nano Energy* **2013**, *2*, 826.

(19) Liu, D.; Zhang, Y.; Xiao, P.; Garcia, B. B.; Zhang, Q.; Zhou, X.; Jeong, Y.-H.; Cao, G. *Electrochim. Acta* **2009**, *54*, 6816.

(20) Yan, Y.; Hao, B.; Wang, D.; Chen, G.; Markweg, E.; Albrecht, A.; Schaaf, P. *J. Mater. Chem. A* **2013**, *1*, 14507.

(21) Henningsson, A.; Rensmo, H.; Sandell, A.; Siegbahn, H.; Södergren, S.; Lindström, H.; Hagfeldt, A. *J. Chem. Phys.* **2003**, *118*, 5607.

(22) Olson, C. L.; Nelson, J.; Islam, M. S. *J. Phys. Chem. B* **2006**, *110*, 9995.

(23) Koudriachova, M. V.; Harrison, N. M.; de Leeuw, S. W. *Phys. Rev. B* **2002**, *65*, 235423.

(24) Koudriachova, M. V.; de Leeuw, S. W.; Harrison, N. M. *Phys. Rev. B* **2004**, *69*, 054106.

(25) Takai, S.; Kamata, M.; Fujine, S.; Yoneda, K.; Kanda, K.; Esaka, T. *Solid State Ionics* **1999**, *123*, 165.

(26) Hu, Y. S.; Kienle, L.; Guo, Y. G.; Maier, J. *Adv. Mater.* **2006**, *18*, 1421.

(27) Howard, C.; Sabine, T.; Dickson, F. *Acta Crystallogr., Sect. B: Struct. Sci* **1991**, *47*, 462.

(28) Arico, A. S.; Bruce, P.; Scrosati, B.; Tarascon, J.-M.; Van Schalkwijk, W. *Nature Mater.* **2005**, *4*, 366.

(29) Reddy, M. A.; Pralong, V.; Varadaraju, U.; Raveau, B. *Electrochem. Solid-State Lett.* **2008**, *11*, A132.

(30) Mackrodt, W. *J. Solid State Chem.* **1999**, *142*, 428.

(31) Navrotsky, A.; Kleppa, O. *J. Am. Ceram. Soc.* **1967**, *50*, 626.

(32) Das, S. K.; Darmakolla, S.; Bhattacharyya, A. J. *J. Mater. Chem.* **2010**, *20*, 1600.

(33) Chen, J. S.; Liu, H.; Qiao, S. Z.; Lou, X. W. D. *J. Mater. Chem.* **2011**, *21*, 5687.

(34) Eder, D.; Windle, A. H. *J. Mater. Chem.* **2008**, *18*, 2036.

(35) Liu, M.; Qiu, X.; Miyauchi, M.; Hashimoto, K. *Chem. Mater.* **2011**, *23*, 5282.

(36) Deng, D.; Kim, M. G.; Lee, J. Y.; Cho, J. *Energy Environ. Sci.* **2009**, *2*, 818.

(37) Liu, H.; Bi, Z.; Sun, X. G.; Unocic, R. R.; Paranthaman, M. P.; Dai, S.; Brown, G. M. *Adv. Mater.* **2011**, *23*, 3450.

(38) Shen, L.; Uchaker, E.; Yuan, C.; Nie, P.; Zhang, M.; Zhang, X.; Cao, G. *ACS Appl. Mater. Interfaces* **2012**, *4*, 2985.

(39) Moriguchi, I.; Hidaka, R.; Yamada, H.; Kudo, T.; Murakami, H.; Nakashima, N. *Adv. Mater.* **2006**, *18*, 69.

(40) Guo, Y. G.; Hu, Y. S.; Sigle, W.; Maier, J. *Adv. Mater.* **2007**, *19*, 2087.

(41) Borghols, W.; Wagemaker, M.; Lafont, U.; Kelder, E.; Mulder, F. *J. Am. Chem. Soc.* **2009**, *131*, 17786.

(42) Cheng, L.; Yan, J.; Zhu, G.-N.; Luo, J.-Y.; Wang, C.-X.; Xia, Y.-Y. *J. Mater. Chem.* **2010**, *20*, 595.

(43) Eom, J.-Y.; Lim, S.-J.; Lee, S.-M.; Ryu, W.-H.; Kwon, H.-S. *J. Mater. Chem. A* **2015**, *3*, 11183.

(44) Chen, X.; Liu, L.; Peter, Y. Y.; Mao, S. S. *Science* **2011**, *331*, 746.

(45) Hu, Y. H. *Angew. Chem. Int. Ed.* **2012**, *51*, 12410.

(46) Zhang, C.; Yu, H.; Li, Y.; Fu, L.; Gao, Y.; Song, W.; Shao, Z.; Yi, B. *Nanoscale* **2013**, *5*, 6834.

(47) Shen, L.; Yuan, C.; Luo, H.; Zhang, X.; Chen, L.; Li, H. *J. Mater. Chem.* **2011**, *21*, 14414.

(48) Xia, T.; Zhang, W.; Murowchick, J.; Liu, G.; Chen, X. *Nano Lett.* **2013**, *13*, 5289.

(49) Nowotny, M.; Bak, T.; Nowotny, J. *J. Phy. Chem. B* **2006**, *110*, 16270.

(50) Liu, Y.; Liu, C.; Li, J. *J. Mater. Chem. A* **2014**, *2*, 15746.

(51) Li, G.; Zhang, Z.; Peng, H.; Chen, K. *RSC Adv.* **2013**, *3*, 11507.

(52) Ren, Y.; Li, J.; Yu, J. *Electrochim. Acta* **2014**, *138*, 41.

(53) Zhang, Z.; Zhou, Z.; Nie, S.; Wang, H.; Peng, H.; Li, G.; Chen, K. *J. Power Sources* **2014**, *267*, 388.

(54) Qiu, J.; Li, S.; Gray, E.; Liu, H.; Gu, Q.-F.; Sun, C.; Lai, C.; Zhao, H.; Zhang, S. *J. Phys. Chem. C* **2014**, *118*, 8824.

(55) Zhu, K.; Wang, Q.; Kim, J.-H.; Pesaran, A. A.; Frank, A. J. *J. Phys. Chem. C* **2012**, *116*, 11895.

(56) Wang, J.; Polleux, J.; Lim, J.; Dunn, B. *J. Phys. Chem. C* **2007**, *111*, 14925.

(57) Kang, Q.; Cao, J.; Zhang, Y.; Liu, L.; Xu, H.; Ye, J. *J. Mater. Chem. A* **2013**, *1*, 5766.

(58) Xia, T.; Zhang, W.; Murowchick, J. B.; Liu, G.; Chen, X. *Adv. Energy Mater.* **2013**, *3*, 1516.

(59) Shi, Y.; Zhang, D.; Chang, C.; Huang, K.; Holze, R. *J. Alloy. Compd.* **2015**, *639*, 274.

(60) Qiu, J.; Dawood, J.; Zhang, S. *Chin. Sci. Bull.* **2014**, *59*, 2144.

(61) Park, M.; Zhang, X.; Chung, M.; Less, G. B.; Sastry, A. M. *J. Power Sources* **2010**, *195*, 7904.

(62) Shen, L.; Uchaker, E.; Zhang, X.; Cao, G. *Adv. Mater.* **2012**, *24*, 6502.

(63) Lu, Z.; Yip, C. T.; Wang, L.; Huang, H.; Zhou, L. *ChemPlusChem* **2012**, *77*, 991.

(64) Diwald, O.; Thompson, T. L.; Zubkov, T.; Goralski, E. G.; Walck, S. D.; Yates, J. T. *J. Phys. Chem. B* **2004**, *108*, 6004.

(65) Oropeza, F. E.; Harmer, J.; Egdell, R.; Palgrave, R. G. *Phys. Chem. Chem. Phys.* **2010**, *12*, 960.

(66) Samiee, M.; Luo, J. *J. Power Sources* **2014**, *245*, 594.

(67) Zhang, Y.; Du, F.; Yan, X.; Jin, Y.; Zhu, K.; Wang, X.; Li, H.; Chen, G.; Wang, C.; Wei, Y. *ACS Appl. Mater. Interfaces* **2014**, *6*, 4458.

(68) Yoon, S.; Bridges, C. A.; Unocic, R. R.; Paranthaman, M. P. *J. Mater. Sci.* **2013**, *48*, 5125.

(69) Yang, Y.; Ji, X.; Jing, M.; Hou, H.; Zhu, Y.; Fang, L.; Yang, X.; Chen, Q.; Banks, C. E. *J. Mater. Chem. A* **2015**, *3*, 5648.

(70) Wang, B.; Wang, J.; Cao, J.; Ge, H.; Tang, Y. *J. Power Sources* **2014**, *266*, 150.

(71) Kim, J.-G.; Shi, D.; Kong, K.-J.; Heo, Y.-U.; Kim, J. H.; Jo, M. R.; Lee, Y. C.; Kang, Y.-M.; Dou, S. X. *ACS Appl. Mater. Interfaces* **2013**, *5*, 691.

(72) Han, H.; Song, T.; Bae, J.-Y.; Nazar, L. F.; Kim, H.; Paik, U. *Energy Environ. Sci.* **2011**, *4*, 4532.

(73) Liu, G.; Yin, L.-C.; Wang, J.; Niu, P.; Zhen, C.; Xie, Y.; Cheng, H.-M. *Energy Environ. Sci.* **2012**, *5*, 9603.

(74) Jiao, W.; Li, N.; Wang, L.; Wen, L.; Li, F.; Liu, G.; Cheng, H.-M. *Chem. Commun.* **2013**, *49*, 3461.

CHAPTER ELEVEN

Black TiO$_2$ Nanomaterials for Lithium–Sulfur Batteries

Zheng Liang, *Xinyong Tao*[†] *and* *Yi Cui*[*]

[*]*Department of Materials Science and Engineering,*
Stanford University, Stanford, CA, USA
[†]*College of Chemical Engineering and Materials Science,*
Zhejiang University of Technology, Hangzhou, China
[‡]*yicui@stanford.edu*

11.1 Introduction

High-performance energy storage devices are crucial for meeting the ever-growing demand for high energy density applications such as portable electronics and electric-vehicles. Among emerging advanced rechargeable batteries, lithium–sulfur (Li–S) system has raised intense attention due to its high energy density (2500 Wh/kg), which is dramatically higher than conventional $LiNi_{1/3}Mn_{1/3}Co_{1/3}O_2$ — graphite system as well as the high energy $LiNi_{1/3}Mn_{1/3}Co_{1/3}O_2$ — silicon system (Figure 11.1(a)).[1] Sulfur cathodes could deliver a gravimetric specific capacity of as high as 1675 mAh/g (volumetric capacity: 3467 mAh/cm^3)[2–5] when pairing with lithium anode. In addition, the natural abundance, low cost, and environmental friendliness further contribute to the potential of sulfur as ideal lithium battery cathodes.

Nevertheless, the practical application of Li–S batteries has been hindered by a series of technical obstacles, including low active material utilization, limited rate capability, and poor cycle

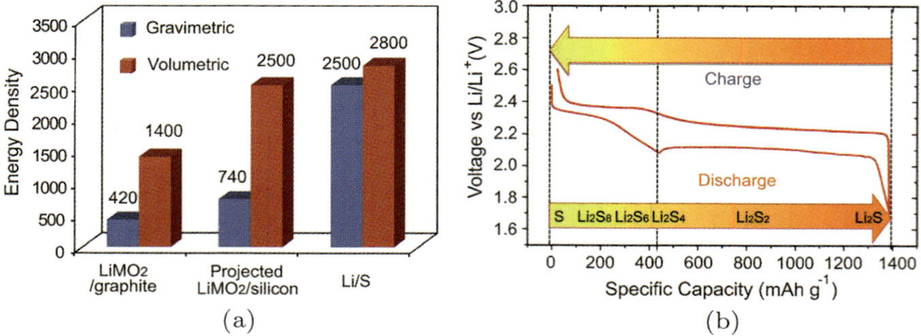

Figure 11.1. An introduction of Li–S battery. (a) Theoretical gravimetric and volumetric energy density of several rechargeable battery systems. M = Ni$_{1/3}$Mn$_{1/3}$Co$_{1/3}$. Units for gravimetric and volumetric energy densities are Wh/kg and Wh/L, respectively. (b) A typical charge–discharge voltage profile of Li–S battery. Reprinted with permission from Ref. [1]. Copyright: The Royal Society of Chemistry.

life. These issues could be attributed to a notorious phenomenon named "polysulfides dissolution and shuttle effect". According to the Li–S battery chemistry shown in Figure 11.1(b), during discharge process, elemental sulfur transforms through a series of intermediate products known as lithium polysulfides (Li$_2$S$_n$, 3 ≤ n < 8), into solid final products (Li$_2$S$_2$ and Li$_2$S).[1] In the subsequent charge process, these discharge products are converted back reversibly into elemental sulfur. Polysulfides possess a high solubility in ether-based electrolyte and tend to diffuse into the electrolyte.[6–9] The dissolved lithium polysulfides migrate back and forth between battery electrodes continuously, with structures switching between higher-order long chain polysulfides and lower-order polysulfides.[1] This effect significantly reduces the cycling stability, active material utilization and Coulombic efficiency of Li–S batteries.

To tackle the aforementioned issues, a variety of sulfur cathode designs were proposed.[2, 10–31] In general, the hydrophobic carbonaceous host materials for sulfur cathodes with closed structures are capable of inhibiting polysulfides diffusion through physical entrapment of polysulfides. However, in this way, there is only physical barrier retarding the polysulfides shuttling. The chemical binding between ionic polysulfides species and non-polar carbon-based host is weak in nature which makes carbonaceous host materials less efficient in polysulfides immobilization.[15,31–34] Therefore one of the most common and dominating strategies is to introduce nanostructured metal oxide additives to provide chemical interaction with polysulfides, including SiO$_2$,[34] MnO$_2$,[35] La$_2$O$_3$,[36] Al$_2$O$_3$,[37] Mg$_{0.6}$Ni$_{0.4}$O,[38] Mg$_{0.8}$Cu$_{0.2}$O,[39] and especially, TiO$_2$.[15,18,40] The effectiveness of metal oxide additives as ideal polysulfides absorbents is mainly attributed to the electrostatic interaction between polar polysulfides species and polar metal oxide. Moreover, nanostructured metal oxides with high surface-to-volume ratio increase the contact area between electrode and electrolyte, offering a high active material utilization and effective polysulfides absorption. Among various metal oxides, TiO$_2$ is especially attractive due to its strong polysulfides absorption ability.[15,40,41] However, as a large bandgap semi-conductor, the poor electrical conductivity of TiO$_2$ (bandgap of 3.0–3.2 eV) limits its widespread implementation in energy-related fields.[42–47] In addition, the chemical interaction (S–Ti–O) needs to be reinforced to achieve a perfect polysulfides confinement.[15]

Thus, researchers are seeking for an effective method to chemically tune the TiO$_2$ to achieve an improved electrical conductivity and polysulfides binding ability. Recently, nanostructured oxygen-deficient titanium dioxide or black TiO$_2$ has opened up a new direction for effectively solving the intrinsic problems of Li–S batteries. Annealing TiO$_2$ under reducing atmosphere or vacuum could lead to the darkening of TiO$_2$ due to the presence of Ti^{3+} and/or oxygen vacancies.[48,49] The resulting black TiO$_2$ could serve as a bifunctional material employed in Li–S batteries integrating high electrical conductivity and enhanced chemical interactions with lithium polysulfides. In this chapter, we will discuss the recent research progress of black TiO$_2$ nanomaterials in the field of Li–S energy storage system. Specifically, we focus on the synthesis of oxygen-deficient or black TiO$_2$ nanomaterials, the chemical properties as well as several application examples. We believe the present chapter will contribute significantly to the energy research field and inspire the research in other areas.

11.2   Black TiO$_2$ Nanomaterials in Li–S System

TiO$_2$ has raised considerable attention due to its applications in energy-related fields. Normal TiO$_2$ (anatase, rutile, or brookite) with large electronic bandgap of 3.0–3.2 eV[42,46] exhibits a white color since no visible light is absorbed. Black TiO$_2$ is usually defined as modified titanium dioxide with a dark color through high degree reduction. When heated under vacuum or reducing atmosphere, TiO$_2$ shows a series of color change from white (pristine), yellow to gray and finally black material depending on the reduction degree.[49] The darkening in the appearance suggests visible light absorption resulting mainly from the bandgap narrowing effect associated with the reduction treatment.[48,50] When applied in the Li–S batteries, these metallic black TiO$_2$ possess highly polar chemical bonds and hydrophilic surfaces that promote strong chemical interaction with polysulfides. In this way, the polysulfides produced during cycling is chemically rather than physically trapped.

In this section, Magneli-phase Ti$_n$O$_{2n-1}$ $(3 < n < 10)$,[51,52] first solved by Magneli in the 1950s, is introduced here as a typical family of hydrogenated black TiO$_2$. And we will study one representative example, Ti$_4$O$_7$, together with a lightly reduced titanium oxide (TiO$_{2-x}$), briefly covering the synthesis methods as well as chemical/electrochemical properties.

11.2.1   *Fabrication of Black TiO$_2$ Nanomaterials and the Composite Electrode for Li-S Batteries*

There is a large variety of strategies regarding the formation of black TiO$_2$ nanomaterials. These fabrication methods include hydrogen thermal treatment, hydrogen plasma treatment, chemical reduction, and electrochemical reduction of pristine TiO$_2$. Among the various synthesis methods, hydrogen thermal treatment is undoubtedly the dominating choice since it is less likely to introduce contaminations.

In the following application examples in this chapter, slightly hydrogen reduced TiO$_{2-x}$, partially reduced Ti$_6$O$_{11}$, and heavily reduced Ti$_4$O$_7$ are all prepared via annealing under pure hydrogen

Figure 11.2. Photographic images of commercial TiO_2 powder (anatase/rutile: 80/20) under hydrogen reduction (35 bar) at room temperature for different reduction time. Reprinted with permission from Ref. [53]. Copyright: The Royal Society of Chemistry.

atmosphere. Several factors could influence the degree of hydrogen reduction such as time, annealing temperature, and pressure of hydrogen. Figure 11.2 summarizes the appearance change as a function of hydrogen treatment time under room temperature.[53] It is clearly shown that TiO_2 powder transforms from white (pristine) to yellow, gray, and finally to black with increasing time as we discussed above. Under constant temperature, TiO_2 powder exhibits a darker color with a longer treatment time. Figure 11.3 illustrates the color evolution of TiO_2 as a function of annealing temperature. The pristine TiO_2 was first calcined in air for 3 h, and then subjected to hydrogen annealing for an additional 30 min at various annealing temperatures. It is found that a higher annealing temperature triggers a darker color. Hence in general, TiO_2 prepared

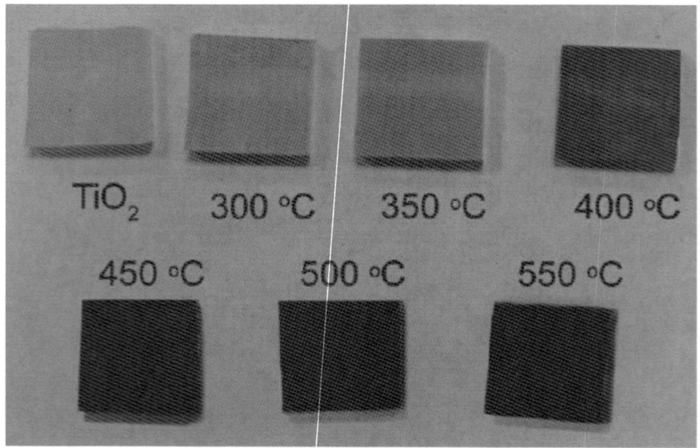

Figure 11.3. Digital images of pristine TiO$_2$ and hydrogen reduced titanium dioxide at various annealing temperatures (300°C, 350°C, 400°C, 450°C, 500°C, and 550°C) Reprinted with permission from Ref. [54]. Copyright: American Chemical Society.

with a longer annealing time and higher annealing temperature would have a higher degree of hydrogen reduction (more oxygen vacancies).[54]

There have been significant efforts placed in utilizing black TiO$_2$ nanomaterials in Li–S batteries. Both Cui group from Stanford University and Nazar group from University of Waterloo pioneer the study.

Cui *et al.* first proposed a sulfur cathode with hydrogen reduced TiO$_{2-x}$ inverse opal structure in 2014. TiO$_{2-x}$ refers to the reduced anatase phase with a low reduction degree. The pristine titania inverse opal was prepared by a template-assisted method as discussed in Ref. [15] and it was slightly reduced into a gray TiO$_2$ during the subsequent hydrogen treatment. The calcination was conducted under pure hydrogen environment with a flow rate of 100 sccm and 80 Torr of inner pressure. The calcination was held at 500°C for 2 h with a heating rate of 5°C/min. To construct an advanced sulfur cathode with this TiO$_{2-x}$ inverse opal, sulfur solution (sulfur dissolved in toluene) was infiltrated from the top openings into the entire porous structure under capillary forces. After toluene was evaporated, the as-prepared mixture was held at 155°C for 12 h to allow a uniform sulfur deposition. The TiO$_{2-x}$/S cathode exhibits excellent cycling stability and rate capability originating from the facile electron transport and enhanced polysulfides binding capability.

In addition to the slightly reduced anatase titania (TiO$_{2-x}$), another work from Cui group related to heavily reduced TiO$_2$ with Magneli phase (Ti$_n$O$_{2n-1}$) has also drawn tremendous attention. Tao *et al.* annealed the white TiO$_2$ at 950°C and 1050°C under pure hydrogen for 4 h respectively to synthesize Ti$_6$O$_{11}$ and Ti$_4$O$_7$. Then they prepared the Ti$_4$O$_7$–Sulfur (Ti$_4$O$_7$–S) and Ti$_6$O$_{11}$–Sulfur (Ti$_6$O$_{11}$–S) composite electrodes through a sulfur submission-deposition method which is facile and cost-effective. Afterwards, they added carbon disulfide (CS$_2$) solution into sublimed sulfur/Ti$_n$O$_{2n-1}$ mixture at an appropriate weight ratio. The resulting mixture was pressed into a cylindrical mold and annealed under vacuum at 155°C for 24 h. The Ti$_4$O$_7$–S cathode demonstrates a superior electrochemical performances in contrast to Ti$_6$O$_{11}$–S cathode, suggesting that a higher reduction degree leads to a superior performance.

Alternatively, Nazar group showed that black Ti$_4$O$_7$ nanopowder could also be prepared by argon treatment at nearly the same time. They anneal a yellow TiO$_2$ gel at ∼950°C under argon atmosphere in a tube furnace with a heating rate of 4°C/min. Then certain amount of sulfur was impregnated into the black powder by a simple melt-diffusion at 155°C described above to build the composite electrode. The resulting cathode shows a decent capacity retention ascribed to the unique properties of Ti$_4$O$_7$.

11.2.2 *Properties of Black TiO$_2$ Nanomaterials*

The conditions and synthesis methods of black TiO$_2$ nanomaterials vary from one another in the literature. Therefore, various black TiO$_2$ nanomaterials exhibit different properties.[47,48] In general, the reduction degree enables fine-tuning of the electrical conductivity, showing either semiconductor (Ti$_6$O$_{11}$) or metallic (Ti$_4$O$_7$) behavior at room temperature.[55] By simply controlling the reduction time and temperature, various reduced titanium oxides with different crystal structures and electrical/chemical/electrochemical properties could be obtained.

11.2.2.1   The presence of Ti^{3+} species and oxygen vacancies

In spite of the effectiveness of titanium oxides for polysulfides absorption, the highly insulating nature of these oxides impedes the overall electrochemical performance. It is proposed that the presence of Ti^{3+} ions and oxygen vacancies greatly enhances the electrical conductivity of titanium oxides. Ti trivalent ions and oxygen vacancies are constantly reported in black TiO$_2$ nanomaterials with hydrogen reduction.[48, 54, 56–65] X-ray photoelectron spectroscopy (XPS) is an effective way to probe the electronic property of titanium oxides. For example, Ti^{3+} ions were detected by Lu *et al.* from the red shift of the Ti 2p spectra (Figure 11.4). They subtract the Ti 2p spectra for unreduced and black TiO$_2$, resulting in a clear observation of a strong characteristic peak at 457.9 eV, corresponding to the existence of Ti^{3+}.

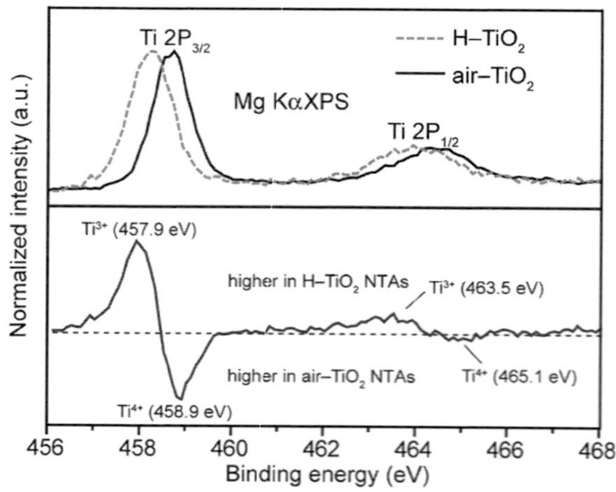

Figure 11.4. Top part is the overlay of normalized Ti 2p core level XPS spectra of unreduced TiO$_2$ (black solid line) and hydrogen reduced titania (red dashed line). Bottom part is the related subtraction spectrum. Reprinted with permission from Ref. [56]. Copyright: American Chemical Society.

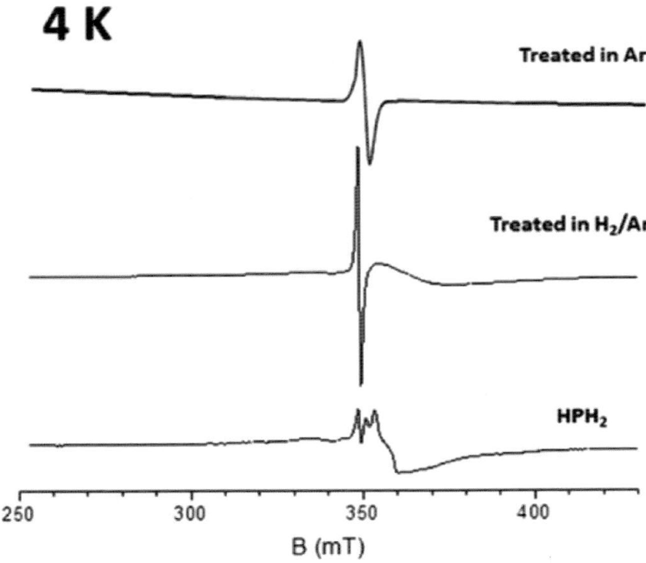

Figure 11.5. EPR spectra of TiO$_2$ nanomaterial annealed in argon, hydrogen/argon mixture, and high pressure hydrogen. Reprinted with permission from Ref. [66]. Copyright: American Chemical Society.

The existence of oxygen vacancies in hydrogen reduced titanium oxides has been confirmed by Schmuki group[66] according to the electron paramagnetic resonance (EPR) study. 4K X-band EPR spectra of the hydrogen treated titanium oxides (Figure 11.5) displays clearly a pronounced oxygen vacancy signal.[66]

Liang *et al.* has performed a simple Current–Voltage measurement to study the electrical property.[15] The experiment setup is shown in Figure 11.6(d). The TiO$_{2-x}$ nanomaterial exhibits a metallic property with linear *I–V* curve and a remarkable through-plane resistance (Figure 11.6(a)) in the proper voltage range. The current across the control anatase TiO$_2$ film is in the order of nA which suggests a high resistance (Figure 11.6(b)). When compared in a log scale, the through-plane current for TiO$_{2-x}$ is several orders of magnitude larger than that of TiO$_2$ under the same applied voltage, indicating a superior electrical conductivity after introducing the Ti^{3+} and oxygen vacancies by hydrogen annealing (Figure 11.6(c)).

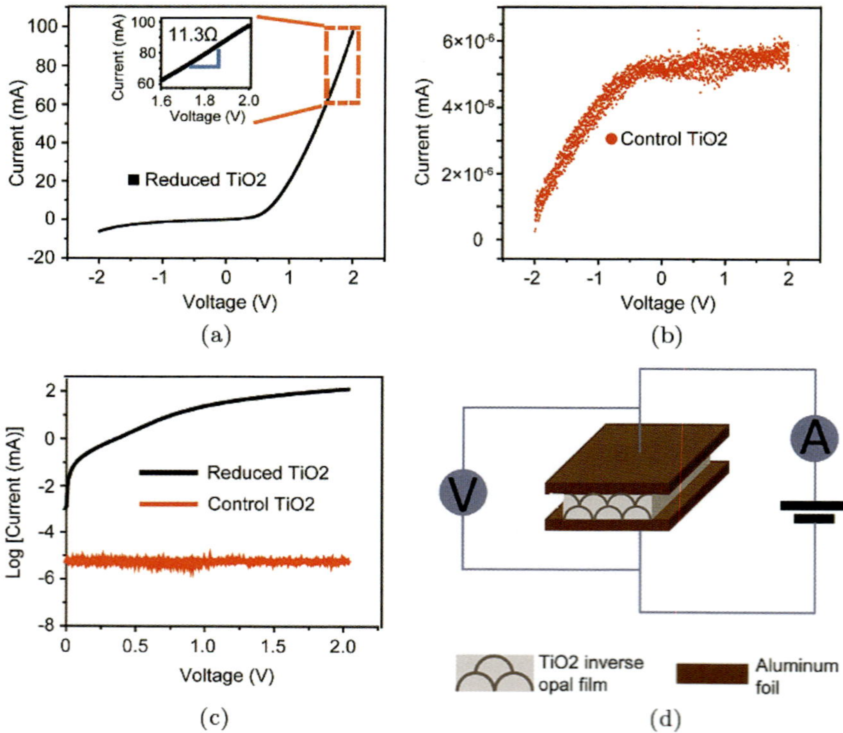

Figure 11.6. Current–Voltage characteristics of the reduced TiO$_{2-x}$ and control anatase TiO$_2$ inverse opal structure. Current–Voltage curve for the TiO$_{2-x}$ (a) and control TiO$_2$ (b). (c) Comparison of current–voltage curve for TiO$_{2-x}$ and TiO$_2$ plotted in a log scale. (d) Schematics of the experiment setup. Reprinted with permission from Ref. [15]. Copyright: American Chemical Society.

11.2.2.2   The structural disorder on surfaces of black TiO$_2$ and surface absorption of sulfur/polysulfides species

Many researchers have proved the disordered structural features at black TiO$_2$ surfaces recently.[53, 60, 63, 65, 68–73] Tao *et al.* from Cui group at Stanford University compared the structural properties of Ti$_4$O$_7$ and unreduced rutile TiO$_2$. Figure 11.7 displays the crystal structures of the most stable planes of Ti$_4$O$_7$ and pristine rutile TiO$_2$. Ti$_{4c}$, Ti$_{5c}$, and Ti$_{6c}$ refer to the titanium atoms with coordination numbers of 4, 5, and 6, respectively. Percentage of Ti$_{6c}$ in the most

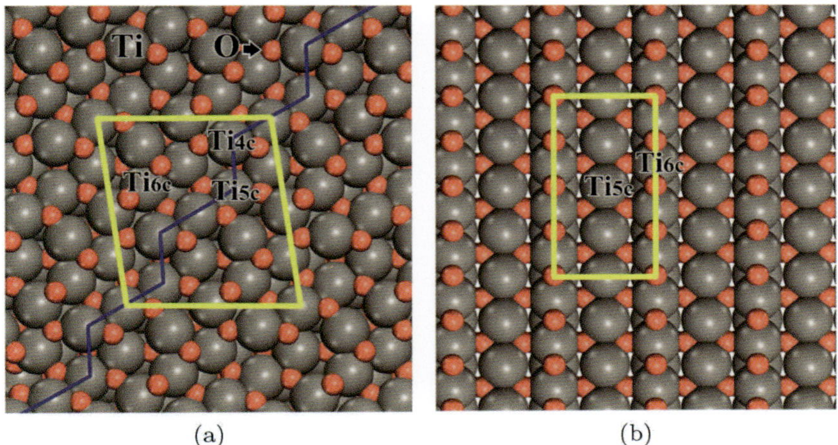

(a)   (b)

Figure 11.7. The schematic of most stable surfaces of (a) heavily reduced black TiO$_2$ (Ti$_4$O$_7$) and (b) unreduced rutile TiO$_2$ (110) planes (white TiO$_2$). Gray atoms: titanium; pink atoms: oxygen. Reprinted with permission from Ref. [67]. Copyright: American Chemical Society.

stable surface of Ti$_4$O$_7$ is 37.5%, implying 62.5% of the surface Ti atoms have unsaturated chemical bonds. For rutile TiO$_2$ (110) planes, ratio for Ti$_{5c}$ to Ti$_{6c}$ is 1:1, indicating a much smaller amount of "active" Ti atoms. More important, the majority of the unsaturated Ti atoms in Ti$_4$O$_7$ surfaces are located along the step sites (blue line in Figure 11.7(a)), which are easily accessible for external molecules/atoms. In contrast, "active" Ti atoms on rutile TiO$_2$ (110) surfaces are mainly spatially distributed on terrace sites with a lower exposure degree. Therefore, structural characteristics of this Ti$_4$O$_7$ would greatly affect the binding ability with sulfur and polysulfides species.

They also used density functional theory (DFT) to analyze the binding property of titanium oxides. They studied interactions of sulfur species (S$_x$, $x = 1$ to 4) and polysulfides species (Li$_2$S$_x$, $x = 1$ to 4) on the most stable titanium oxides surfaces. As illustrated in Figures 11.8(a) and (c), low coordination Ti sites (Ti$_{4c}$ and Ti$_{5c}$) act as favorable absorption sites for small sulfur clusters. The resulting highly polar Ti–S interaction facilitates the strong bonding effect.

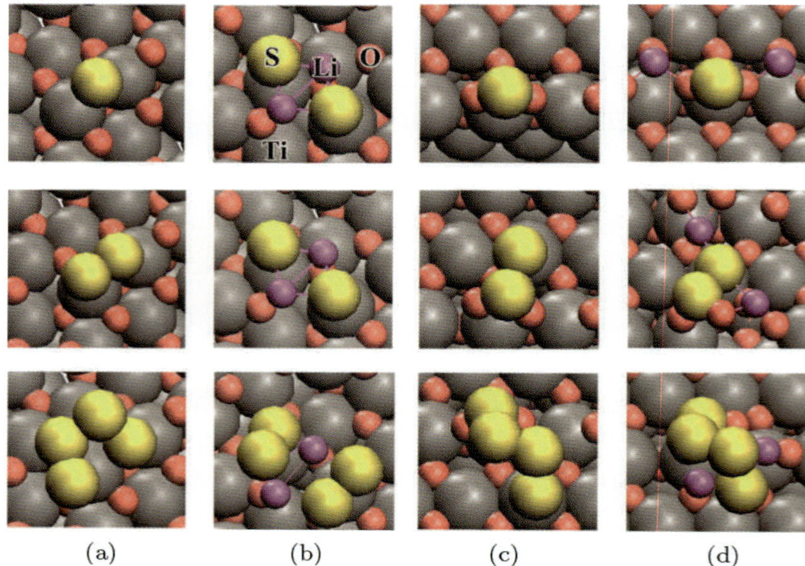

(a)   (b)   (c)   (d)

Figure 11.8. DFT analysis of sulfur species on most stable Ti_4O_7 surfaces ((a) and (b)) and unreduced rutile TiO_2 (110) surfaces ((c) and (d)). (a) Optimized geometries for S_x on Ti_4O_7 surfaces. (b) Optimized geometries for polysulfides on Ti_4O_7 surfaces. (c) Optimized geometries for S_x on rutile TiO_2 surfaces. (d) Optimized geometries for polysulfides on rutile TiO_2 surfaces. Gray atoms: titanium; pink atoms: oxygen; yellow atoms: sulfur; purple atoms: lithium. Reprinted with permission from Ref. [67]. Copyright: American Chemical Society.

While on rutile TiO_2 (110) surfaces, elemental sulfur is bonded with two bridging oxygen atoms, forming a SO_2 like species. Small sulfur molecules have weak bonding because of the lack of unsaturated Ti sites and oxygen vacancies. This also applies to the absorption of polysulfides species. Defective Ti_4O_7 with abundant low coordinated Ti ions provides active sites for Ti–S interactions (Figure 11.8(b)) while solely SO_2 like binding effect exists for polysulfides on rutile TiO_2 surfaces (Figure 11.8(d)).

The mechanism of formation of the earlier mentioned Ti–S bonds is related to the binding ability of defective titanium oxides with oxygen species. The interaction between defective titanium oxides and oxygen species has been investigated extensively during the past decades.[49,67,74] Many researchers have reported that surface low coordination Ti and oxygen vacancies on reduced titanium oxides could function as absorb sites for oxygen species or oxides.[49,74] Tao *et al.* believes electrostatic attractions exist for defective titanium oxides with sulfur species in a similar way since both oxygen and sulfur belong to Group 16 in the periodic table with similar electron configuration as well as chemical bonding properties.[15,67,75]

11.2.2.3 The electrochemical and mechanical stability of black TiO$_2$ nanomaterials in Li–S batteries

The electrochemical stability of black TiO$_2$ nanomaterials plays a key role towards stable battery cycling. During the battery charge stage, sulfur cathodes could be charged to a high voltage of ~2.5 V vs. Li$^+$/Li. Under this highly oxidizing environment, it is possible for the Ti trivalent species to be re-oxidized into Ti^{4+}. The composite electrode would lose its outstanding electrical conductivity and this would lead to a deterioration of the battery performance. According to the chemistry handbook,[76] the standard redox potential of Ti^{4+}/Ti^{3+} pair is 2.5 V vs. Li$^+$/Li which is close to the common cut-off voltage for Li–S battery operations. Due to the overpotential during battery cycling, the re-oxidation of Ti^{3+} could be considered beyond the scope.[15] Therefore, the re-oxidation of black TiO$_2$ is ruled out within the operation voltage of Li–S batteries.

The mechanical stability of the black TiO$_2$ nanomaterials mainly originates from its ceramic nature. In the next section, we will study several application examples, and the structural stability after long-term battery cycling is confirmed by the microscopic characterizations.

11.3 Case Studies of Black TiO$_2$ in Li–S Batteries

In this chapter, we study four examples of black TiO$_2$ or oxygen-deficient TiO$_2$ nanomaterials used in Li–S batteries. Section 11.3.1 focuses on a work conducted by Cui group at Stanford University using slightly hydrogen reduced TiO$_2$ (gray) inverse opal structure to trap sulfur and polysulfides. Sections 11.3.2 and 11.3.3, done by Tao *et al.* and Nazar group recently, show the employment of highly polar and metallic Magneli-phase Ti$_4$O$_7$ nanomaterials (black TiO$_2$) as effective additives and/or stable conductive matrix which enables an impressive electrochemical performance. Section 11.3.4 demonstrates a work studying the facets dependence in the interaction of hydrogen reduced titania with polysulfides molecules. Based on these application examples, we want to introduce and demonstrate how these nanostructured black or oxygen-deficient TiO$_2$ are effectively involved in Li–S batteries.

11.3.1   *Application Example-1: "Sulfur Cathodes with Hydrogen Reduced Titanium Dioxide Inverse Opal Structure"*

Hydrogen reduced TiO$_2$ has been explored as advanced anode materials for rechargeable lithium ion batteries due to its ultrafast ion transport and excellent electrical conductivity.[77–79] Here, Liang *et al.* at Stanford University led by Professor Yi Cui has now identified a new application of this material as conductive matrix for sulfur cathodes. They took the advantages of the excellent electrical conductivity and polysulfides binding property of hydrogen reduced titania and developed a novel composite electrode with a remarkable electrochemical performance based on this material.

For the preparation of the inverse opal structure, as illustrated in Figure 11.9(a), polystyrene (PS) spheres were drop-casted to form a close-packed template. Amorphous TiO$_2$ with a 30 nm thickness was then deposited uniformly through the low temperature atomic layer deposition (ALD). The sacrificial PS template was removed by the subsequent heat treatment. Sulfur was infiltrated into the porous three-dimensional structure by a melt-diffusion process. Figure 11.9(b) presents the scanning electron microscopy (SEM) images of the homogeneous distributed sulfur particles inside the inverse opal structure. Nanostructured TiO$_{2-x}$ inverse opal with highly ordered nanopores (780 nm in diameter) ensures a high sulfur utilization and accommodates the sulfur volume change during cycling.

Liang *et al.* probed the electronic and chemical environments of this TiO$_{2-x}$ by XPS. The red shift in binding energy indicates that

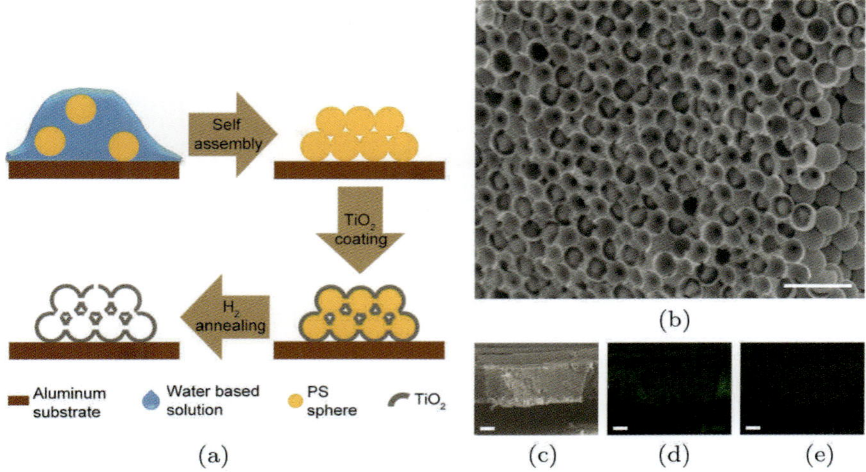

Figure 11.9. Fabrication process of the TiO_{2-x} and the corresponding microscopic characterization. (a) Schematics of the fabrication. (b) SEM image of the TiO_{2-x}/S composite. ((c)–(e)) Elemental mapping for sulfur and titanium. Scale bars are $2\,\mu m$ in (b) and $5\,\mu m$ in ((c)–(e)). Reprinted with permission from Ref. [15]. Copyright: American Chemical Society.

hydrogen reduction of TiO_2 could generate Ti^{3+} trivalent species and oxygen vacancies (Figure 11.10(a)), leading to a dramatically improved electrical conductivity.[56,80,81] Moreover, oxygen vacancies created could promote the chemical interaction of S–Ti–O, which enhances polysulfides absorption on the TiO_{2-x} surfaces, as clearly indicated by the Ti–S shoulder peak (Figure 11.10(b)).[31,82] This enhanced binding ability originates from the strong absorption of Group 16 elements on defective titania. The phase of this slightly reduced titania is confirmed from the X-ray diffraction pattern (XRD) which mainly consists of reduced anatase and rutile (Figure 11.10(c)).

They also demonstrated the structural stability of this TiO_{2-x} matrix after long-term cycling. The TiO_{2-x} inverse opal scaffold exhibits little morphology change before and after lithiation. Liang *et al.* proposed three attributions of this structurally intact TiO_{2-x} as follows: (1) the empty space of the inverse opal structure provides

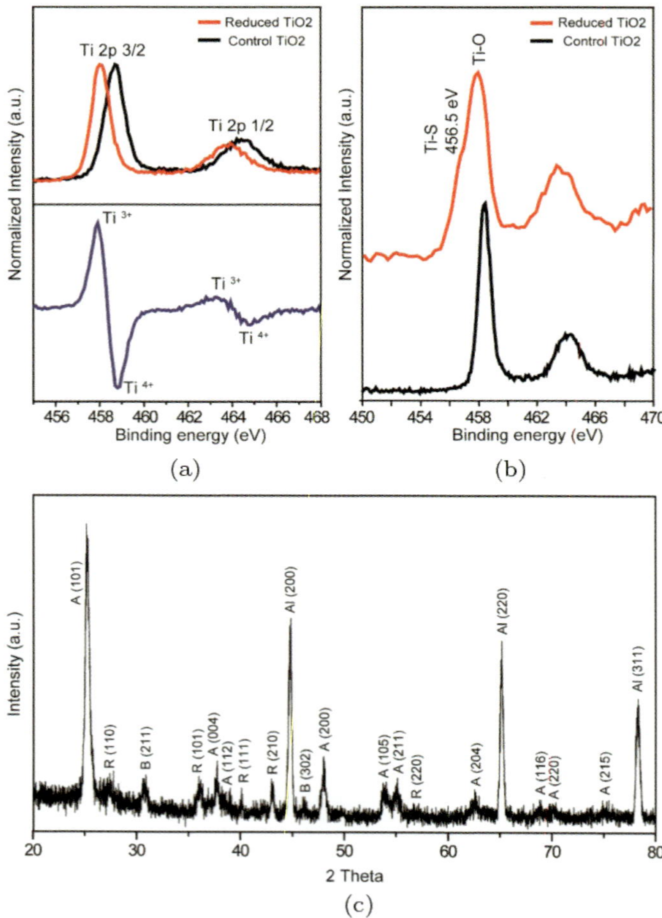

Figure 11.10. XPS and XRD characterizations. (a) Top part is the normalized Ti 2p spectra of TiO_{2-x} (red solid line) and anatase TiO_2 (black solid line). Bottom part is the corresponding subtraction spectrum. (b) Normalized Ti 2p XPS spectra of polysulfides treated anatase TiO_2 (black) and TiO_{2-x} (red). (c) XRD pattern collected from the TiO_{2-x}. Reprinted with permission from Ref. [15]. Copyright: American Chemical Society.

facile accommodation of the sulfur volume expansion (2) TiO_{2-x} is electrochemically inactive within the voltage range of Li–S battery operations and (3) crystalline TiO_{2-x} is mechanically strong owing to its ceramic nature.

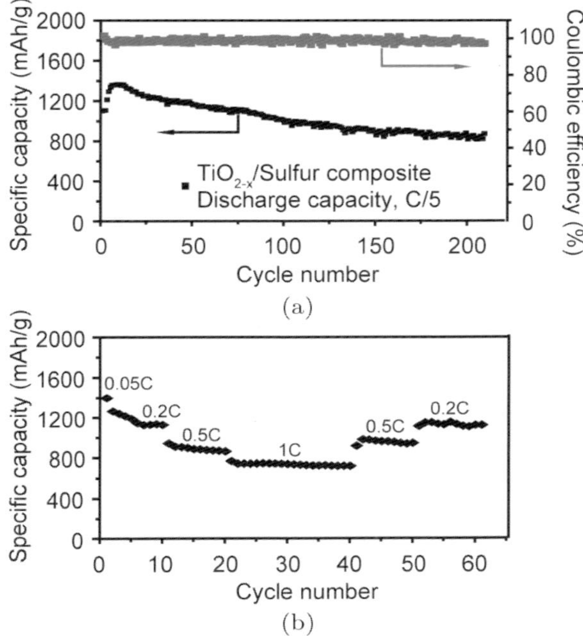

Figure 11.11. Cycling performance and rate capability of the as-prepared TiO$_{2-x}$/S composite cathode. (a) Black curve: cycling performance. Blue: Coulombic efficiency. The current rate is C/5. (d) Rate capability of TiO$_{2-x}$/S composite electrode from 0.05 C to 1 C. Reprinted with permission from Ref. [15]. Copyright: American Chemical Society.

By using the TiO$_{2-x}$ to fabricate the sulfur cathodes, they have achieved a good battery performance, with an initial specific capacity of ∼1100 mAh/g and an 81% capacity retention over 200 cycles at a C/5 rate. The corresponding Coulombic efficiency reaches 99.5% (Figure 11.11(a)). They also present an impressive rate capability, related to the facile electron/ion conduct of this material (Figure 11.11(b)).

This hydrogen reduced titania developed by Cui group shows great promise as a favorable host material for Li–S batteries. Three-dimensional nanoporous scaffold facilitates the physical confinement of polysulfides as well as remarkable kinetics associated with electrons and lithium ions. The resulting TiO$_{2-x}$/S composite cathode exhibits an impressive electrochemical behavior with good electrical conductivity, fast-ion transport, stable structure during cycling and the ability to bind polysulfides both physically and chemically. Therefore, hydrogen reduced titania inverse opal would have the potential use as a novel electrode structure for Li–S system. And this concept opens up a new avenue for constructing sulfur cathodes with oxygen-deficient metal oxides. However, the fabrication process of the composite electrode is still relatively complicated due to the introduction of sacrificial template. The low temperature ALD process requires a long-synthesis period and is not cost-effective. This strategy would become more promising if the ALD process could be replaced with other simple coating methods.

11.3.2　*Application Example-2: "Strong Sulfur Binding with Conducting Magneli-Phase Ti$_n$O$_{2n-1}$ Nanomaterials for Improving Lithium–Sulfur Batteries"*

Substoichiometric titanium dioxide, referred to as Magneli-phase, exhibits a greatly improved polar and metallic behavior compared to normal TiO$_2$ or even the slightly reduced TiO$_2$ (TiO$_{2-x}$).[33,67] Recently Tao *et al.* has selected from several Magneli-phase titanium oxides and fabricated a novel Ti$_4$O$_7$–sulfur (Ti$_4$O$_7$–S) composite electrode. Tao *et al.* demonstrated the promise of this Ti$_4$O$_7$ as a highly effective absorbent for polysulfides binding. This is one of the most representative examples of a Li–S battery using nanostructured black TiO$_2$.

Tao *et al.* constructed Type-CR2025 coin cells to evaluate the electrochemical property of the composite sulfur cathode based on this black titania. The electrochemical performance is shown in Figure 11.13. They have successfully constructed three different sulfur composite cathodes based on pristine TiO$_2$, partially reduced Ti$_6$O$_{11}$ and highly reduced Ti$_4$O$_7$. They revealed that black titania with a higher reduction degree offers a superior electrochemical property. Among the three electrodes, sulfur cathodes with highly reduced metallic Ti$_4$O$_7$ displays an impressive discharge

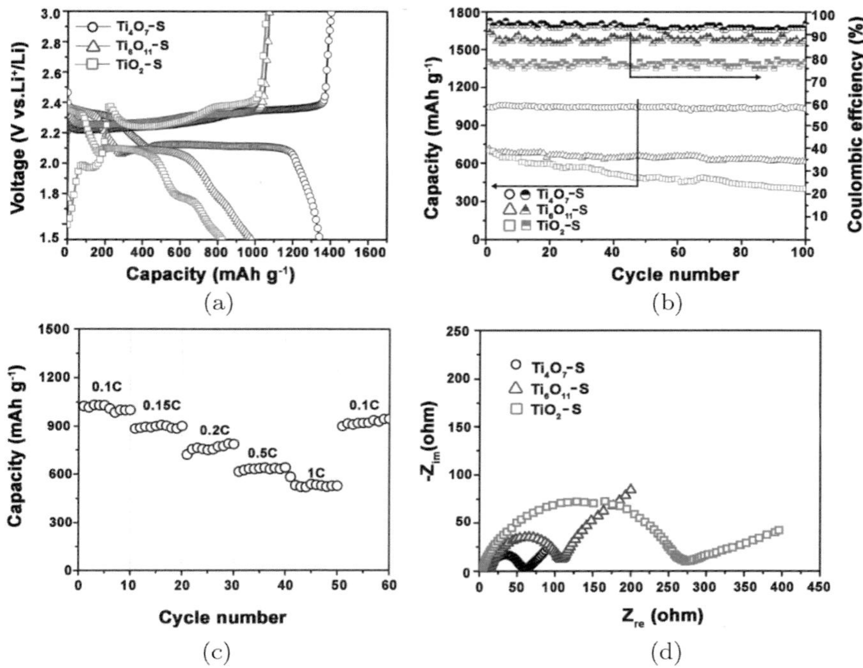

Figure 11.12. Electrochemical properties of Ti$_4$O$_7$–S composite electrode. (a) Initial charge–discharge voltage profiles of the composite electrode at C/50 for different titanium oxides. (b) Long-term cycling performance and the related Coulombic efficiency for various composite electrodes at C/10. (c) Rate capability of the Ti$_4$O$_7$–S composite. (d) Nyquits plots for composite electrodes with various titanium oxides. Reprinted with permission from Ref. [67]. Copyright: American Chemical Society.

capacity of ~1100 mAh/g with a remarkable 99% capacity retention at C/10 over 100 cycles with a relatively high sulfur loading (1–3 mg/cm^2) and a relatively high sulfur content (65–75 wt.%). The high Coulombic efficiency of 96% (Figure 11.12(b)), excellent rate stability (Figure 11.12(c)) and minimum charge-transfer resistance (Figure 11.12(d)) of this Ti$_4$O$_7$–S electrode further confirm its high electrical conductivity and improved chemical binding ability with migrating polysulfides.

The attribution of the outstanding battery performance of sulfur cathodes with this black titania lies in the strong interaction between

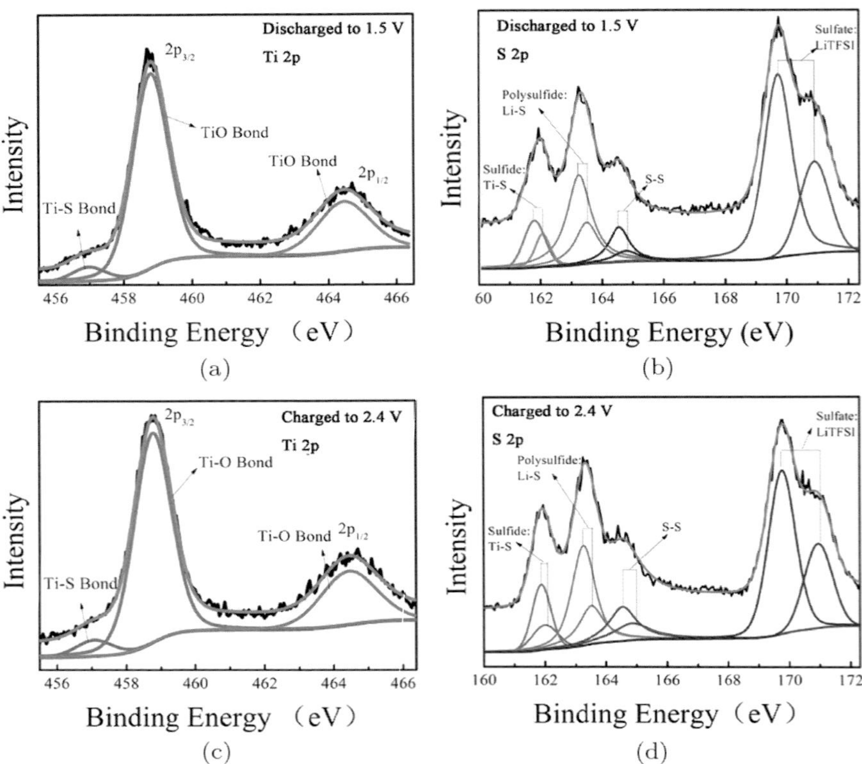

Figure 11.13. XPS analysis of titanium oxide–sulfur composites. (a) Ti 2p and (b) S 2p XPS spectra of the Ti_4O_7–S composite electrode at discharge state. (c) Ti 2p and (d) S 2p XPS spectra of Ti_4O_7–S composite electrode at charge state. Reprinted with permission from Ref. [67]. Copyright: American Chemical Society.

Ti_4O_7 surfaces and polysulfides species. Tao *et al.* proposed that the low coordination Ti could promote a highly polar Ti–S chemical bond. To confirm this hypothesis, they performed XPS study on Ti_4O_7–S composite electrode at both charge and discharge state. For Ti 2p spectra, at both fully charge and discharge state, there is existence of Ti–S bond (Figures 11.13(a) and (c)). They also show the evidence of Ti–S bond from the S 2p spectra (Figures 11.13(b) and (d)). Hence, Tao *et al.* claimed that the presence of Ti–S bond triggers strong sulfur and polysulfides interactions.

In summary, Tao *et al.* from Cui group successfully developed the high-performance Li–S batteries based on Magneli-phase Ti$_4$O$_7$, which serves as an ideal matrix for sulfur owning to its unique sulfur/polysulfides trapping property and high electrical conductivity. They also fabricated Li–S batteries with unreduced TiO$_2$ and partially reduced Ti$_6$O$_{11}$ for comparison. They found that titanium oxides with a higher reduction degree facilitates a superior battery performance. For practical battery assembly, conducting agent such as carbon black and polymer binder such as polyvinylidene fluoride (PVDF) are still required for fabricating the cells. Despite the fact that (1) introduction of metal oxides additive compromises the high specific capacity of the composite sulfur cathodes and (2) the demand for hydrogen treatment further increases the complexity of the fabrication process, this new finding has great impact on Li–S battery field and it will stimulate the potential applications of other Magneli-phase metal oxides (Mo$_n$O$_{2n-1}$, W$_n$O$_{2n-1}$, V$_n$O$_{2n-1}$, etc.)[33,67] with abundant unsaturated metal atoms on the surface.

11.3.3 *Application Example-3: "Surface-Enhanced Redox Chemistry of Polysulfides on a Metallic and Polar Host for Lithium–Sulfur Batteries"*

Traditional carbonaceous materials rely solely on the closed structure to physically trap polysulfides, Nazar group from University of Waterloo created a bifunctional nanomaterial for sulfur cathodes combining inherent metallic conductivity with chemical interactions with polysulfides. They discovered the unique property of this Ti$_4$O$_7$ nanomaterial at nearly the same time with Cui group. Nazar *et al.* reported similar outstanding electrochemical performances (Figure 11.14). They showed that at a high cycling rate, the nanostructured Ti$_4$O$_7$ electrode could deliver a high initial capacity of 800 mAh/g with a negligible capacity fade of 0.06% per cycle (Figure 11.14(d)). For comparison, Nazar *et al.* also constructed a composite electrode using carbonaceous material (VC). The VC/S composite electrode displays an inferior cycling stability (Figure 11.14(b)) indicating the lack of polysulfides confinement.

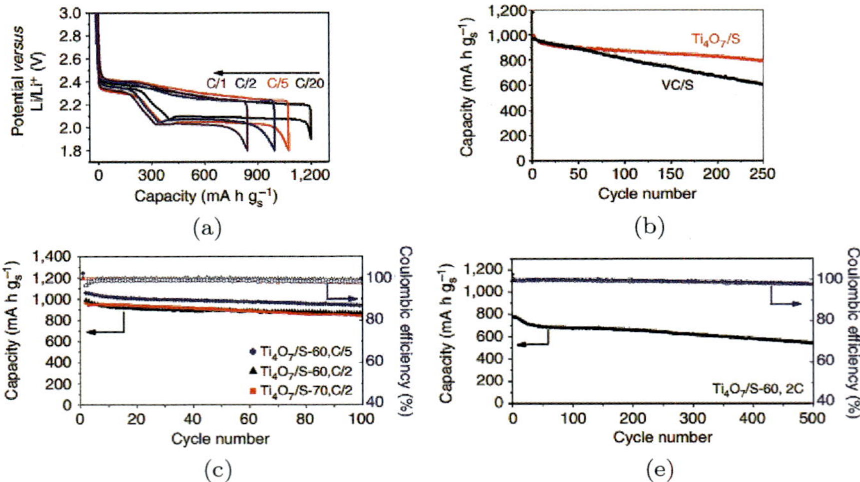

Figure 11.14. Electrochemical performance of Ti_4O_7/S and VC/S electrodes. (a) Voltage profiles of Ti_4O_7/S positive electrodes at various rates. (b) Cycling performance of Ti_4O_7/S and VC/S at C/2. (c) Cycling performance of Ti_4O_7/S composites with different sulfur fractions and the corresponding Coulombic efficiency. (d) High-rate cycling performance of Ti_4O_7/S with corresponding Coulombic efficiency. Reprinted with permission from Ref. [33]. Copyright: Nature Publishing Group.

Nazar *et al.* designed an innovative visual study to demonstrate the strong polysulfides binding effect of Ti_4O_7. A most representative polysulfide species (Li_2S_4) is dissolved in tetrahydrofuran (THF) to form a yellow solution. Various absorbents were put in contact with the polysulfides solution. They observed an almost completely colorless solution after the addition of Ti_4O_7 followed by the subsequent 1 h stirring (Figures 11.15(b) and (c)). This implies a superior intrinsic capability for Ti_4O_7 to absorb polysulfides, while polysulfides solutions remained intense yellow upon contact with VC or graphite, indicative of no evidence for chemical interactions.

This conclusion is further supported by XPS analysis. For a typical Li_2S_4 molecule, S_T^{-1} and S_B^0 are defined as terminal sulfur atom and bridging sulfur atom, respectively. For the Ti_4O_7/Li_2S_4 mixture, the polarization of electrons away from the electronegative

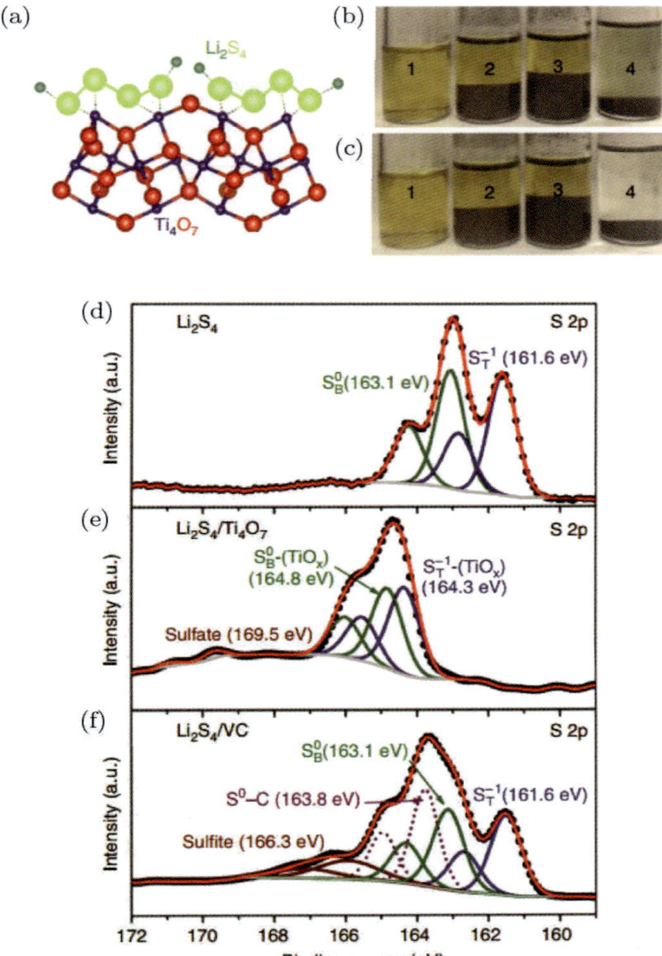

Figure 11.15. Demonstration of the strong interaction of polysulfides with Ti$_4$O$_7$. (a) A schematic showing the electron density transfer between Li$_2$S$_4$ and Ti$_4$O$_7$ (yellow = S, green = Li, blue = Ti, red = O). ((b),(c)) Sealed vials of a Li$_2$S$_4$/THF solution (1), and after contact with graphite (2), VC carbon (3) and Ti$_4$O$_7$ (4), immediately upon contact (b) and after 1 h stirring (c). ((d)–(f)) High-resolution XPS S 2p spectra of (d) Li$_2$S$_4$, (e) Li$_2$S$_4$/Ti$_4$O$_7$ and (f) Li$_2$S$_4$/VC. Reprinted with permission from Ref. [33]. Copyright: Nature Publishing Group.

sulfur atoms to the electropositive titanium and oxygen vacancies is attributed to the interaction of both terminal and bridging sulfur in Li$_2$S$_4$ with Ti$_4$O$_7$ surfaces (Figures 11.15(d) and (e)). Whereas the S 2p spectra of VC/S mixture exhibits two dominating characteristic peaks identical to that of bare Li$_2$S$_4$. This unchanged Li$_2$S$_4$ spectral feature together with the intense yellow color of VC/Li$_2$S$_4$ solutions implies that no chemical absorption occurs for carbonaceous materials.

Based on the two highly innovative studies of Ti$_4$O$_7$ by Tao *et al.* and Nazar *et al.*, it is expected that Magneli-phase titanium oxides could be utilized for construction of practical high performance sulfur cathodes.

11.3.4   *Application Example-4: "Hierarchical Sulfur-Impregnated Hydrogenated TiO$_2$ Mesoporous Spheres Comprising Anatase Nanosheets with Highly Exposed (001) Facets for Advanced Li–S Batteries"*

It is believed that the absorption behavior of polysulfides species on black TiO$_2$ molecules should have some dependence on the different titanium oxide crystal facets.[83–85] Lin group from Xiamen University reported a strong interaction between sulfur species and titanium oxides on (001) facets and they have conducted an effective investigation on sulfur cathodes based on hydrogenated titanium oxide nanosheets with highly exposed (001) planes.

Lin *et al.* investigated sulfur species absorption on several crystal planes of hydrogenated titanium oxide and concluded that (001) facets possess the very strong bonding effect.[83, 85] Then they compared the chemical interactions between S$_4^{2-}$ specie and (001) surfaces for both hydrogenated titanium oxide and the unreduced anatase. According to their VASP computation, on the (001) surface of hydrogenated titanium oxide, sulfur species tends to form Ti–S bond while for (001) surface of unreduced anatase TiO$_2$, it forms S–O bond, which agrees with Tao *et al.*[67] Moreover the average bond length for reduced oxide (001) surface with S$_4^{2-}$ is smaller than that for the unreduced anatase surfaces, indicating a larger bond energy of reduced (001) surfaces.

11.4 Challenges and Outlooks

We have demonstrated the effectiveness of black TiO$_2$ nanomaterials in Li–S batteries based on several innovative works done by different groups. To target on large-scale practical battery applications, there is still room for improvement.

(1) The fabrication process of this nanostructured black TiO$_2$ requires some complex and highly demanding procedures such as the hydrogen treatment and thermal annealing. This means the cost of sulfur cathodes based on this material is still a concern for practical batteries. The scale-up is also challenging. Taking the material expense of titanium oxides into consideration, the total cost of batteries based on this material could be a disadvantage for its potential commercial development.

(2) Introducing this black TiO$_2$ into current Li–S batteries would compromise the high energy density of the entire composite electrode. Traditional battery additives such as the conducting agent (carbon black) possess a small molecular weight and reduce the overall specific capacity only to a small degree. The weight percentage of this black TiO$_2$ in the entire composite should be further optimized.

(3) Although it is shown that electrochemical re-oxidation of the reduced black TiO$_2$ should not be a concern within normal applied voltage range (1.7–2.5 V vs. Li$^+$/Li) for Li–S batteries, under some extreme conditions such as over charge, it is still possible for re-oxidation of the Ti^{3+} into Ti^{4+}. In addition, the intercalation between Li and black TiO$_2$ is also likely to occur which may affect the unique electronic properties of this material. Therefore, the working voltage of Li–S batteries based on black TiO$_2$ should be under careful control.

(4) Despite various hypothesis proposed by Cui group and Nazar group, the underlying mechanism of the enhanced polysulfides binding effect of this material is still not clearly elucidated. Many

more research on fully understanding how polysulfides species interacts with black TiO$_2$ molecules are required.

11.5 Summary

Li–S battery is one of the leading candidates for next-generation energy storage systems. Traditional carbonaceous host material for sulfur cathodes is non-polar in nature, not being able to provide favorable chemical binding with the ionic sulfides species. Metal oxides as host or supplemental material facilitate electrostatic binding effect because of the absorption of electronegative polysulfides species on metal cations. However, the inherent insulating property of oxides blocks electron and Li ion conductive pathway, therefore leading to a low active material utilization and poor rate capability.

Black TiO$_2$ nanomaterials are therefore acting as a two-in-one strategy to tackle the aforementioned issues. On one hand, the reduced titanium oxide with metallic behavior could electrically wire-up the sulfur and dramatically improve the Li-storage performance of the composite electrode.[86] On the other hand, utilization of chemical interactions between polysulfides molecules and lower coordinated Ti species effectively hinders the migration of dissolved polysulfides. Furthermore, nanoporous black TiO$_2$ with a high surface area enables a high capacity, improved electrolyte accessibility as well as facile volume change toleration, which realizes excellent electrochemical performance.

To conclude, though several issues need to be considered to achieve the widespread implementation of black TiO$_2$ nanomaterials in practical Li–S batteries, the utilization of this material as additive or matrix in sulfur cathodes is very attractive and of great significance.

References

(1) Yang, Y.; Zheng, G.; Cui, Y. *Chem. Soc. Rev.* **2013**, *7*, 3018–3032.
(2) Ji, X.; Lee, K. T.; Nazar, L. F. *Nat Mater.* **2009**, *6*, 500–506.
(3) Manthiram, A.; Fu, Y.; Su, Y.-S. *Acc. Chem. Res.* **2013**, *5*, 1125–1134.
(4) Evers, S.; Nazar, L. F. *Acc. Chem. Res.* **2013**, *5*, 1135–1143.

(5) Bruce, P. G.; Freunberger, S. A.; Hardwick, L. J.; Tarascon, J.-M. *Nat. Mater.* **2012**, *1*, 19–29.

(6) Mikhaylik, Y. V.; Akridge, J. R. *J. Electrochem. Soc.* **2004**, *11*, A1969–A1976.

(7) Suo, L.; Hu, Y.-S.; Li, H.; Armand, M.; Chen, L. *Nat. Commun.* **2013**, *4*, 1481.

(8) Yamin, H.; Gorenshtein, A.; Penciner, J.; Sternberg, Y.; Peled, E. *J. Electrochem. Soc.* **1988**, *5*, 1045–1048.

(9) Barchasz, C.; Leprêtre, J.-C.; Alloin, F.; Patoux, S. *J. Power Sources* **2012**, 322–330.

(10) Jayaprakash, N.; Shen, J.; Moganty, S. S.; Corona, A.; Archer, L. A. *Angew. Chem. Int. Ed.* **2011**, *26*, 5904–5908.

(11) He, G.; Evers, S.; Liang, X.; Cuisinier, M.; Garsuch, A.; Nazar, L. F. *ACS Nano* **2013**, *12*, 10920–10930.

(12) Evers, S.; Nazar, L. F. *Chem. Comm.* **2012**, *9*, 1233–1235.

(13) Wang, H.; Yang, Y.; Liang, Y.; Robinson, J. T.; Li, Y.; Jackson, A.; Cui, Y.; Dai, H. *Nano Lett.* **2011**, *7*, 2644–2647.

(14) Xiao, L.; Cao, Y.; Xiao, J.; Schwenzer, B.; Engelhard, M. H.; Saraf, L. V.; Nie, Z.; Exarhos, G. J.; Liu, J. *Adv. Mater.* **2012**, *9*, 1176–1181.

(15) Liang, Z.; Zheng, G.; Li, W.; Seh, Z. W.; Yao, H.; Yan, K.; Kong, D.; Cui, Y. *ACS Nano* **2014**, *5*, 5249–5256.

(16) Su, Y.-S.; Fu, Y.; Cochell, T.; Manthiram, A. *Nat. Commun.* **2013**, *4*.

(17) Yao, H.; Zheng, G.; Li, W.; McDowell, M. T.; Seh, Z.; Liu, N.; Lu, Z.; Cui, Y. *Nano Lett.* **2013**, *7*, 3385–3390.

(18) Wei Seh, Z.; Li, W.; Cha, J. J.; Zheng, G.; Yang, Y.; McDowell, M. T.; Hsu, P.-C.; Cui, Y. *Nat. Commun.* **2013**, 1331.

(19) Seh, Z. W.; Yu, J. H.; Li, W.; Hsu, P.-C.; Wang, H.; Sun, Y.; Yao, H.; Zhang, Q.; Cui, Y. *Nat. Commun.* **2014**, *5*.

(20) Li, W.; Zheng, G.; Yang, Y.; Seh, Z. W.; Liu, N.; Cui, Y. *Proc. Nat. Acad. Sci.* **2013**, *18*, 7148–7153.

(21) Yao, H.; Zheng, G.; Hsu, P.-C.; Kong, D.; Cha, J. J.; Li, W.; Seh, Z. W.; McDowell, M. T.; Yan, K.; Liang, Z.; Narasimhan, V. K.; Cui, Y. *Nat. Commun.* **2014**.

(22) Li, W.; Zhang, Q.; Zheng, G.; Seh, Z. W.; Yao, H.; Cui, Y. *Nano Lett.* **2013**, *11*, 5534–5540.

(23) Guo, J.; Yang, Z.; Yu, Y.; Abruña, H. D.; Archer, L. A. *J. Am. Chem. Soc.* **2013**, *2*, 763–767.

(24) Zheng, G.; Yang, Y.; Cha, J. J.; Hong, S. S.; Cui, Y. *Nano Lett.* **2011**, *10*, 4462–4467.

(25) Ji, L.; Rao, M.; Zheng, H.; Zhang, L.; Li, Y.; Duan, W.; Guo, J.; Cairns, E. J.; Zhang, Y. *J. Amer. Chem. Soc.* **2011**, *46*, 18522–18525.

(26) Wang, H.; Zhang, Q.; Yao, H.; Liang, Z.; Lee, H.-W.; Hsu, P.-C.; Zheng, G.; Cui, Y. *Nano Lett.* **2014**, *12*, 7138–7144.

(27) Guo, J.; Xu, Y.; Wang, C. *Nano Lett.* **2011**, *10*, 4288–4294.

(28) Kim, J.; Lee, D.-J.; Jung, H.-G.; Sun, Y.-K.; Hassoun, J.; Scrosati, B. *Adv. Funct. Mater.* **2013**, *8*, 1076–1080.

(29) Schuster, J.; He, G.; Mandlmeier, B.; Yim, T.; Lee, K. T.; Bein, T.; Nazar, L. F. *Angew. Chem. Int. Ed.* **2012**, *51*, 3591–3595.

(30) Su, Y.-S.; Manthiram, A. *Nat. Commun.* **2012**, *3*, 1166.

(31) Evers, S.; Yim, T.; Nazar, L. F. *J. Phys. Chem. C* **2012**, *37*, 19653–19658.

(32) Pope, M. A.; Aksay, I. A. *Adv. Energy Mater.* **2015**, *16*, n/a-n/a.

(33) Pang, Q.; Kundu, D.; Cuisinier, M.; Nazar, L. F. *Nat. Commun.* **2014**, *5*.

(34) Ji, X.; Evers, S.; Black, R.; Nazar, L. F. *Nat. Commun.* **2011**, *2*, 325.

(35) Li, Z.; Zhang, J.; Lou, X. W. *Angew. Chem. Int. Ed.* **2015**, *54*, 12886–12890.

(36) Zheng, W.; Hu, X. G.; Zhang, C. F. *Electrochem. Solid State Lett.* **2006**, *7*, A364–A367.

(37) Dong, K.; Wang, S.; Zhang, H.; Wu, J. *Mater. Res. Bull.* **2013**, *6*, 2079–2083.

(38) Zhang, Y.; Zhao, Y.; Yermukhambetova, A.; Bakenov, Z.; Chen, P. *J. Mater. Chem. A* **2013**, *2*, 295–301.

(39) Zhang, Y.; Wu, X.; Feng, H.; Wang, L.; Zhang, A.; Xia, T.; Dong, H. *Int. J. Hydrogen Energ.* **2009**, *3*, 1556–1559.

(40) Ding, B.; Shen, L.; Xu, G.; Nie, P.; Zhang, X. *Electrochim. Acta* **2013**, *107*, 78–84.

(41) Zhang, Z.; Li, Q.; Jiang, S.; Zhang, K.; Lai, Y.; Li, J. *Chem. Eur. J.* **2015**, *3*, 1343–1349.

(42) Liu, L.; Chen, X. *Chem. Rev.* **2014**, *19*, 9890–9918.

(43) Lin, C.-H.; Chao, J.-H.; Liu, C.-H.; Chang, J.-C.; Wang, F.-C. *Langmuir* **2008**, *17*, 9907–9915.

(44) Kuo, H.-L.; Kuo, C.-Y.; Liu, C.-H.; Chao, J.-H.; Lin, C.-H. *Catal. Lett.* **2007**, *1-2*, 7–12.

(45) Lin, C. H.; Lee, C. H.; Chao, J. H.; Kuo, C. Y.; Cheng, Y. C.; Huang, W. N.; Chang, H. W.; Huang, Y. M.; Shih, M. K. *Catal. Lett.* **2004**, *1*, 61–66.

(46) Ma, Y.; Wang, X.; Jia, Y.; Chen, X.; Han, H.; Li, C. *Chem. Rev.* **2014**, *19*, 9987–10043.

(47) Chen, X.; Li, C.; Gratzel, M.; Kostecki, R.; Mao, S. S. *Chem. Soc. Rev.* **2012**, *23*, 7909–7937.

(48) Chen, X.; Liu, L.; Huang, F. *Chem. Soc. Rev.* **2015**, *7*, 1861–1885.

(49) Diebold, U. *Surfa. Sci. Rep.* **2003**, *5–8*, 53–229.

(50) Pan, H.; Zhang, Y.-W.; Shenoy, V. B.; Gao, H. *J. Phys. Chem. C* **2011**, *24*, 12224–12231.

(51) Andersson, S. C., Bengt; Kruuse, Georg; Kuylenstierna, Ulf; Magnéli, Arne; Pestmalis, Herbert; Åsbrink, Stig. *Acta Chem. Scand.* **1957**, *11*, 1653–1657.

(52) Andersson, S. C., Bengt; Kuylenstierna, Ulf; Magnéli, Arne. *Acta Chem. Scand.* **1957**, *11*, 1641–1652.

(53) Lu, H.; Zhao, B.; Pan, R.; Yao, J.; Qiu, J.; Luo, L.; Liu, Y. *RSC Advances* **2014**, *3*, 1128–1132.

(54) Wang, G.; Wang, H.; Ling, Y.; Tang, Y.; Yang, X.; Fitzmorris, R. C.; Wang, C.; Zhang, J. Z.; Li, Y. *Nano Lett.* **2011**, *7*, 3026–3033.

(55) Bartholomew, R. F.; Frankl, D. R. *Phys. Rev.* **1969**, *3*, 828–833.

(56) Lu, X.; Wang, G.; Zhai, T.; Yu, M.; Gan, J.; Tong, Y.; Li, Y. *Nano Lett.* **2012**, *3*, 1690–1696.

(57) Zhang, Q.; Re Ko, N.; Kwon Oh, J. *RSC Adv.* **2012**, *21*, 8079–8086.

(58) Rekoske, J. E.; Barteau, M. A. *J. Phys. Chem. B* **1997**, *7*, 1113–1124.

(59) Li, S.; Qiu, J.; Ling, M.; Peng, F.; Wood, B.; Zhang, S. *ACS Appl. Mater. Interfaces* **2013**, *21*, 11129–11135.

(60) Jiang, X.; Zhang, Y.; Jiang, J.; Rong, Y.; Wang, Y.; Wu, Y.; Pan, C. *J. Phys. Chem. C* **2012**, *42*, 22619–22624.

(61) Khader, M. M.; Kheiri, F. M. N.; El-Anadouli, B. E.; Ateya, B. G. *J. Phys. Chem.* **1993**, *22*, 6074–6077.

(62) Haerudin, H.; Bertel, S.; Kramer, R. Surface stoichiometry of "titanium suboxide" Part I Volumetric and FTIR study. *J. Chem. Soc. Faraday Trans.* **1998**, *10*, 1481–1487.

(63) Chen, X.; Liu, L.; Yu, P. Y.; Mao, S. S. *Science* **2011**, *6018*, 746–750.

(64) Zhang, C.; Yu, H.; Li, Y.; Gao, Y.; Zhao, Y.; Song, W.; Shao, Z.; Yi, B. *ChemSusChem* **2013**, *4*, 659–666.

(65) Naldoni, A.; Allieta, M.; Santangelo, S.; Marelli, M.; Fabbri, F.; Cappelli, S.; Bianchi, C. L.; Psaro, R.; Dal Santo, V. *J. Am. Chem. Soc.* **2012**, *18*, 7600–7603.

(66) Liu, N.; Schneider, C.; Freitag, D.; Hartmann, M.; Venkatesan, U.; Müller, J.; Spiecker, E.; Schmuki, P. *Nano Lett.* **2014**, *6*, 3309–3313.

(67) Tao, X.; Wang, J.; Ying, Z.; Cai, Q.; Zheng, G.; Gan, Y.; Huang, H.; Xia, Y.; Liang, C.; Zhang, W.; Cui, Y. *Nano Lett.* **2014**, *9*, 5288–5294.

(68) Xia, T.; Chen, X. *J. Mater. Chem. A* **2013**, *9*, 2983–2989.

(69) Chen, X.; Liu, L.; Liu, Z.; Marcus, M. A.; Wang, W.-C.; Oyler, N. A.; Grass, M. E.; Mao, B.; Glans, P.-A.; Yu, P. Y.; Guo, J.; Mao, S. S. *Sci. Rep.* **2013**, 1510.

(70) Wei, W.; Yaru, N.; Chunhua, L.; Zhongzi, X. *RSC Adv.* **2012**, *2*, 8286–8288.

(71) Zheng, Z.; Huang, B.; Lu, J.; Wang, Z.; Qin, X.; Zhang, X.; Dai, Y.; Whangbo, M.-H. *Chem. Commun.* **2012**, *46*, 5733–5735.

(72) Wang, Z.; Yang, C.; Lin, T.; Yin, H.; Chen, P.; Wan, D.; Xu, F.; Huang, F.; Lin, J.; Xie, X.; Jiang, M. *Adv. Funct. Mat.* **2013**, *43*, 5444–5450.

(73) Wang, Z.; Yang, C.; Lin, T.; Yin, H.; Chen, P.; Wan, D.; Xu, F.; Huang, F.; Lin, J.; Xie, X.; Jiang, M. *Energ. Environ. Sci.* **2013**, *10*, 3007–3014.

(74) Setvín, M.; Aschauer, U.; Scheiber, P.; Li, Y.-F.; Hou, W.; Schmid, M.; Selloni, A.; Diebold, U. *Science* **2013**, *6149*, 988–991.

(75) Huang, J.-Q.; Zhang, Q.; Peng, H.-J.; Liu, X.-Y.; Qian, W.-Z.; Wei, F. *Energ. Environ. Sci.* **2014**, *1*, 347–353.

(76) Lide, D. R., *CRC Handbook of Chemistry and Physics.* 94 ed.; CRC Press: Boca Raton, FL, 2013.

(77) Myung, S.-T.; Kikuchi, M.; Yoon, C. S.; Yashiro, H.; Kim, S.-J.; Sun, Y.-K.; Scrosati, B. *Energ. Environ. Sci.* **2013**, *9*, 2609–2614.

(78) Wang, J.; Shen, L.; Nie, P.; Xu, G.; Ding, B.; Fang, S.; Dou, H.; Zhang, X. *J. Materials Chem. A* **2014**, *24*, 9150–9155.

(79) Jeong, G.; Kim, J.-G.; Park, M.-S.; Seo, M.; Hwang, S. M.; Kim, Y.-U.; Kim, Y.-J.; Kim, J. H.; Dou, S. X. *ACS Nano* **2014**, *3*, 2977–2985.

(80) Shin, J.-Y.; Joo, J. H.; Samuelis, D.; Maier, J. *Chem. Mater.* **2012**, *3*, 543–551.

(81) Stefik, M.; Heiligtag, F. J.; Niederberger, M.; Grätzel, M. *ACS Nano* **2013**, *10*, 8981–8989.

(82) Song, T.; Paik, U. *J. Mater. Chem. A* **2016**, *1*, 14–31.

(83) Changzhou, Y.; Siqi, Z.; Hui, C.; Linrui, H.; Jingdong, L. *Nanotechnology* **2016**, *4*, 045403.

(84) Zhang, Z.; Li, Q.; Zhang, K.; Chen, W.; Lai, Y.; Li, J. *J. Power Sources* **2015**, *290*, 159–167.

(85) Chen, J. S.; Tan, Y. L.; Li, C. M.; Cheah, Y. L.; Luan, D.; Madhavi, S.; Boey, F. Y. C.; Archer, L. A.; Lou, X. W. *J. Am. Chem. Soc.* **2010**, *17*, 6124–6130.

(86) Liang, X.; Kwok, C. Y.; Lodi-Marzano, F.; Pang, Q.; Cuisinier, M.; Huang, H.; Hart, C. J.; Houtarde, D.; Kaup, K.; Sommer, H.; Brezesinski, T.; Janek, J.; Nazar, L. F. *Adv. Energy Mater.* **2015**, *5*, n/a-n/a.

Index